CAMBRIDGE TRACTS IN MATHEMATICS
AND MATHEMATICAL PHYSICS

GENERAL EDITORS
H. BASS, J. F. C. KINGMAN, F. SMITHIES
J. A. TODD & C. T. C. WALL

63. *Contiguity of probability measures: some applications in statistics*

GEORGE G. ROUSSAS

Professor of Statistics
University of Wisconsin

Contiguity of probability measures: some applications in statistics

CAMBRIDGE
AT THE UNIVERSITY PRESS
1972

CAMBRIDGE UNIVERSITY PRESS
Cambridge, New York, Melbourne, Madrid, Cape Town, Singapore, São Paulo, Delhi

Cambridge University Press
The Edinburgh Building, Cambridge CB2 8RU, UK

Published in the United States of America by Cambridge University Press, New York

www.cambridge.org
Information on this title: www.cambridge.org/9780521083546

First published 1972
This digitally printed version 2008

A catalogue record for this publication is available from the British Library

Library of Congress Catalogue Card Number: 71–171682

ISBN 978-0-521-08354-6 hardback
ISBN 978-0-521-09095-7 paperback

To

HERCULES

ANGELIQUE

DEMETRA

STELLA

Contents

Acknowledgement   *page* x

Preface   xi

1 On the concept of contiguity and related theorems   1
    Summary   1
    1 Some preliminary definitions and results   2
    2 Contiguity and its relation to other concepts of
       'nearness' of sequences of probability measures   7
    3 Alternative characterizations of contiguity   10
    4 Some auxiliary results   18
    5 Proof of Proposition 3.1   25
    6 An additional characterization of contiguity   31
    7 Some results following from contiguity   33
    Exercises   39

2 Asymptotic expansion and asymptotic distribution
   of likelihood functions   41
    Summary   41
    1 Preliminaries   42
    2 Assumptions   45
    3 Some examples   47
    4 Asymptotic expansion and asymptotic normality
       of likelihood functions   52
    5 Some lemmas   54
    6 Proof of theorems of Section 4   63
    Exercise   66

Contents

3 Approximation of a given family of probability measures by an exponential family – asymptotic sufficiency   *page* 67

    Summary   67

  1 Formulation of the problem and some preliminary results   68

  2 Some auxiliary results   72

  3 The proof of the theorem   76

  4 Differential equivalence of sequences of probability measures and differential sufficiency   79

  5 Some statistical implications of Theorem 1.1   81

    Exercises   84

4 Some statistical applications: AUMP and AUMPU tests for certain testing hypotheses problems   85

    Summary   85

  1 Additional assumptions – Examples   86

  2 Some lemmas   96

  3 Testing a simple hypothesis against one-sided alternatives   99

  4 AUMP tests for the examples of Section 1   105

  5 Testing a simple hypothesis against two-sided alternatives   107

  6 Testing a one-sided hypothesis against one-sided alternatives   123

    Exercises   127

5 Some statistical applications: asymptotic efficiency of estimates   128

    Summary   128

  1 W-efficiency – preliminaries   128

  2 Some lemmas   131

  3 A representation theorem   135

Contents

4 W-efficiency of estimates: upper bounds via
   Theorem 3.1   *page* 141
5 W-efficiency of estimates: upper bounds   147
6 Asymptotic efficiency of estimates: the classical
   approach   157
7 Classical efficiency of estimates: the multiparameter
   case   160
   Exercise   166

6   Multiparameter asymptotically optimal tests   167
1 Some notation and preliminary results   167
2 Formulation of some of the main results   169
3 Restriction to the class of tests $\mathscr{F}$ 171
4 Proof of the first main result   177
5 Proof of the second main result   184
6 Formulation and proof of the third main result   188
7 Behaviour of the power under non-local
   alternatives   196
   Exercises   197

Appendix
1 Some theorems employed in Chapter 1   199
2 Some theorems employed in Chapter 2   204
   Exercise   223
3 A theorem employed in Chapter 5   224
4 Some theorems employed in Chapter 6   225
   Exercises   239
Bibliography   240
Index   246

Acknowledgement

Grateful thanks are due to the University of California Press for permission to use the material in Chapter 6, which is a version of a talk given by the author at the Sixth Berkeley Symposium on Mathematical Statistics and Probability. A paper based on this address has appeared in *Sixth Symposium of Mathematical Statistics and Probability*, edited by Lucien LeCam and published by the University of California Press.

Preface

'Although this may seem a paradox, all science is dominated by
the idea of approximation.'

Bertrand Russell

This monograph represents a modest attempt on my part
to introduce the concept of contiguity, elaborate on the mathematical theory behind it, and also indicate some of its statistical
applications. It lays no claim in containing an exhaustive discussion of results pertaining to contiguity. In fact, there are
already new results available which, however, could not have
been included in this book. It is simply the result of an attempt
to make the concept of contiguity and some of its statistical
applications more familiar to several kinds of research workers.
These include Theoretical Statisticians, Probabilists, Mathe-
maticians whose primary interest lies in measure theory or
approximation theory, and perhaps to practitioners of Statistics
as well. It is my belief that contiguity deserves more attention
than it has received. I hope that this monograph will be a step
in that direction, anticipating the appearance of a more com-
prehensive treatise on the subject.

The concept of contiguity was introduced by Professor Lucien
LeCam as a criterion of nearness of sequences of probability
measures. In addition to its purely mathematical interest, contiguity is a powerful and very useful tool in Statistics, when one is
concerned with asymptotic theory, leading to elegant derivations
of the asymptotic properties of tests and estimates under less
restrictive assumptions than usual. Of course, this presupposes
that one subscribes to the usefulness of large sample theory
results. Frequently, however, this is the only alternative open
to the Statistician in an unfriendly real world.

Preface

The concept of contiguity is introduced in Chapter 1 and its relationship to some other modes of nearness of sequences of probability measures is investigated. Some general results which are based on contiguity and which form the backbone of the later chapters are also derived here. In Chapter 2, the statistical model to be employed throughout this monograph is introduced. This is that of a Markov process satisfying certain mild regularity conditions. Of course, the important independent identically distributed case is included in the Markovian model as a special case. Among the assumptions made here the most important one is that of differentiability in quadratic mean of a certain random function. This replaces, in effect, certain of the classical assumptions made in the literature. We refer to those assumptions which deal with the existence of pointwise derivatives up to the third order, and their boundedness by integrable random variables, of essentially the same random function referred to earlier. Under the regularity conditions imposed on the model some general results regarding the asymptotic expansion, in the probability sense, and the asymptotic normality of the log-likelihood function are derived under both a fixed and a moving sequence of parameter points. Incidentally, these derivations attest to the powerfulness and the elegance of the contiguity approach. The essence of Chapter 3 rests on an exponential approximation to the likelihood function with a view to taking advantage of the optimal inference procedures available for the exponential family in a wide variety of problems. The subsequent Chapters 4 and 5 exhibit some statistical applications. Some testing hypotheses problems for real-valued parameters are discussed in Chapter 4. In Chapter 5, the problem of asymptotic efficiency of sequences of estimates is discussed to a certain extent both from the classical point of view as well as along the lines suggested by Wolfowitz. Section 3 of Chapter 4, may be studied directly after Chapter 2. The results included in Sections 5 and 6, however, presuppose knowledge of the material in Chapter 3. In Chapter 6, a multi-parameter testing hypothesis problem is discussed and, finally, in the Appendix various results used in the body of the monograph are gathered together. Most of them are also proved.

Preface

The first serious thought and attempt to organize the results of this monograph in their present form was made when I was visiting the Mathematics Institute of Aarhus University, Denmark, during the spring semester of 1969. The financial support of the Institute is gratefully acknowledged here. Some financial support was also provided by the University of Wisconsin Research Committee and the National Science Foundation when many of the results included here were written up as Technical Reports. This is also gratefully acknowledged.

Some of the material of this monograph has been discussed in seminars both in Aarhus University and the University of Wisconsin in Madison. Comments of the participants in these seminars helped in modifying the original proofs of some theorems. In connection with this, I wish to thank O. Barndorff-Nielsen, R. A. Johnson, E. Spjøtvoll and G. K. Bhattacharyya. Many thanks are due to B. Lind for his invaluable contribution in helping clarify and organize various parts of this monograph. Thanks also go to A. Philippou and A. Soms for their many helpful suggestions and constructive comments. W. Davis is also to be thanked for a number of useful comments as well as all those who attended my lectures and contributed their comments. Professor J. F. C. Kingman read the first draft of this monograph on behalf of the Cambridge University Press. His thoughtful comments have contributed to improve the presentation of the material in the monograph. Also the Cambridge University Press itself has been very helpful throughout the editorial and publication process. I extend my sincere thanks to both. Last, but not least, I am grateful to Professor Lucien LeCam who introduced me to the subject of contiguity.

Madison, Wisconsin　　　　　　　GEORGE G. ROUSSAS
February 1972

1. *On the concept of contiguity and related theorems*

Summary

The main purpose of this chapter is to present the concept of contiguity (see Definition 2.1) introduced by LeCam [4] and study some alternative characterizations of it (see Theorem 6.1). In the process of doing so, some auxiliary concepts such as weak convergence, relative compactness and tightness of a sequence of probability measures are needed. These concepts are introduced in this chapter, as we go along, and also some of their relationships are stated and/or proved. For the omitted proofs, the reader is always referred to appropriate sources. The various characterizations of contiguity provide alternative methods one may employ in establishing the presence (or absence) of contiguity in a given case. Some concrete examples are used for illustrative purposes.

Contiguity is a concept of 'nearness' of sequences of probability measures. It would then be appropriate to relate it to other more familiar concepts of the same nature such as 'nearness' of two sequences of probability measures expressed by the norm (L_1-norm) associated with convergence in variation. By means of examples, it is shown, as one would expect, that 'nearness' of two sequences of probability measures expressed by contiguity is weaker than that expressed by the L_1-norm.

Some attention is also focused to possible relationships between contiguity on the one hand, and mutual absolute continuity and tightness on the other. In connection with this, it is shown, by means of examples, that mutual absolute continuity of the (corresponding) measures in two sequences of probability measures need not imply contiguity of the sequences. Although the converse is not true either, it is always possible to replace two contiguous sequences of probability measures by two other contiguous sequences whose (corresponding) members are absolutely continuous with respect to one another and these latter sequences lie close to the given ones in the L_1-norm sense (see Theorem 5.1). Also it is shown that tightness need not imply contiguity and contiguous sequences of probability measures need not be tight.

Finally, a number of important theorems (see Theorems 7.1, 7.2 and their corollaries), based on the assumption of contiguity, are formulated and proved. As will be seen in later chapters, these results are very essential for the statistical applications to be discussed in this monograph. Loosely speaking, these results provide the asymptotic distribution of a sequence of statistics, under a given sequence of probability measures $\{P_n'\}$, if the same sequence of statistics has a limiting distribution, under a sequence of probability measures $\{P_n\}$ contiguous to $\{P_n'\}$. For example, $\{P_n\}$ may correspond to a hypothesis being tested and $\{P_n'\}$ to ' close ' alternatives.

In order to avoid undue repetition, we should like to mention at the outset that in this chapter, as well as in the subsequent ones, all limits are taken as $\{n\}$, or subsequences thereof, converges to infinity through the positive (or non-negative) integers unless otherwise specified. Also, integrals without limits are understood to be taken over the entire (appropriate) space.

1 Some preliminary definitions and results

In all that follows, $(S, \mathscr{T})$ is a topological space, where the topology $\mathscr{T}$ is defined by a metric on S. $\mathscr{S}$ is the topological Borel σ-field generated by $\mathscr{T}$.

Let $\{\mathscr{L}_n\}$ and $\mathscr{L}$ be a sequence of probability measures and a probability measure, respectively, defined on $\mathscr{S}$. Then

DEFINITION 1.1 We say that $\{\mathscr{L}_n\}$ converges *weakly* to $\mathscr{L}$ and we write $\mathscr{L}_n \Rightarrow \mathscr{L}$ if $\int f \mathrm{d}\mathscr{L}_n \to \int f \mathrm{d}\mathscr{L}$ for all real-valued, bounded and continuous functions f defined on S.

If
$$(S, \mathscr{S}) = (R^m, \mathscr{B}^m) = \prod_{j=1}^{m} (R_j, \mathscr{B}_j) \quad (m = 1, 2, \ldots),$$
where $(R_j, \mathscr{B}_j) = (R, \mathscr{B})$, the Borel real line, and if F_n and F are the distribution functions (d.f.s) corresponding to $\mathscr{L}_n$ and $\mathscr{L}$, re-

spectively, then weak convergence is equivalent to the convergence $F_n(x) \to F(x)$, $x \in C(F)$, the set of continuity points of F (this is a generalization of the Helly–Bray theorem discussed, e.g. in Loève [1], p. 182; see also Billingsley [4], p. 18).

Theorems 1.1 A, 1.2 A in the appendix (and also Theorem 1.1 below along with Remark 1.2) provide alternative characterizations of weak convergence.

THEOREM 1.1 Suppose $(S, \mathscr{S}) = (R^m, \mathscr{B}^m)$ and let $\{\mathscr{L}_n\}$ and $\mathscr{L}$ be a sequence of probability measures and a probability measure, respectively, defined on $\mathscr{S}$ such that $\mathscr{L}_n \Rightarrow \mathscr{L}$. Let f be a real-valued, bounded function defined on S such that its restriction to a compact set K is continuous, f vanishes outside K and f need not vanish on the boundary ∂K of K, provided $\mathscr{L}(\partial K) = 0$. Then $\int f \mathrm{d}\mathscr{L}_n \to \int f \mathrm{d}\mathscr{L}$.

Proof Of course, if f is continuous, the conclusion holds true whether or not $\mathscr{L}(\partial K) = 0$. Thus it suffices to restrict ourselves to fs which are discontinuous on ∂K, provided $\mathscr{L}(\partial K) = 0$. Since

$$\int_{K^c} f \mathrm{d}\mathscr{L}_n = \int_{K^c} f \mathrm{d}\mathscr{L} = 0,$$

it suffices to show that

$$\int_K f \mathrm{d}\mathscr{L}_n \to \int_K f \mathrm{d}\mathscr{L}. \tag{1.1}$$

By Theorem 1.1 A, $\mathscr{L}_n(K) \to \mathscr{L}(K)$ since $\mathscr{L}(\partial K) = 0$. Therefore, if $\mathscr{L}(K) = 0$, then

$$\left| \int_K f \mathrm{d}\mathscr{L}_n \right| \leqslant M \mathscr{L}_n(K) \to 0 = \int_K f \mathrm{d}\mathscr{L},$$

where M is an upper bound for $|f|$ on K, and hence (1.1) is true. Thus we assume that $\mathscr{L}(K) > 0$ and on $\mathscr{S} \cap K$, define $\mathscr{L}_n^*$ for all sufficiently large n (so that $\mathscr{L}_n(K) > 0$), and $\mathscr{L}^*$ by

$$\mathscr{L}_n^*(B) = \mathscr{L}_n(B)/\mathscr{L}_n(K), \quad \mathscr{L}^*(B) = \mathscr{L}(B)/\mathscr{L}(K).$$

Let $(\partial B)_r$ be the boundary of a set B with respect to the induced (in K) topology. Then it is clear that $\partial B \subseteq (\partial B)_r \cup \partial K$, so that

$\mathscr{L}^*[(\partial B)_r] = 0$ implies $\mathscr{L}(\partial B) = 0$. Thus $\mathscr{L}^*$-continuity sets are also $\mathscr{L}$-continuity sets. Therefore $\mathscr{L}_n^* \Rightarrow \mathscr{L}^*$. It follows that

$$\int_K f \mathrm{d}\mathscr{L}_n = \mathscr{L}_n(K) \int_K f \mathrm{d}\mathscr{L}_n^* \to \mathscr{L}(K) \int_K f \mathrm{d}\mathscr{L}^* = \int_K f \mathrm{d}\mathscr{L}.$$

This establishes (1.1) and hence the theorem itself. ▮

COROLLARY 1.1   Let f be as in the theorem except that f is equal to a constant c on K^c and f need not be equal to c on ∂K, provided $\mathscr{L}(\partial K) = 0$. Then $\int f \mathrm{d}\mathscr{L}_n \to \int f \mathrm{d}\mathscr{L}$.

Proof   Replace f by $f - c$ and apply the theorem. ▮

REMARK 1.1   By Theorem 1.2 A, it follows that the converse of Theorem 1.1 is also true. That is, if $\int f \mathrm{d}\mathscr{L}_n \to \int f \mathrm{d}\mathscr{L}$ for each f as described in Theorem 1.1, then $\mathscr{L}_n \Rightarrow \mathscr{L}$.

We recall that $\{\mathscr{L}_n\}$ is a sequence of probability measures defined on $\mathscr{S}$. Then

DEFINITION 1.2   The sequence $\{\mathscr{L}_n\}$ is said to be *relatively compact* if for every subsequence $\{n'\} \subseteq \{n\}$ there exists a further subsequence $\{n''\} \subseteq \{n'\}$ such that $\{\mathscr{L}_{n''}\}$ converges weakly to a probability measure (which, in general, depends on $\{n''\}$).

The following definition will also be useful.

DEFINITION 1.3   The sequence $\{\mathscr{L}_n\}$ is said to be *tight* if for every $\epsilon > 0$ there is a compact set $K = K(\epsilon)$ such that

$$\mathscr{L}_n(K) > 1 - \epsilon$$

for all n.

If $(S, \mathscr{S}) = (R^m, \mathscr{B}^m)$, this condition can be replaced by:

$$\mathscr{L}_n([a, b]) > 1 - \epsilon$$

for all n, where $[a, b]$ is a closed interval in R^m. This is so because any compact set in R^m can be enclosed in a bounded, closed interval, and bounded, closed intervals are compact.

The concepts of relative compactness and tightness are related as follows.

THEOREM 1.2   Let $(S, \mathscr{S}) = (R^m, \mathscr{B}^m)$. Then $\{\mathscr{L}_n\}$ is relatively compact if and only if it is tight.

By Theorem 1.3 A, tightness always implies relative compactness. The reverse implication is also true, by the same theorem, provided S is separable and complete. This requirement is, clearly, satisfied in the present situation, where, of course, $\mathscr{T}$ is assumed to be the usual topology in R^m. So Theorem 1.2 is a special case of Theorem 1.3 A. However, a simple proof of Theorem 1.2 can be presented along the following lines.

Proof of Theorem 1.2 Suppose that $\{\mathscr{L}_n\}$ is tight. Then, for every $\epsilon > 0$, there exists a closed interval in R^m, $[a, b]$, depending on ϵ, such that $\mathscr{L}_n([a, b]) > 1 - \epsilon$ for all n. Let F_n be the d.f. corresponding to $\mathscr{L}_n$. Then by the weak compactness theorem for d.f.s, for every $\{n'\} \subset \{n\}$ there exists $\{n''\} \subset \{n'\}$ such that $F_{n''}(x) \to F^*(x)$ for all $x \in C(F^*)$, where F^* is a d.f. in all other respects except that it might fail to have variation equal to one. We shall, actually, show that this does not occur. To this end, let $\mathscr{L}^*$ be the measure induced by F^*. Then $\mathscr{L}^*$ is a probability measure. In fact, $\mathscr{L}_{n''}([a, b]) > 1 - \epsilon$ for all n'' and a, b can be taken to be continuity points of F^*. Taking the limit as $n'' \to \infty$, we have $\mathscr{L}^*([a, b]) \geqslant 1 - \epsilon$. Thus, for every $\epsilon > 0$, there exists a closed interval $[a, b]$ in R^m, depending on ϵ, such that

$$\mathscr{L}^*([a, b]) \geqslant 1 - \epsilon,$$

and this implies that $\mathscr{L}^*$ is a probability measure. Hence, tightness implies relative compactness.

Now assume that $\{\mathscr{L}_n\}$ is relatively compact. Then, for every $\epsilon > 0$, we claim that there exists a closed interval $[a, b]$ in R^m, depending on ϵ, such that $\mathscr{L}_n([a, b]) > 1 - \epsilon$ for all n. In fact, if this were not true, then for some $\epsilon > 0$ and for every $[a, b]$ in R^m there would exist a k, depending on ϵ and $[a, b]$, such that $\mathscr{L}_k([a, b]) \leqslant 1 - \epsilon$. Apply this argument for $a_n, b_n \in R^m$, where $a_n = (-n, ..., -n)'$, $b_n = (n, ..., n)'$ and $'''$ denotes transpose. Thus, for every n, there exists n' such that $\mathscr{L}_{n'}([a_n, b_n]) \leqslant 1 - \epsilon$, where $\{n'\} \subseteq \{n\}$ and $n' \to \infty$. Then there cannot exist $\{n''\} \subseteq \{n'\}$ such that $\{\mathscr{L}_{n''}\}$ converges (weakly) to a probability measure $\mathscr{L}$, say. This is so because, if F is the d.f. corresponding to $\mathscr{L}$, then for every $x, y \in C(F)$ with $x < y$ (to be understood in the coordinatewise sense), we have

$$\mathscr{L}((x, y]) = \lim \mathscr{L}_{n''}((x, y]) \leqslant \liminf \mathscr{L}_{n''}((a_{n*}, b_{n*}]) \leqslant 1 - \epsilon,$$

where n^* is the subscript of the interval associated with n'' and the limits are taken as $n^* \to \infty$ which implies $n'' \to \infty$. Thus $\mathscr{L}((x,y]) \leqslant 1 - \epsilon$ for every $x, y \in C(F)$, which implies that $\mathscr{L}$ is not a probability measure. This contradicts our assumption of $\{\mathscr{L}_n\}$ being relatively compact. The proof of the theorem is completed. ∎

Now let $\{(\mathscr{X}, \mathscr{A}_n, P_n)\}$ be a sequence of probability spaces and let $\{T_n\}$ be a sequence of m-dimensional random vectors such that T_n is $\mathscr{A}_n$-measurable. Set $\mathscr{L}_n = \mathscr{L}(T_n | P_n)$. Then the following proposition will prove useful on many occasions.

PROPOSITION 1.1   The sequence $\{\mathscr{L}_n\}$ just defined is relatively compact (equivalently, tight) if and only if, for every $\epsilon > 0$, there exists $b = b(\epsilon) > 0$ such that $P_n(\|T_n\| > b) < \epsilon$ for all n, where $\|\cdot\|$ denotes the usual Euclidean norm in R^m.

Proof   By Theorem 1.2, $\{\mathscr{L}_n\}$ is relatively compact if and only if it is tight. On the other hand, tightness of $\{\mathscr{L}_n\}$ is equivalent to the existence of an interval $[a, b]$ in R^m, depending on ϵ, such that $\mathscr{L}_n([a, b]) > 1 - \epsilon$. This is so by the comments following Definition 1.3. But $\mathscr{L}_n([a, b]) = P_n(T_n \in [a, b])$. Thus $\{\mathscr{L}_n\}$ is tight if and only if $P_n(T_n \in [a, b]) > 1 - \epsilon$ for all n. This, however, is equivalent to the existence of a $c(\epsilon) = c = (c_1, ..., c_m)'$ with positive coordinates such that $P_n(T_n \in [-c, c]) > 1 - \epsilon$ for all n. Taking $b(\epsilon) = \|c(\epsilon)\|$, we get the desired result. ∎

Let now P and Q be two probability measures on the σ-field $\mathscr{A}$ of subsets of $\mathscr{X}$. Then the L_1-*norm* of $P - Q$, denoted by $\|P - Q\|$, is defined by $\|P - Q\| = 2 \sup\{|P(A) - Q(A)|; A \in \mathscr{A}\}$. Next, if f and g are densities of P and Q, respectively, relative to a dominating σ-finite measure μ (e.g. $\mu = P + Q$), then

$$\|P - Q\| = \int |f - g| \, \mathrm{d}\mu.$$

This is shown in Theorem 1.4A.

The following result is isolated here for convenient reference.

PROPOSITION 1.2   For each n, let P_n and Q_n be probability measures on $\mathscr{A}_n$ and let $\{T_n\}$ be a sequence of m-dimensional random vectors such that T_n is $\mathscr{A}_n$-measurable. Set

$$\mathscr{L}_n^P = \mathscr{L}(T_n | P_n) \quad \text{and} \quad \mathscr{L}_n^Q = \mathscr{L}(T_n | Q_n),$$

and suppose that $\|P_n - Q_n\| \to 0$. Then we have

(i) $$\|\mathscr{L}_n^P - \mathscr{L}_n^Q\| \to 0.$$

(ii) If $\mathscr{L}_n^P \Rightarrow \mathscr{L}$, a probability measure, then $\mathscr{L}_n^Q \Rightarrow \mathscr{L}$ and conversely.

Proof (i) For $B \in \mathscr{B}^m$, one has

$$\left|\mathscr{L}_n^P(B) - \mathscr{L}_n^Q(B)\right| = \left|P_n(T_n \in B) - Q_n(T_n \in B)\right|,$$

so that

$$\|\mathscr{L}_n^P - \mathscr{L}_n^Q\| = 2 \sup\left[\left|\mathscr{L}_n^P(B) - \mathscr{L}_n^Q(B)\right|; B \in \mathscr{B}^m\right]$$

$$\leqslant 2 \sup\left[\left|P_n(A) - Q_n(A)\right|; A \in \mathscr{A}_n\right] = \|P_n - Q_n\| \to 0.$$

(ii) Let B be a continuity set of $\mathscr{L}$. Then we have

$$\left|\mathscr{L}_n^Q(B) - \mathscr{L}(B)\right| \leqslant \left|\mathscr{L}_n^Q(B) - \mathscr{L}_n^P(B)\right| + \left|\mathscr{L}_n^P(B) - \mathscr{L}(B)\right|.$$

If $\mathscr{L}_n^P \Rightarrow \mathscr{L}$, then the second term on the right-hand side of the above inequality tends to zero, by Theorem 1.1 A. The first term also converges to zero by (i). Therefore, $\mathscr{L}_n^Q \Rightarrow \mathscr{L}$. That $\mathscr{L}_n^Q \Rightarrow \mathscr{L}$ implies $\mathscr{L}_n^P \Rightarrow \mathscr{L}$ follows by symmetry. The proof of the proposition is concluded. ∎

2 Contiguity and its relation to other concepts of 'nearness' of sequences of probability measures

In this section, the concept of contiguity is introduced and some alternative characterizations of it are studied. Its relationship to other familiar concepts of 'nearness' of sequences of probability measures is investigated, and, finally, a number of illustrative examples are discussed.

Let $\{(\mathscr{X}, \mathscr{A}_n)\}$ be a sequence of measurable spaces, and let P_n, P_n' be probability measures on $\mathscr{A}_n$. Also, let $\{T_n\}$ be a sequence of random variables (r.v.s) such that T_n is $\mathscr{A}_n$-measurable.

DEFINITION 2.1 The sequences of probability measures $\{P_n\}$ and $\{P_n'\}$ are said to be *contiguous* if the following is true: for any $\mathscr{A}_n$-measurable r.v.s T_n, $T_n \to 0$ in P_n-probability if and only if $T_n \to 0$ in P_n'-probability.

The concept of contiguity is a concept expressing 'closeness' or 'nearness' between the sequences of probability measures $\{P_n\}$ and $\{P'_n\}$ in the sense of the definition just given. Some more light is shed on this concept by the fact that, if $A_n \in \mathscr{A}_n$, then $P_n(A_n) \to 0$ if and only if $P'_n(A_n) \to 0$, provided $\{P_n\}$ and $\{P'_n\}$ are contiguous, as will be shown below. Also some concrete examples will further illustrate the point.

REMARK 2.1   It is worth noticing that contiguity is transitive. That is, if $\{P_n\}$, $\{Q_n\}$ and $\{Q_n\}$, $\{R_n\}$ are contiguous, then $\{P_n\}$, $\{R_n\}$ are contiguous. This is an immediate consequence of Definition 2.1.

PROPOSITION 2.1   The sequences of probability measures $\{P_n\}$ and $\{P'_n\}$ are contiguous if and only if, for $A_n \in \mathscr{A}_n$, $P_n(A_n) \to 0$ if and only if $P'_n(A_n) \to 0$.

Proof   Assume $\{P_n\}$ and $\{P'_n\}$ to be contiguous, and let $P_n(A_n) \to 0$ with $A_n \in \mathscr{A}_n$. Set $T_n = I_{A_n}$. Then, for every

$$(1 >) \epsilon > 0, \quad P_n(|T_n| > \epsilon) = P_n(A_n),$$

so that $T_n \to 0$ in P_n-probability. This implies $T_n \to 0$ in P'_n-probability by Definition 2.1. Since $P'_n(|T_n| > \epsilon) = P'_n(A_n)$, we have then $P'_n(A_n) \to 0$. The fact that $P'_n(A_n) \to 0$ implies $P_n(A_n) \to 0$ is treated entirely symmetrically. For the converse, we have: let $T_n \to 0$ in P_n-probability and, for $\epsilon > 0$, set

$$A_n = (|T_n| > \epsilon).$$

Then $P_n(A_n) \to 0$ and this implies $P'_n(A_n) \to 0$. However, this last convergence is equivalent to the convergence $T_n \to 0$ in P'_n-probability. That $T_n \to 0$ in P'_n-probability implies $T_n \to 0$ in P_n-probability follows by symmetry. ∎

Of course, if $\|P_n - P'_n\| \to 0$, then the sequences $\{P_n\}$ and $\{P'_n\}$ are as close together as they can be, apart from having identical elements. One would then expect that convergence in L_1-norm would imply contiguity. That this is, actually, the case is seen in the following lemma.

LEMMA 2.1   If $\|P_n - P'_n\| \to 0$, then $\{P_n\}$ and $\{P'_n\}$ are contiguous.

Proof The convergence $\|P_n - P'_n\| \to 0$ implies that, for every $\epsilon > 0$, $|P_n(A_n) - P'_n(A_n)| < \epsilon$ for every $A_n \in \mathscr{A}_n$ and all sufficiently large n, $n \geqslant n_1$, say. Thus $P_n(A_n) < \epsilon$ implies $P'_n(A_n) < 2\epsilon$ and $P'_n(A_n) < \epsilon$ implies $P_n(A_n) < 2\epsilon$ for every $A_n \in \mathscr{A}_n$ and all $n \geqslant n_1$. Then Proposition 2.1 applies and gives the desired result. ∎

REMARK 2.2 It is demonstrated by means of examples (see, e.g. Example 3.1 (i)) that the converse of Lemma 2.1 is not true. Thus contiguity is a weaker 'measure of nearness' of sequences of probability measures than that expressed by convergence in L_1-norm.

We now consider the following example as a simple application of Lemma 2.1.

EXAMPLE 2.1 Let $(\mathscr{X}, \mathscr{A}_n) = (R, \mathscr{B})$, $P_n = U(-1/n, 1)$, the uniform measure over $(-1/n, 1)$ and $P'_n = U(0, 1 + 1/n)$, the uniform measure over $(0, 1 + 1/n)$.

Then, if $$f_n = dP_n/dl \quad \text{and} \quad g_n = dP'_n/dl,$$

where l is the Lebesgue measure in R, we have

$$\|P_n - P'_n\| = \int |f_n - g_n| \, dl = 2/(n+1) \to 0.$$

Thus $\{P_n\}$ and $\{P'_n\}$ are contiguous, by Lemma 2.1.

REMARK 2.3 Example 2.1 also shows that contiguity need not imply mutual absolute continuity (either for all n or only for all sufficiently large n), as one might be tempted to (wrongly) infer by Proposition 2.1. It will be shown later, however, that any given pair of contiguous sequences of probability measures can always be replaced by another pair of contiguous sequences whose members are mutually absolutely continuous and lie arbitrarily close to the given ones (for this, see Theorem 5.1). On the other hand, mutual absolute continuity need not imply contiguity either, as is demonstrated by the following example.

EXAMPLE 2.2 Let

$$(\mathscr{X}, \mathscr{A}_n) = (R, \mathscr{B}), \quad P_n = N(\mu_n, 1) \quad \text{and} \quad P'_n = N(\mu'_n, 1),$$

where $\mu_n \to -\infty$ and $\mu'_n \to \infty$.

Then, clearly, $P_n \approx P'_n$ for all n. (As usual, by $P \approx Q$ we express the fact that $P \ll Q$ and $Q \ll P$.) Consider the set A_n defined by

$A_n = (\mu_n - 1, \mu_n + 1)$. Then $P_n(A_n) = c$, a positive constant ($c \approx 0.68$), so that $P_n(A_n)$ does not tend to zero. But $P'_n(A_n) \to 0$, clearly. Thus, by Proposition 2.1, $\{P_n\}$ and $\{P'_n\}$ cannot be contiguous.

REMARK 2.4 From this example, we might (wrongly) conclude that the absence of contiguity of $\{P_n\}$, $\{P'_n\}$ is due to the lack of tightness. (It is clear that the sequences are not tight.) That tightness is not a necessary condition for contiguity is shown in Example 3.1(i), where the sequences involved are not tight (for a special choice of the parameters) and yet contiguous. In the same example and for another choice of the parameters, it is also shown that tightness is not a sufficient condition for contiguity.

Summarizing some of the results obtained so far, we have:

If $\|P_n - P'_n\| \to 0$, then $\{P_n\}$, $\{P'_n\}$ are contiguous. The converse need not be true.

Contiguity of $\{P_n\}$, $\{P'_n\}$ need not imply $P_n \approx P'_n$ for all or sufficiently large n (see, however, Theorem 5.1). The converse is also true.

Contiguity of $\{P_n\}$, $\{P'_n\}$ need not imply their tightness. The converse is also true.

The following remark is relevant here.

REMARK 2.5 If $\{P_n\}$ and $\{P'_n\}$ are contiguous, but $P_n \not\approx P'_n$, then it follows from Proposition 2.1 that the support of the singular part of P_n with respect to P'_n has P_n-probability tending to zero, and the support of the singular part of P'_n with respect to P_n has P'_n-probability tending to zero.

3 Alternative characterizations of contiguity

Proposition 2.1 provides an alternative characterization of contiguity. Additional ones will be given in the sequel but for their formulation and proof, we need some more notation and also some auxiliary results. To this end, for each n, let P_n, P'_n be probability measures on $\mathscr{A}_n$ and let μ_n be a σ-finite measure dominating them. Furthermore, let

$$f_n = \mathrm{d}P_n/\mathrm{d}\mu_n, \quad g_n = \mathrm{d}P'_n/\mathrm{d}\mu_n. \tag{3.1}$$

Define the sets B_n, C_n, D_n, E_n by

$$\left.\begin{array}{l} B_n = \{x \in \mathscr{X}; f_n(x)\, g_n(x) > 0\} \\ C_n = \{x \in \mathscr{X}; f_n(x) > 0, g_n(x) = 0\} \\ D_n = \{x \in \mathscr{X}; f_n(x) = 0, g_n(x) > 0\} \\ E_n = \{x \in \mathscr{X}; f_n(x) = g_n(x) = 0\}, \end{array}\right\} \tag{3.2}$$

and let the r.v. (log-likelihood) Λ_n be defined as follows

$$\Lambda_n = \log\,(g_n/f_n) \tag{3.3}$$

on B_n and arbitrary (but measurable) on $C_n \cup D_n \cup E_n$.

Finally, *for each determination of Λ_n by (3.3)*, let

$$\mathscr{L}_n = \mathscr{L}(\Lambda_n|P_n), \quad \mathscr{L}'_n = \mathscr{L}(\Lambda_n|P'_n). \tag{3.4}$$

Let S_i $(i = 1, 2, 3)$, stand for the following statements:

(S_1) $\{P_n\}$ and $\{P'_n\}$ are contiguous.

(S_2) $\{\mathscr{L}_n\}$ and $\{\mathscr{L}'_n\}$ are relatively compact (for each determination of Λ_n).

(S_3) $\{\mathscr{L}_n\}$ is relatively compact (for each determination of Λ_n) and for any $\{m\} \subseteq \{n\}$ for which $\mathscr{L}_m \Rightarrow \mathscr{L}$, a probability measure (depending, in general, on $(\mathscr{L}_n\})$, we have $\int \exp \lambda\, d\mathscr{L} = 1$, where λ is a dummy variable.

$$\hspace{10cm} (3.5)$$

In terms of the notation introduced so far, we may formulate the following result.

PROPOSITION 3.1   The statements S_i $(i = 1, 2, 3)$ given by (3.5), are equivalent.

The proof of the proposition is found in Sections 3 and 5. In the meantime, we consider some illustrative examples.

EXAMPLE 3.1   Let $(\mathscr{X}, \mathscr{A}_n) = (R, \mathscr{B})$.

(i) Here we take $P_n = N(\mu_n, \sigma_n^2)$ and $P'_n = N(\mu'_n, \sigma_n^2)$. Then we have

$$\Lambda_n = \frac{\mu'_n - \mu_n}{\sigma_n^2}\, X + \frac{\mu_n^2 - \mu_n'^2}{2\sigma_n^2},$$

so that $\mathscr{L}_n = \mathscr{L}(\Lambda_n | P_n) = N\left[-\dfrac{(\mu_n'-\mu_n)^2}{2\sigma_n^2}, \dfrac{(\mu_n'-\mu_n)^2}{\sigma_n^2}\right],$

$$\mathscr{L}_n' = \mathscr{L}(\Lambda_n | P_n') = N\left[\dfrac{(\mu_n'-\mu_n)^2}{2\sigma_n^2}, \dfrac{(\mu_n'-\mu_n)^2}{\sigma_n^2}\right].$$

Suppose that $\left\{\dfrac{\mu_n'-\mu_n}{\sigma_n}\right\}$ is bounded and let the subsequence $\left\{\dfrac{\mu_m'-\mu_m}{\sigma_m}\right\}$ converge to σ, say. Then, clearly,

$$\mathscr{L}_m \Rightarrow \mathscr{L} = N(-\tfrac{1}{2}\sigma^2, \sigma^2) \quad \text{and} \quad \mathscr{L}_m' \Rightarrow \mathscr{L}' = N(\tfrac{1}{2}\sigma^2, \sigma^2).$$

It follows that $\{\mathscr{L}_n\}$, $\{\mathscr{L}_n'\}$ are relatively compact and hence $\{P_n\}$, $\{P_n'\}$ are contiguous on account of the implication $S_2 \Rightarrow S_1$. The contiguity of $\{P_n\}$, $\{P_n'\}$ may also be concluded, on the basis of the implication $S_3 \Rightarrow S_1$, from the relative compactness of $\{\mathscr{L}_n\}$ and the fact that

$$\int \exp \lambda \, \mathrm{d}\mathscr{L} = 1,$$

as is easily seen.

As a special case, let $\sigma_n = 1$, $\mu_n \to \mu$, $\mu_n' \to \mu'$, where both μ, μ' are finite and $\mu \neq \mu'$. Then $\{P_n\}$, $\{P_n'\}$ are contiguous but, clearly, $\|P_n - P_n'\| \nrightarrow 0$.

Also, let $\sigma_n = 1$ as above and let $\mu_n' = \mu_n + c$, where c is a constant $\neq 0$ and $\mu_n \to -\infty$. Then again $\{P_n\}$, $\{P_n'\}$ are contiguous, but, clearly, they are not tight.

Finally, let $\mu_n' = \mu_n + c$ with c as above, let $\mu_n \to \mu$, finite, and let $\sigma_n \to 0$. Then $\{P_n\}$, $\{P_n'\}$ are not contiguous and yet they are tight.

(ii) Here we take P_n and P_n' to be the probability measures corresponding to one-parameter exponential densities; that is,

$$\frac{\mathrm{d}P_n}{\mathrm{d}l} = \lambda_n \exp(-\lambda_n x), \quad \frac{\mathrm{d}P_n'}{\mathrm{d}l} = \lambda_n' \exp(-\lambda_n' x) \quad (x > 0).$$

Suppose that $\left\{\dfrac{\lambda_n'}{\lambda_n}\right\}$ is bounded from above and also away from 0.

Without loss of generality, we may assume that it is bounded away from 1 since if $\{m\} \subseteq \{n\}$ is such that $\lambda_m'/\lambda_m \to 1$, then $\|P_m - P_m'\| \to 0$ (see Exercise 1). Then we have

$$\Lambda_n = (\lambda_n - \lambda_n') X + \log(\lambda_n'/\lambda_n),$$

so that

$$\mathscr{L}_n = \mathscr{L}(\Lambda_n|P_n): \frac{\lambda_n}{|\lambda_n - \lambda_n'|}\exp\left(\frac{\lambda_n}{\lambda_n - \lambda_n'}\log\frac{\lambda_n'}{\lambda_n}\right)\exp\left(-\frac{\lambda_n}{\lambda_n - \lambda_n'}y\right),$$

$$\mathscr{L}_n' = \mathscr{L}(\Lambda_n|P_n'): \frac{\lambda_n'}{|\lambda_n - \lambda_n'|}\exp\left(\frac{\lambda_n'}{\lambda_n - \lambda_n'}\log\frac{\lambda_n'}{\lambda_n}\right)\exp\left(-\frac{\lambda_n'}{\lambda_n - \lambda_n'}y\right).$$

The domain of y is $\left(\log\dfrac{\lambda_n'}{\lambda_n}, \infty\right)$ if $\lambda_n > \lambda_n'$ and $\left(-\infty, \log\dfrac{\lambda_n'}{\lambda_n}\right)$ if $\lambda_n < \lambda_n'$.

Let $\{m\}$ be a subsequence of $\{n\}$ such that $\lambda_m'/\lambda_m \to \lambda$. Then

$$\mathscr{L}_m \Rightarrow \mathscr{L}: \frac{1}{|1-\lambda|}\exp\left(\frac{1}{1-\lambda}\log\lambda\right)\exp\left(-\frac{1}{1-\lambda}y\right)$$

and

$$\mathscr{L}_m' \Rightarrow \mathscr{L}': \frac{\lambda}{|1-\lambda|}\exp\left(\frac{\lambda}{1-\lambda}\log\lambda\right)\exp\left(-\frac{\lambda}{1-\lambda}y\right).$$

The domain of y is $(\log\lambda, \infty)$ if $\lambda < 1$ and $(-\infty, \log\lambda)$ if $\lambda > 1$. Thus we have that $\{\mathscr{L}_n\}$, $\{\mathscr{L}_n'\}$ are relatively compact and hence $\{P_n\}$, $\{P_n'\}$ are contiguous since $S_2 \Rightarrow S_1$. The same conclusion can be reached on the basis of the implication $S_3 \Rightarrow S_1$, since

$$\int_{\log\lambda}^{\infty} \exp y \, d\mathscr{L} = 1$$

for the case that $\lambda < 1$ and $\displaystyle\int_{-\infty}^{\log\lambda} \exp y \, d\mathscr{L} = 1$ for the case that $\lambda > 1$, as is seen by integration.

(iii) Let now P_n and P_n' be Poisson measures; i.e.

$$P_n(\{x\}) = \exp\left(-\lambda_n\right)\frac{\lambda_n^x}{x!},$$

$$P_n'(\{x\}) = \exp\left(-\lambda_n'\right)\frac{\lambda_n'^x}{x!}; \quad x = 0, 1, \ldots.$$

Then we have $\Lambda_n = a_n X + b_n$, where

$$a_n = \log\left(\lambda_n'/\lambda_n\right), \qquad b_n = \lambda_n - \lambda_n',$$

so that

$$\mathscr{L}_n = \mathscr{L}(\Lambda_n|P_n): P_n(\Lambda_n = y_n)$$

$$= \exp\left(-\lambda_n\right)\frac{\lambda_n^x}{x!}; \, y_n = a_n x + b_n \quad (x = 0, 1, \ldots),$$

$$\mathcal{L}'_n = \mathcal{L}(\Lambda_n | P'_n) : P'_n(\Lambda_n = y_n)$$

$$= \exp(-\lambda'_n) \frac{\lambda_n'^x}{x!} : y_n = a_n x + b_n \quad (x = 0, 1, \ldots).$$

Suppose that $\{\lambda_n\}$, $\{\lambda'_n\}$ are bounded from above and away from 0. Then, if $\{m\}$ is a subsequence of $\{n\}$ such that $\lambda_m \to \lambda$, $\lambda'_m \to \lambda'$, it follows that $\mathcal{L}_m \Rightarrow \mathcal{L}$, $\mathcal{L}'_m \Rightarrow \mathcal{L}'$, where $\mathcal{L}, \mathcal{L}'$ are probability distributions over the points

$$y = ax + b, \quad x = 0, 1, \ldots, a = \log(\lambda'/\lambda), \quad b = \lambda - \lambda'.$$

Therefore $\{\mathcal{L}_n\}$, $\{\mathcal{L}'_n\}$ are relatively compact and hence $\{P_n\}$, $\{P'_n\}$ contiguous on account of the fact that $S_2 \Rightarrow S_1$. The implication $S_3 \Rightarrow S_1$ could also be used since

$$\int \exp y \, d\mathcal{L} = \sum_{x=0}^{\infty} \exp(ax+b) \exp(-\lambda) \frac{\lambda^x}{x!}$$

$$= \exp(-\lambda') \sum_{x=0}^{\infty} \frac{1}{x!} (\lambda \exp a)^x$$

$$= \exp(-\lambda') \exp(\lambda \exp a) = 1.$$

(For another example on contiguity of sequences of probability measures, see Exercise 2.)

We recall that the main objective of this section is to establish the equivalences stated in Propostion 3.1. The implication $S_2 \Rightarrow S_1$ will be shown here (see Corollary 3.2). In the process of doing so, some auxiliary results useful in the sequel will be established. The remaining parts of the proposition are dealt with in Section 5.

LEMMA 3.1   Let $\{\mathcal{L}_n\}$, $\{\mathcal{L}'_n\}$ and C_n, D_n be defined by (3.4) and (3.2), respectively. Then if $\{\mathcal{L}_n\}$ is relatively compact (equivalently, tight), it follows that $P_n(C_n) \to 0$. Also, if $\{\mathcal{L}'_n\}$ is relatively compact (equivalently, tight), it follows that $P'_n(D_n) \to 0$.

Proof   If $\{\mathcal{L}_n\}$ is relatively compact for any determination of Λ_n, then it is so for $\Lambda_n = -n$ on C_n. This being the case, suppose that $P_n(C_n) \nrightarrow 0$. Then there exists $\{m\} \subseteq \{n\}$ such that

$$P_m(C_m) \to 2\delta > 0$$

and therefore for $m \geqslant$ some m_1,

$$\delta < P_m(C_m) \leqslant P_m(\Lambda_m = -m).$$

From this it follows by Proposition 1.1 that $\{\mathcal{L}_m\}$, and hence $\{\mathcal{L}_n\}$, is not relatively compact, a contradiction. Thus the relative compactness of $\{\mathcal{L}_n\}$ implies that $P_n(C_n) \to 0$. The relative compactness of $\{\mathcal{L}_n'\}$ implies that $P_n(D_n) \to 0$ by taking $\Lambda_n = n$ on D_n and arguing as above. ∎

COROLLARY 3.1 If $\{\mathcal{L}_n\}$ is relatively compact (equivalently, tight), it follows that $P_n(B_n) \to 1$, where B_n is given by (3.2). Also, if $\{\mathcal{L}_n'\}$ is relatively compact (equivalently, tight), then $P_n'(B_n) \to 1$.

Proof It follows from the lemma and the equations

$$1 = P_n(B_n) + P_n(C_n) = P_n'(B_n) + P_n'(D_n). ∎$$

LEMMA 3.2 Let $\mathcal{L}_n$, $\mathcal{L}_n'$ be defined by (3.4) and suppose that $\{\mathcal{L}_n\}$ is relatively compact (equivalently, tight). Then for any $A_n \in \mathcal{A}_n$ for which $P_n'(A_n) \to 0$, it follows that $P_n(A_n) \to 0$. Also, if $\{\mathcal{L}_n'\}$ is relatively compact (equivalently, tight), then for any $A_n \in \mathcal{A}_n$ for which $P_n(A_n) \to 0$, it follows that $P_n'(A_n) \to 0$.

Proof Let $P_n'(A_n) \to 0$ and suppose that $P_n(A_n) \nrightarrow 0$. Then there exists $\{m\} \subseteq \{n\}$ and $\delta > 0$ such that

$$P_m(A_m) > 3\delta, \quad m \geqslant \text{some } m_2, \quad P_m'(A_m) \to 0. \quad (3.6)$$

The relative compactness of $\{\mathcal{L}_n\}$ implies that of $\{\mathcal{L}_m\}$. Thus there exists $c = c(\delta) > 0$ such that $P_m(|\Lambda_m| < c) > 1 - \delta$ for all m, or $P_m(F_m) > 1 - \delta$ for all m, where

$$F_m = (|\Lambda_m| < c). \quad (3.7)$$

The relative compactness of $\{\mathcal{L}_n\}$ also implies that $P_m(B_m) \to 1$ by Corollary 3.1. Thus

$$P_m(B_m) > 1 - \delta \quad (m \geqslant \text{some } m_3). \quad (3.8)$$

From (3.6) and (3.8), it follows that

$$P_m(G_m) > 2\delta \quad (m \geqslant \text{some } m_4),$$

where $G_m = A_m \cap B_m$, and this, together with (3.7), gives that

$$P_m(H_m) > \delta \quad (m \geqslant m_4),$$

where $\qquad H_m = F_m \cap G_m = A_m \cap B_m \cap F_m.$ $\qquad$ (3.9)

On H_m, $\Lambda_m = \log (g_m/f_m)$ and also $|\Lambda_m| < c$ which is equivalent to

$$f_m \exp(-c) < g_m < f_m \exp c.$$

Integrating all members of the above inequalities with respect to μ_m over H_m, one has

$$P_m(H_m) \exp(-c) \leqslant P'_m(H_m) \leqslant P_m(H_m) \exp c. \qquad (3.10)$$

From (3.9) and (3.10), $P'_m(H_m) > \delta \exp(-c)$, whereas by assumption, $P'_m(H_m) \to 0$ which is a contradiction. Thus, by assuming that $P'_n(A_n) \to 0$ and utilizing the relative compactness of $\{\mathscr{L}_n\}$ alone, we were able to conclude that $P_n(A_n) \to 0$. The relative compactness of $\{\mathscr{L}'_n\}$ implies, by symmetry, that $P'_n(A_n) \to 0$ whenever $P_n(A_n) \to 0$.]

COROLLARY 3.2 If $\{\mathscr{L}_n\}$ and $\{\mathscr{L}'_n\}$ are relatively compact (equivalently, tight), it follows that $\{P_n\}$ and $\{P'_n\}$ are contiguous.

Proof This follows from the lemma and Proposition 2.1.]

The proof of the converse of this corollary, i.e. the proof of the implication $S_1 \Rightarrow S_2$, as well as that of the equivalence $S_2 \Leftrightarrow S_3$ is considerably easier if $P_n \approx P'_n$ for all sufficiently large n. This, however, need not be true. Our programme then is to replace P_n, P'_n by the probability measures Q_n, Q'_n such that $Q_n \approx Q'_n$ for all sufficiently large n and which are close to P_n, P'_n, in the sense that $\|P_n - Q_n\| + \|P'_n - Q'_n\| \to 0$. In connection with Q_n, Q'_n, one establishes the equivalence and the implication cited above and then one concludes that $S_2 \Leftrightarrow S_3$ and $S_2 \Rightarrow S_1$. For the implementation of these assertions, we need additional notation and also some auxiliary results.

For each n, let Q_n, Q'_n be probability measures on $\mathscr{A}_n$ dominated by the σ-finite measure μ_n mentioned in connection with (2.1) (e.g. $\mu_n = P_n + P'_n + Q_n + Q'_n$), and set

$$\bar{f}_n = \frac{\mathrm{d}Q_n}{\mathrm{d}\mu_n}, \quad \bar{g}_n = \frac{\mathrm{d}Q'_n}{\mathrm{d}\mu_n}. \qquad (3.11)$$

Next, define the sets $\bar{B}_n$, $\bar{C}_n$, $\bar{D}_n$, $\bar{E}_n$ as in (3.2); i.e.

$$\left.\begin{aligned}
\bar{B}_n &= \{x \in \mathscr{X}; \bar{f}_n(x)\,\bar{g}_n(x) > 0\} \\
\bar{C}_n &= \{x \in \mathscr{X}; \bar{f}_n(x) > 0, \bar{g}_n(x) = 0\} \\
\bar{D}_n &= \{x \in \mathscr{X}; \bar{f}_n(x) = 0, \bar{g}_n(x) > 0\} \\
\bar{E}_n &= \{x \in \mathscr{X}; \bar{f}_n(x) = \bar{g}_n(x) = 0\},
\end{aligned}\right\} \tag{3.12}$$

and let the r.v. (log-likelihood) $\bar{\Lambda}_n$ be defined as follows

$$\bar{\Lambda}_n = \log(\bar{g}_n/\bar{f}_n) \text{ on } \bar{B}_n \tag{3.13}$$

and arbitrary (but measurable) on $\bar{C}_n \cup \bar{D}_n \cup \bar{E}_n$. Finally, *for each determination* of $\bar{\Lambda}_n$ by (3.13), let

$$\bar{\mathscr{L}}_n = \mathscr{L}(\bar{\Lambda}_n | Q_n), \quad \bar{\mathscr{L}}'_n = \mathscr{L}'_n(\bar{\Lambda}_n | Q'_n). \tag{3.14}$$

Let $\bar{S}_i$ ($i = 1, 2, 3$), stand for statements similar to S_i ($i = 1, 2, 3$), namely,

($\bar{S}_1$) $\{Q_n\}$ and $\{Q'_n\}$ are contiguous.

($\bar{S}_2$) $\{\bar{\mathscr{L}}_n\}$ and $\{\bar{\mathscr{L}}'_n\}$ are relatively compact (for each determination of $\bar{\Lambda}_n$).

($\bar{S}_3$) $\{\bar{\mathscr{L}}_n\}$ is relatively compact (for each determination of $\bar{\Lambda}_n$) and for any $\{m\} \subseteq \{n\}$ for which $\bar{\mathscr{L}}_m \Rightarrow \bar{\mathscr{L}}$, a probability measure (depending, in general, on $\{\bar{\mathscr{L}}_m\}$), we have $\int \exp \lambda \, d\bar{\mathscr{L}} = 1$,

$$\qquad\qquad\qquad\qquad\qquad\qquad\qquad\qquad\qquad (3.15)$$

where λ is a dummy variable.

In terms of the notation introduced so far, we may then formulate the following result.

PROPOSITION 3.2   For each n, let P_n, P'_n and Q_n, Q'_n be probability measures defined on $\mathscr{A}_n$, and let $\mathscr{L}_n$, $\mathscr{L}'_n$ and $\bar{\mathscr{L}}_n$, $\bar{\mathscr{L}}'_n$ be defined by (3.4) and (3.14), respectively. Suppose that

$$\|P_n - Q_n\| + \|P'_n - Q'_n\| \to 0. \tag{3.16}$$

Then for $i = 1, 2, 3$, S_i is fulfilled if and only if $\bar{S}_i$ is, where S_i and $\bar{S}_i$ are defined by (3.5) and (3.15), respectively.

The proof of this result will follow after some lemmas and propositions have been established in the next section.

4 Some auxiliary results

The following lemma is a variant of Corollary 3.1 in that the relative compactness of $\{\mathscr{L}'_n\}$ is replaced by the assumption that $\int \exp \lambda \, d\mathscr{L} = 1$. The conclusion is the same. More precisely, one has

LEMMA 4.1 Let $\{\mathscr{L}_n\}$, defined by (3.4), be relatively compact (equivalently, tight) and suppose that for any $\{m\} \subseteq \{n\}$ for which $\mathscr{L}_m \Rightarrow \mathscr{L}$, a probability measure, one has $\int \exp \lambda \, d\mathscr{L} = 1$. Then $P_n(B_n) \to 1$ and $P'_n(B_n) \to 1$, where B_n is given by (3.2).

Proof In the first place, the relative compactness of $\{\mathscr{L}_n\}$ implies that $P_n(B_n) \to 1$ by Corollary 3.1. We next show that $P'_n(B_n) \to 1$. We have

$$\int \exp \Lambda_n \, dP_n = \int_{B_n} \exp \Lambda_n \, dP_n + \int_{C_n} \exp \Lambda_n \, dP_n$$

$$= \int_{B_n} \frac{g_n}{f_n} \, dP_n + \int_{C_n} \exp \Lambda_n \, dP_n$$

$$= P'_n(B_n) + \int_{C_n} \exp \Lambda_n \, dP_n.$$

That is,
$$\int \exp \Lambda_n \, dP_n = P'_n(B_n) + \int_{C_n} \exp \Lambda_n \, dP_n. \tag{4.1}$$

This is true for any determination of Λ_n on C_n and therefore, in particular, for $\Lambda_n = -n$. One has then, from (4.1)

$$\int \exp \lambda \, d\mathscr{L}_n = \int \exp \Lambda_n \, dP_n = P'_n(B_n) + \exp(-n) P_n(C_n). \tag{4.2}$$

In order to show that $P'_n(B_n) \to 1$, it suffices to show that for every $\{n'\} \subseteq \{n\}$ there exists $\{m\} \subseteq \{n'\}$ such that $P'_m(B_m) \to 1$. Let $\{n'\}$ be an arbitrary subsequence of $\{n\}$. Then there exists $\{r\} \subseteq \{n'\}$ such that $P'_r(C_r) \to \alpha$ ($0 \leqslant \alpha \leqslant 1$). By the relative compactness of $\{\mathscr{L}_n\}$, there exists $\{m\} \subseteq \{r\}$ such that $\mathscr{L}_m \Rightarrow \mathscr{L}$, a probability measure. For any $-\infty < x < y < \infty$ with x, y being continuity points of $\mathscr{L}$, we have

$$\int \exp \lambda \, d\mathscr{L}_m \geqslant \int_{[x, y]} \exp \lambda \, d\mathscr{L}_m,$$

so that

$$\liminf \int \exp \lambda \, d\mathscr{L}_m \geqslant \liminf \int_{[x,\,y]} \exp \lambda \, d\mathscr{L}_m$$

$$= \lim \int_{[x,\,y]} \exp \lambda \, d\mathscr{L}_m = \int_{[x,\,y]} \exp \lambda \, d\mathscr{L}.$$

The last equality holds because of Theorem 1.1. Thus we have

$$\liminf \int \exp \lambda \, d\mathscr{L}_m \geqslant \int_{[x,\,y]} \exp \lambda \, d\mathscr{L}. \qquad (4.3)$$

Letting now $x \to -\infty$, $y \to \infty$ through continuity points of $\mathscr{L}$ (see also Exercise 3) and utilizing the dominated convergence theorem, we obtain

$$\int_{[x,\,y]} \exp \lambda \, d\mathscr{L} \to \int \exp \lambda \, d\mathscr{L} = 1. \qquad (4.4)$$

Relations (4.3) and (4.4) yield

$$\liminf \int \exp \lambda \, d\mathscr{L}_m \geqslant 1. \qquad (4.5)$$

Now replacing n by m in (4.2) and taking the limits, we obtain on account of (4.5), $1 \leqslant \liminf P'_m(B_m)$, so that $P'_m(B_m) \to 1$. Therefore $P'_n(B_n) \to 1$, as was to be seen. ▮

COROLLARY 4.1 The conclusion of the lemma holds with $\mathscr{L}_n$, P_n, P'_n and B_n being replaced by $\bar{\mathscr{L}}_n$, Q_n, Q'_n and $\bar{B}_n$, respectively.

Proof Immediate from the lemma. ▮

The lemma below will be needed in the sequel.

LEMMA 4.2 Let f_n, g_n and $\bar{f}_n$, $\bar{g}_n$ be given by (3.1) and (3.11), respectively, and for any $\epsilon > 0$, define $A_n(\epsilon)$, $A'_n(\epsilon)$

$$A_n(\epsilon) = A_n = \left\{ x \in B_n \cap \bar{B}_n; \left| \frac{\bar{f}_n(x)}{f_n(x)} - 1 \right| > \epsilon \right\},$$

$$A'_n(\epsilon) = A'_n = \left\{ x \in B_n \cap \bar{B}_n; \left| \frac{\bar{g}_n(x)}{g_n(x)} - 1 \right| > \epsilon \right\},$$

where B_n and $\bar{B}_n$ are given by (3.2) and (3.12), respectively. Then if $\|P_n - Q_n\| + \|P_n' - Q_n'\| \to 0$, it follows that $P_n(A_n) + P_n'(A_n') \to 0$ and $Q_n(A_n) + Q_n'(A_n') \to 0$.

Proof   Let

$$F_n = \left\{ x \in B_n \cap \bar{B}_n; \frac{\bar{f}_n(x)}{f_n(x)} - 1 > \epsilon \right\}.$$

Then

$$Q_n(F_n) = \int_{F_n} \bar{f}_n \, \mathrm{d}\mu_n \geqslant (1+\epsilon) \int_{F_n} f_n \, \mathrm{d}\mu_n = (1+\epsilon) P_n(F_n). \quad (4.6)$$

Now suppose that $P_n(F_n) \nrightarrow 0$. Then there exists $\{m\} \subseteq \{n\}$ such that $P_m(F_m) \to 2\delta$, some $\delta > 0$. Thus for all sufficiently large m, $m \geqslant m_1$, say, $P_m(F_m) > \delta$. Then one has from (4.6)

$$Q_m(F_m) - P_m(F_m) \geqslant \epsilon P_m(F_m) \geqslant \epsilon\delta \quad (m \geqslant m_1).$$

However, this contradicts the assumption that $\|P_n - Q_n\| \to 0$. Next, let

$$G_n = \left\{ x \in B_n \cap \bar{B}_n; \frac{\bar{f}_n(x)}{f_n(x)} - 1 < -\epsilon \right\}.$$

Then it is shown in a similar fashion that $P_n(G_n) \to 0$, so that $P_n(A_n) \to 0$. That $P_n'(A_n') \to 0$ follows by symmetry on account of $\|P_n' - Q_n'\| \to 0$. Finally, the last assertion of the lemma is immediate. ▎

The following result asserts that for probability measures which are close together statements S_2 and $\bar{S}_2$ are equivalent. More precisely, one has

PROPOSITION 4.1   Consider the statements S_2 and $\bar{S}_2$ given by (3.5) and (3.15), respectively. Then if

$$\|P_n - Q_n\| + \|P_n' - Q_n'\| \to 0,$$

it follows that S_2 and $\bar{S}_2$ are equivalent.

Proof   We first show that the relative compactness of $\{\mathscr{L}_n'\}$ implies that of $\{\bar{\mathscr{L}}_n'\}$. The proof is by contradiction. In fact, if $\{\bar{\mathscr{L}}_n'\}$ is not relatively compact, there exists an

$$\epsilon > 0 \quad \text{and} \quad \{m_n\} \subseteq \{n\}$$

with $m_n \uparrow \infty$ such that $Q_{m_n}'(|\bar{\Lambda}_{m_n}| > n) > 8\epsilon, n \geqslant 1$. Setting

$$A_{m_n} = (|\bar{\Lambda}_{m_n}| > n), \quad (4.7)$$

we have then that

$$|\overline{\Lambda}_{m_n}| > n \text{ on } A_{m_n} \quad \text{and} \quad Q'_{m_n}(A_{m_n}) > 8\epsilon \quad (n \geqslant 1). \quad (4.8)$$

Let $\qquad A^+_{m_n} = (\overline{\Lambda}_{m_n} > n), \quad A^-_{m_n} = (\overline{\Lambda}_{m_n} < -n).$

Then at least one of the inequalities

$$Q'_{m_n}(A^+_{m_n}) > 4\epsilon, \quad Q'_{m_n}(A^-_{m_n}) > 4\epsilon$$

holds true for infinitely many ns, say $n_i \uparrow \infty$. For simplicity, set $r_i = m_{n_i}$ and let F_{r_i} $(i \geqslant 1)$, be sets for which one of the inequalities (always the same) $\overline{\Lambda}_{r_i} > n_i$, $\overline{\Lambda}_{r_i} < -n_i$ holds. Of course, $Q'_{r_i}(F_{r_i}) > 4\epsilon$. In other words, one has $Q'_{r_i}(F_{r_i}) > 4\epsilon$, and on

$$F_{r_i}, \overline{\Lambda}_{r_i} > n_i \quad \text{or} \quad \overline{\Lambda}_{r_i} < -n_i \quad \text{(always the same)}, \quad (i \geqslant 1). \quad (4.9)$$

Now $\qquad 4\epsilon < Q'_{r_i}(F_{r_i}) = Q'_{r_i}(F_{r_i} \cap \overline{B}_{r_i}) + Q'_{r_i}(F_{r_i} \cap \overline{D}_{r_i}).$

Hence at least one of the inequalities

$$Q'_{r_i}(F_{r_i} \cap \overline{B}_{r_i}) > 2\epsilon, \quad Q'_{r_i}(F_{r_i} \cap \overline{D}_{r_i}) > 2\epsilon$$

holds true for infinitely many is, say, $i_j \uparrow \infty$. For simplicity, set $r_{i_j} = k_j$ and consider the following two cases.

Case 1 $\qquad Q'_{k_j}(F_{k_j} \cap \overline{D}_{k_j}) > 2\epsilon \quad (j \geqslant 1).$

That is, $\qquad Q'_{k_j}(G_{k_j}) > 2\epsilon \quad (j \geqslant 1), \quad G_{k_j} = F_{k_j} \cap \overline{D}_{k_j}. \quad (4.10)$

From (4.10) and $\|P'_n - Q'_n\| \to 0$, it follows that for some j_1,

$$P'_{k_j}(G_{k_j}) > \epsilon \quad (j \geqslant j_1). \quad (4.11)$$

Now $G_{k_j} \subseteq \overline{D}_{k_i}$ and $Q_{k_j}(\overline{D}_{k_j}) = 0$. Thus $Q_{k_j}(G_{k_.}) = 0$ and then $\|P_n - Q_n\| \to 0$ implies that $P_{k_j}(G_{k_j}) \to 0$. That is, for $G_{k_j} \in \mathscr{A}_{k_j}$, one has $P_{k_j}(G_{k_j}) \to 0$, whereas $P'_{k_j}(G_{k_j}) > \epsilon \; (j \geqslant j_1)$. However, this cannot happen because of Lemma 3.2.

Case 2 $\qquad Q'_{k_j}(F_k \cap \overline{B}_{k_j}) > 2\epsilon \quad (j \geqslant 1).$

That is $\qquad Q'_{k_j}(\overline{G}_{k_j}) > 2\epsilon \quad (j \geqslant 1), \quad \overline{G}_{k_j} = F_{k_j} \cap \overline{B}_{k_j}. \quad (4.12)$

From (4.12) and $\|P'_n - Q'_n\| \to 0$, it follows that for some j_2,

$$P'_k(\overline{G}_k) > \epsilon \quad (j \geqslant j_2). \quad (4.13)$$

Next, from (4.9), it follows that on F_{k_j}, $\overline{\Lambda}_{k_j} > n'_j$ or $\overline{\Lambda}_{k_j} < -n'_j$ (always the same inequality), where $n'_j = n_{i_j} \uparrow \infty$. Suppose first that $\overline{\Lambda}_{k_j} > n'_j$. Then since $\bar{G}_{k_j} \subseteq \bar{B}_{k_j}$ on $\bar{G}_{k_j}$, we have

$$\overline{\Lambda}_{k_j} = \log \frac{\bar{g}_{k_j}}{\bar{f}_{k_j}},$$

so that $\overline{\Lambda}_{k_j} > n'_j$ is equivalent to $\bar{g}_{k_j} > \bar{f}_{k_j} \exp n'_j$. Integrating both members of this inequality with respect to μ_{k_j} over $\bar{G}_{k_j}$, one obtains $Q'_{k_j}(\bar{G}_{k_j}) \geq Q_{k_j}(\bar{G}_{k_j}) \exp n'_j$ and hence $Q_{k_j}(\bar{G}_{k_j}) \leq \exp(-n'_j)$. Letting $j \to \infty$, we have $Q_{k_j}(\bar{G}_{k_j}) \to 0$. This result, together with $\|P_n - Q_n\| \to 0$, implies that $P_{k_j}(\bar{G}_{k_j}) \to 0$. That is, for $\bar{G}_{k_j} \in \mathscr{A}_{k_j}$, one has $P_{k_j}(\bar{G}_{k_j}) \to 0$, whereas $P'_{k_j}(\bar{G}_{k_j}) > \epsilon \; (j \geq j_2)$, by (4.13). This is, however, a contradiction on the basis of Lemma 3.2. Next we show that in the present case, the inequality $\overline{\Lambda}_{k_j} < -n'_j$ may not occur except possibly for finitely many js. In fact, if $\overline{\Lambda}_{k_j} < -n'_j$, then on $\bar{G}_{k_j}$, one has as above that

$$Q'_{k_j}(\bar{G}_{k_j}) \leq Q_{k_j}(\bar{G}_{k_j}) \exp(-n'_j);$$

or, by means of (4.12), $2\epsilon \exp n'_j \leq Q_{k_j}(\bar{G}_{k_j})$. Since $n'_j \to \infty$ as $j \to \infty$, this last inequality, and therefore the original one, $\overline{\Lambda}_{k_j} < -n'_j$, cannot be true except possibly for finitely many js. Thus it has been established that (4.8) cannot occur and hence $\{\overline{\mathscr{L}}'_n\}$ is relatively compact. That relative compactness of $\{\mathscr{L}_n\}$ implies that of $\{\overline{\mathscr{L}}'_n\}$ follows similarly. This shows that S_2 implies $\bar{S}_2$. The converse follows by symmetry.▮

It is to be noted that the result just established is part of Proposition 3.2. The following results establish another part of the same proposition.

PROPOSITION 4.2 Consider the statements S_3 and $\bar{S}_3$ given by (3.5) and (3.15), respectively. Then if

$$\|P_n - Q_n\| + \|P'_n - Q'_n\| \to 0,$$

it follows that S_3 and $\bar{S}_3$ are equivalent.

Proof From the proof of the previous proposition, it follows that $\{\mathscr{L}_n\}$ is relatively compact if and only if $\{\overline{\mathscr{L}}_n\}$ is so. Let $\{\overline{\mathscr{L}}_n\}$ be relatively compact and let $\{m\} \subseteq \{n\}$ be such that $\overline{\mathscr{L}}_m \Rightarrow \overline{\mathscr{L}}$, a probability measure. Then there exists $\{r\} \subseteq \{m\}$ such that

$\mathscr{L}_r \Rightarrow \mathscr{L}$, a probability measure. We shall actually show that $\mathscr{L} = \bar{\mathscr{L}}$. The proof will then be completed by symmetry. Now in order to show that $\mathscr{L} = \bar{\mathscr{L}}$, it suffices to prove that

$$\Lambda_r - \bar{\Lambda}_r \to 0 \quad \text{in } P_r\text{-probability.} \tag{4.14}$$

This is so because $\mathscr{L}_r \Rightarrow \mathscr{L}$ and this, together with (4.14), implies that $\mathscr{L}(\bar{\Lambda}_r | P_r) \Rightarrow \mathscr{L}$ by the standard Slutsky's theorems (see also Exercise 4). This result, together with $\|P_n - Q_n\| \to 0$ and Proposition 1.2, yields the convergence $\mathscr{L}(\bar{\Lambda}_r | Q_r) \Rightarrow \mathscr{L}$. But

$$\mathscr{L}(\bar{\Lambda}_r | Q_r) = \bar{\mathscr{L}}_r \Rightarrow \bar{\mathscr{L}}.$$

Hence $\mathscr{L} = \bar{\mathscr{L}}$ by the uniqueness of weak convergence limits.

For the proof of (4.14), let $\epsilon > 0$ be arbitrary and define $\bar{A}_r$, $\bar{A}_r'$ by

$$\bar{A}_r = \bar{A}_r(\epsilon) = (\Lambda_r - \bar{\Lambda}_r > \epsilon), \quad \bar{A}_r' = \bar{A}_r'(\epsilon) = (\Lambda_r - \bar{\Lambda}_r < -\epsilon). \tag{4.15}$$

Then we have to show that $P_r(\bar{A}_r) + P_r(\bar{A}_r') \to 0$. It will be shown that $P_r(\bar{A}_r) \to 0$. The proof that $P_r(\bar{A}_r') \to 0$ follows in a similar manner. We have

$$\begin{aligned} P_r(\bar{A}_r) &= P_r(\bar{A}_r \cap B_r \cap \bar{B}_r) + P_r(\bar{A}_r \cap B_r \cap \bar{C}_r) + P_r(\bar{A}_r \cap B_r \cap \bar{D}_r) \\ &\quad + P_r(\bar{A}_r \cap B_r \cap \bar{E}_r) + P_r(\bar{A}_r \cap C_r \cap \bar{B}_r) \\ &\quad + P_r(\bar{A}_r \cap C_r \cap \bar{C}_r) + P_r(\bar{A}_r \cap C_r \cap \bar{D}_r) \\ &\quad + P_r(\bar{A}_r \cap C_r \cap \bar{E}_r). \end{aligned} \tag{4.16}$$

From the relative compactness of $\{\mathscr{L}_n\}$ and Lemma 3.1, one has that $P_r(C_r) \to 0$, so that

$$\begin{aligned} P_r(\bar{A}_r \cap C_r \cap \bar{B}_r) + P_r(\bar{A}_r \cap C_r \cap \bar{C}_r) + P_r(\bar{A}_r \cap C_r \cap \bar{D}_r) \\ + P_r(\bar{A}_r \cap C_r \cap \bar{E}_r) \to 0. \end{aligned} \tag{4.17}$$

Next, $Q_r(\bar{D}_r) = Q_r(\bar{E}_r) = 0$ and hence $P_r(\bar{D}_r) + P_r(\bar{E}_r) \to 0$ since $\|P_n - Q_n\| \to 0$. Thus

$$P_r(\bar{A}_r \cap B_r \cap \bar{D}_r) + P_r(\bar{A}_r \cap B_r \cap \bar{E}_r) \to 0. \tag{4.18}$$

Now apply Lemma 3.1 with $\{\bar{\mathscr{L}}_n\}$ and $\bar{C}_n$ instead of $\{\mathscr{L}_n\}$ and C_n, which can be done since $\{\bar{\mathscr{L}}_n\}$ is relatively compact. We then have $Q_r(\bar{C}_r) \to 0$ and this, together with $\|P_n - Q_n\| \to 0$, implies that $P_r(\bar{C}_r) \to 0$. Therefore

$$P_r(\bar{A}_r \cap B_r \cap \bar{C}_r) \to 0. \tag{4.19}$$

Relations (4.16)–(4.19) imply that $P_r(\bar{A}_r) \to 0$ if

$$P_r(F_r) \to 0, \quad F_r = \bar{A}_r \cap B_r \cap \bar{B}_r. \tag{4.20}$$

We have

$$P_r(F_r) = P_r[(\Lambda_r - \bar{\Lambda}_r > \epsilon) \cap B_r \cap \bar{B}_r] = P_r\left[\left(\frac{\bar{f}_r}{\tilde{f}_r} > \delta\frac{\bar{g}_r}{g_r}\right) \cap B_r \cap \bar{B}_r\right], \tag{4.21}$$

where $\delta = \exp\epsilon\ (> 1)$. Next for $0 < \epsilon_1 < (\delta - 1/\delta)$ arbitrary, let $\epsilon_2 > 0$ be such that $(1 + \epsilon_2/\delta) < 1 - \epsilon_1$. Then

$$P_r\left[\left(\frac{\bar{f}_r}{\tilde{f}_r} > \delta\frac{\bar{g}_r}{g_r}\right) \cap B_r \cap \bar{B}_r\right] = P_r\left[\left(\frac{\bar{f}_r}{\tilde{f}_r} > \delta\frac{\bar{g}_r}{g_r}\right) \cap A_r(\epsilon_2)\right]$$

$$+ P_r\left[\left(\frac{\bar{f}_r}{\tilde{f}_r} > \delta\frac{\bar{g}_r}{g_r}\right) \cap A_r^c(\epsilon_2)\right]$$

$$\leqslant P_r[A_r(\epsilon_2)] + P_r\left[\left(1 + \epsilon_2 > \delta\frac{\bar{g}_r}{g_r}\right) \cap B_r \cap \bar{B}_r\right]$$

$$< \epsilon_1 + P_r\left[\left(\frac{\bar{g}_r}{g_r} < \frac{1 + \epsilon_2}{\delta}\right) \cap B_r \cap \bar{B}_r\right], \tag{4.22}$$

provided r is sufficiently large; this is so by Lemma 4.2. From (4.20)–(4.22), it follows that in order to complete the proof, it suffices to show that

$$P_r(G_r) \to 0, \quad G_r = \left(\frac{\bar{g}_r}{g_r} < \frac{1 + \epsilon_2}{\delta}\right) \cap B_r \cap \bar{B}_r.$$

For contradiction, suppose that $P_r(G_r) \nrightarrow 0$. Then there exists $\{s\} \subseteq \{r\}$ such that $P_s(G_s) \to 2\delta_0$, some $\delta_0 > 0$, so that

$$P_s(G_s) > \delta_0 \quad (s \geqslant s_1) \quad \text{say.} \tag{4.23}$$

Now

$$P_r'(G_r) \leqslant P_r'\left[\left(\frac{\bar{g}_r}{g_r} < 1 - \epsilon_1\right) \cap B_r \cap \bar{B}_r\right]$$

$$\leqslant P_r'\left[\left(\left|\frac{\bar{g}_r}{g_r} - 1\right| > \epsilon_1\right) \cap B_r \cap \bar{B}_r\right] = P_r'[A_r'(\epsilon_1)] \to 0$$

by Lemma 4.2. Therefore $P_r'(G_r) \to 0$ and hence

$$P_s'(G_s) \to 0. \tag{4.24}$$

However, in view of the relative compactness of $\{\mathcal{L}_n\}$, $\{\bar{\mathcal{L}}_n\}$ and Corollary 3.2, relations (4.23) and (4.24) lead to a contradiction. The proof is completed. ∎

R EMARK 4.1   It is to be noted that in the proposition just proved, one has that if $\{r\} \subseteq \{n\}$ is such that $\mathscr{L}_r \Rightarrow \mathscr{L}$, $\bar{\mathscr{L}}_r \Rightarrow \bar{\mathscr{L}}$, where $\mathscr{L}$, $\bar{\mathscr{L}}$ are probability measures, then $\mathscr{L} = \bar{\mathscr{L}}$.

We may now conclude the proof of Proposition 3.2 as follows.

Proof of Proposition 3.2   The equivalence of S_1, $\bar{S}_1$ follows from Lemma 2.1 and the transitive property of contiguity (see Remark 2.1). The equivalence of S_2, $\bar{S}_2$ is established by Proposition 4.1, and, finally, the equivalence of S_3, $\bar{S}_3$ is the content of Proposition 4.2.∎

5   Proof of Proposition 3.1

Consider the statements S_i $(i = 1, 2, 3)$, given by (3.5) and recall that it is our purpose to prove their equivalence, thus providing additional characterizations of contiguity. In proving these equivalences, one may assume that $P_n \approx P_n'$ for all sufficiently large n. This is so because of Proposition 3.2 and also Theorem 5.1 below. For the proof of this last theorem, the following simple lemma is needed.

L EMMA 5.1   Under any one of the statements S_i $(i = 1, 2, 3)$, given by (3.5), it follows that $P_n(B_n) \to 1$, $P_n'(B_n) \to 1$, where B_n is defined in (3.2).

Proof   From (3.2) one has

$$\left.\begin{array}{l} 1 = P_n(B_n) + P_n(C_n) + P_n(D_n) + P_n(E_n) = P_n(B_n) + P_n(C_n), \\ 1 = P_n'(B_n) + P_n'(C_n) + P_n'(D_n) + P_n'(E_n) = P_n'(B_n) + P_n'(D_n). \end{array}\right\} \tag{5.1}$$

Let S_1 hold. Then by Proposition 2.1, the fact that

$$P_n(D_n) = P_n'(C_n) = 0$$

implies that $P_n'(D_n) \to 0$, $P_n(C_n) \to 0$, so that the conclusion follows from (5.1). Under S_2 and S_3, the conclusion follows from Corollary 3.1 and Lemma 4.1, respectively.∎

T HEOREM 5.1   Under any one of the statements

$$S_i \quad (i = 1, 2, 3),$$

as defined by (3.5), there exist sequences of probability measures $\{Q_n\}$ and $\{Q_n'\}$ such that

(i) $Q_n \approx Q_n'$ for all sufficiently large n.

(ii) $\|P_n - Q_n\| + \|P_n' - Q_n'\| \to 0$.

(iii) For $i = 1, 2, 3$, S_i implies $\bar{S}_i$, where the $\bar{S}_i$s are given by (3.15).

Proof Under any one of the statements S_i $(i = 1, 2, 3)$, Lemma 5.1 implies that

$$P_n(B_n) \to 1 \quad \text{and} \quad P_n'(B_n) \to 1. \qquad (5.2)$$

For all sufficiently large n (so that $P_n(B_n)$, $P_n'(B_n) > 0$), define the probability measures Q_n and Q_n' on $\mathscr{A}_n$ as follows:

$$Q_n(A) = \frac{1}{P_n(B_n)} \int_{A \cap B_n} f_n \, \mathrm{d}\mu_n, \quad Q_n'(A) = \frac{1}{P_n'(B_n)} \int_{A \cap B_n} g_n \, \mathrm{d}\mu_n. \qquad (5.3)$$

Then we claim that these probability measures satisfy the requirements of the theorem. In fact,

(i) Clearly, $Q_n(A) = 0$ if and only if $\int_{A \cap B_n} f_n \, \mathrm{d}\mu_n = 0$. Since $f_n > 0$ on B_n, this can happen only if $\mu_n(A \cap B_n) = 0$. But then $\int_{A \cap B_n} g_n \, \mathrm{d}\mu_n = 0$, which implies that $Q_n'(A) = 0$. Thus $Q_n' \ll Q_n$ and by symmetry, $Q_n \ll Q_n'$.

(ii) By (5.3), we have

$$\frac{\mathrm{d}Q_n}{\mathrm{d}\mu_n} = f_n', \quad \text{where} \quad f_n' = \frac{1}{P_n(B_n)} f_n I_{B_n}.$$

Hence by Theorem 1.4 A,

$$\|P_n - Q_n\| = \int |f_n - f_n'| \, \mathrm{d}\mu_n = \frac{1}{P_n(B_n)} \int |f_n P_n(B_n) - f_n I_{B_n}| \, \mathrm{d}\mu_n.$$

By (5.2), it suffices then to show that $\int |f_n P_n(B_n) - f_n I_{B_n}| \, \mathrm{d}\mu_n \to 0$.

We have $\int |f_n P_n(B_n) - f_n I_{B_n}| \, \mathrm{d}\mu_n$

$$= \int_{B_n} f_n |P_n(B_n) - I_{B_n}| \, \mathrm{d}\mu_n + \int_{B_n^c} f_n |P_n(B_n) - I_{B_n}| \, \mathrm{d}\mu_n$$

$$= \int_{B_n} f_n [1 - P_n(B_n)] \, \mathrm{d}\mu_n + \int_{B_n^c} f_n P_n(B_n) \, \mathrm{d}\mu_n$$

$$= P_n(B_n) - P_n^2(B_n) + P_n(B_n) \, P_n(B_n^c)$$

and this converges to $1 - 1^2 + 1 \cdot 0 = 0$; that is, $\|P_n - Q_n\| \to 0$, as was to be seen. The proof of $\|P_n' - Q_n'\| \to 0$ is entirely analogous.

(iii) This is a consequence of (ii) and Proposition 3.2.∎

We may now proceed with the proof of the equivalence of the statements S_i $(i = 1, 2, 3)$. To this effect, we have

PROPOSITION 5.1   Suppose that $P_n \approx P_n'$ for all sufficiently large n. Then the statements S_i $(i = 1, 2, 3)$, given by (3.5), are equivalent.

Proof   (i) $S_1 \Leftrightarrow S_2$. That $S_2 \Rightarrow S_1$ follows from Corollary 3.2 (without making use of the assumption $P_n \approx P_n'$). The converse is shown as follows. For $c \in R$ and all sufficiently large n, we have

$$P_n(\Lambda_n > c) = \int_{(\Lambda_n > c)} \mathrm{d}P_n = \int_{(\Lambda_n > c)} \exp(-\Lambda_n) \, \mathrm{d}P_n'$$

$$= \int_{(-\Lambda_n < -c)} \exp(-\Lambda_n) \, \mathrm{d}P_n'$$

$$\leqslant \exp(-c) \, P_n'(\Lambda_n > c) \leqslant \exp(-c).$$

(The second equality from the left is where the assumption $P_n \approx P_n'$ is used.) Thus, replacing c by $(0 <) c_n \uparrow \infty$, we have

$$P_n(\Lambda_n > c_n) \to 0, \tag{5.4}$$

which, by contiguity and Proposition 2.1, implies

$$P_n'(\Lambda_n > c_n) \to 0. \tag{5.5}$$

In a similar way, we prove that

$$P_n'(\Lambda_n < -c_n) \to 0, \tag{5.6}$$

which, again by contiguity and Proposition 2.1, implies

$$P_n(\Lambda_n < -c_n) \to 0. \tag{5.7}$$

Relations (5.4) and (5.7) imply that

$$P_n(|\Lambda_n| > c_n) \to 0.$$

Therefore, for every $\epsilon > 0$, there exists $b^* = b^*(\epsilon) > 0$ such that

$$P_n(|\Lambda_n| > b^*) < \epsilon \quad \text{for all sufficiently large } n.$$

Increasing b^* to take care of the finitely many exceptional ns and setting $b = b(\epsilon)$, we have that

$$P_n(|\Lambda_n| > b) < \epsilon \quad \text{for all } n.$$

This establishes the relative compactness of $\{\mathscr{L}_n\}$, by Proposition 1.1. The relative compactness of $\{\mathscr{L}'_n\}$ is concluded in a similar fashion, by using relations (5.5) and (5.6).

(ii) $S_2 \Leftrightarrow S_3$. We first show that $S_2 \Rightarrow S_3$. By the tightness of $\{\mathscr{L}'_n\}$, we have that, for every $\epsilon > 0$, there exists a compact set $K = K(\epsilon)$ such that $\mathscr{L}'_n(K) > 1 - \epsilon$ for all n and $\mathscr{L}(\partial K) = 0$ (see Exercise 3). This can be rewritten as $\int_K \exp \lambda \, d\mathscr{L}_n > 1 - \epsilon$ for all n, by the fact that $\mathscr{L}'_n(K) = \int_K \exp \lambda \, d\mathscr{L}_n$. (Here is one instance where the assumption $P_n \approx P'_n$ is employed.) Now $\mathscr{L}_m \Rightarrow \mathscr{L}$ implies $\int_K \exp \lambda \, d\mathscr{L}_m \to \int_K \exp \lambda \, d\mathscr{L}$, by Theorem 1.1. Therefore

$$\int_K \exp \lambda \, d\mathscr{L} \geqslant 1 - \epsilon. \tag{5.8}$$

Next, $\int \exp \lambda \, d\mathscr{L}_n = \int dP'_n = 1$ and hence

$$\int_B \exp \lambda \, d\mathscr{L}_n \leqslant 1 \quad \text{for any } B \in \mathscr{B}. \tag{5.9}$$

Let $\{K_n\}$ be a sequence of compact sets such that $K_n \uparrow R$ and such that $\mathscr{L}(\partial K_n) = 0$ for all n (e.g. take K_n to be closed intervals). Then $0 \leqslant \exp \lambda I_{K_n}(\lambda) \uparrow \exp \lambda$ and the monotone convergence theorem implies

$$\int_{K_n} \exp \lambda \, d\mathscr{L} \to \int \exp \lambda \, d\mathscr{L}. \tag{5.10}$$

But $\int_{K_n} \exp \lambda \, d\mathscr{L} = \lim \int_{K_n} \exp \lambda \, d\mathscr{L}_m$ as $m \to \infty$, and therefore

$\int_{K_n} \exp \lambda \, d\mathscr{L} \leqslant 1$ for all n, by (5.9). Relation (5.10) then gives that $\int \exp \lambda \, d\mathscr{L} \leqslant 1$. We now claim that $\int \exp \lambda \, d\mathscr{L} = 1$. This is a consequence of (5.8). The proof of the implication $S_2 \Rightarrow S_3$ is complete.

Next, we show that $S_3 \Rightarrow S_2$. To this end, let $\{m\}$ be an arbitrary subsequence of $\{n\}$. Then by the relative compactness of $\{\mathscr{L}_n\}$, there exists $\{r\} \subseteq \{m\}$ such that $\mathscr{L}_r \Rightarrow \mathscr{L}$, a probability measure. Since $\mathscr{L}'_r = \exp \lambda \mathscr{L}_r$, we have that for all real-valued, continuous functions f on R each of which vanishes outside a compact set,

$$\int f \, d\mathscr{L}'_r = \int f \exp \lambda \, d\mathscr{L}_r \to \int f \exp \lambda \, d\mathscr{L} = \int f \, d\mathscr{L}',$$

where the probability measure $\mathscr{L}'$ is defined by $\mathscr{L}' = \exp \lambda \mathscr{L}$. So $\mathscr{L}'_r \Rightarrow \mathscr{L}'$ and the proof is completed.]

We may now conclude the proof of Proposition 3.1. Namely,

Proof of Proposition 3.1   As was mentioned in the course of the proof of Proposition 5.1, the implications $S_2 \Rightarrow S_1$ has already been established in Corollary 3.2. Next in discussing the equivalences under consideration, there is no loss of generality by assuming that $P_n \approx P'_n$ for all sufficiently large n. This is so because of Theorem 5.1 and Proposition 3.2. Thus the proof is concluded by Proposition 5.1.

This section is closed with an alternative proof of the implication $S_2 \Rightarrow S_1$ (see Corollary 3.2) by making use of the additional assumption in Proposition 5.1 that $P_n \approx P'_n$ for all sufficiently large n. In the course of the proof we utilize a certain fact which also has an interesting statistical interpretation.

Consider $A_n \in \mathscr{A}_n$ with $0 < P_n(A_n) < 1$. Then there exists a test ϕ_n defined by

$$\phi_n = \begin{cases} 1 & \text{if} \quad L_n > c_n, \\ \gamma_n & \text{if} \quad L_n = c_n, \\ 0 & \text{if} \quad L_n < c_n, \end{cases} \tag{5.11}$$

where $L_n = dP'_n/dP_n$, and $0 \leqslant \gamma_n \leqslant 1$ and c_n are determined so that $\int \phi_n \, dP_n = P_n(A_n)$, with the property that

$$\int \phi_n \, dP'_n \geqslant P'_n(A_n). \tag{5.12}$$

In order to see this, we set $P_n(A_n) = \alpha_n$ and, for each n, we consider the problem of testing the hypothesis H_n^*: the underlying probability measure is P_n against the alternative A_n^*: the underlying probability measure is P_n', at level of significance α_n. Then the test defined by (5.11) is most powerful of level α_n and its power, β_n, is $\int \phi_n \, \mathrm{d}P_n'$. Now, $P_n'(A_n)$ must be $\leqslant \beta_n$ because otherwise the test ϕ_n^* defined as follows

$$\phi_n^* = \begin{cases} 1 & \text{on} \quad A_n, \\ 0 & \text{on} \quad A_n^c \end{cases}$$

is of level $\int \phi_n^* \, \mathrm{d}P_n = P_n(A_n) = \alpha_n$, and has power

$$\int \phi_n^* \, \mathrm{d}P_n' = P_n'(A_n) > \beta_n.$$

But this contradicts the optimality of ϕ_n, by the Neyman–Pearson fundamental lemma. Therefore $\int \phi_n \, \mathrm{d}P_n' \geqslant P_n'(A_n)$, as was to be seen. ∎

We are now in a position to prove the following result.

Alternative proof of Corollary 3.2 $(S_2 \Rightarrow S_1)$ under the additional assumption that $P_n \approx P_n'$ for all sufficiently large n.

Let $A_n \in \mathcal{A}_n$ be such that $P_n(A_n) \to 0$. We will show that $P_n'(A_n) \to 0$. By (5.12), it suffices to show that $\int \phi_n \, \mathrm{d}P_n' \to 0$, where ϕ_n is defined by (5.11). For $1 < c \ (< \infty)$, we have

$$\begin{aligned}
\int \phi_n \, \mathrm{d}P_n' &= \int_{(L_n \leqslant c)} \phi_n \, \mathrm{d}P_n' + \int_{(L_n > c)} \phi_n \, \mathrm{d}P_n' \\
&= \int_{(L_n \leqslant c)} \phi_n L_n \, \mathrm{d}P_n + \int_{(L_n > c)} \phi_n \, \mathrm{d}P_n' \\
&\leqslant c \int \phi_n \, \mathrm{d}P_n + \int_{(L_n > c)} \mathrm{d}P_n' \\
&= c P_n(A_n) + P_n'(\Lambda_n > \log c) \\
&\leqslant c P_n(A_n) + P_n'(|\Lambda_n| > \log c).
\end{aligned}$$

Now, by the relative compactness of $\{\mathscr{L}_n'\}$, we can choose c sufficiently large, so that $P_n'(|\Lambda_n| > \log c) < \epsilon$ for all n. This is so by Proposition 1.1. Since also $P_n(A_n) \to 0$, we get $\int \phi_n \, \mathrm{d}P_n' < 2\epsilon$. Thus $P_n'(A_n) \to 0$. By symmetry, $P_n'(A_n) \to 0$ implies $P_n(A_n) \to 0$. Then Proposition 2.1 applies and completes the proof. ∎

REMARK 5.1   Let $A_n \in \mathscr{A}_n$ with $0 < P_n(A_n) < 1$, and set $\alpha_n = P_n(A_n)$. Now suppose $\alpha_n \to 0$ and, for each n, consider the problem of testing the hypothesis H_n^*: the underlying probability measure is P_n against the alternative A_n^*: the underlying probability measure is P_n', at level of significance α_n. The optimal test is defined by (5.11) and its power, β_n, is $\beta_n = \int \phi_n \, \mathrm{d} P_n'$. Since $\alpha_n \to 0$, it would be reasonable to expect that $\beta_n \to 0$. The presence of contiguity does secure that this happens. To see this, suppose $\{P_n\}$, $\{P_n'\}$ are contiguous. Then $\{\mathscr{L}_n\}$, $\{\mathscr{L}_n'\}$ are relatively compact, as has been shown, and $\beta_n \to 0$, as was seen in the course of this proof.

6   An additional characterization of contiguity

In this section, we establish a fourth characterization of contiguity. We also present the four characterizations of contiguity together for easy reference. In proving the proposition below, it is to be noted that the (unnecessary but convenient) assumption $P_n \approx P_n'$ is not required.

PROPOSITION 6.1   Let $\{T_n\}$ be any sequence of m-dimensional random vectors. Then the sequences $\{P_n\}$ and $\{P_n'\}$ are contiguous if and only if the following happens: whenever $\{\mathscr{L}(T_n|P_n)\}$ is relatively compact (equivalently, tight), so is $\{\mathscr{L}(T_n|P_n')\}$, and conversely.

Proof   Suppose $\{P_n\}$ and $\{P_n'\}$ are contiguous and assume that $\{\mathscr{L}(T_n|P_n)\}$ is tight. Then we have to show that $\{\mathscr{L}(T_n|P_n')\}$ is also tight.

The tightness of $\{\mathscr{L}(T_n|P_n)\}$ implies that for $\epsilon_i \downarrow 0$, there exists $b_i = b_i(\epsilon_i) \uparrow \infty$ such that

$$P_n(\|T_n\| > b_i) \leqslant \epsilon_i \quad \text{for all } n, \quad (i = 1, 2, \ldots). \qquad (6.1)$$

Now suppose that $\{\mathscr{L}(T_n|P_n')\}$ is not tight. Then there exists $\epsilon > 0$ and $\{n_i\} \subseteq \{n\}$ with $n_i \uparrow \infty$ such that

$$P_{n_i}'(\|T_{n_i}\| > b_i) \geqslant \epsilon \quad (i = 1, 2, \ldots). \qquad (6.2)$$

Set $A_{n_i} = (\|T_{n_i}\| > b_i)$. Then relations (6.1) and (6.2) imply that, as $i \to \infty$,

$$P_{n_i}(A_n) \to 0 \quad \text{while} \quad P_{n_i}'(A_{n_i}) \nrightarrow 0. \qquad (6.3)$$

Now, since contiguity of $\{P_n\}$ and $\{P'_n\}$, clearly, implies contiguity of $\{P_{n_i}\}$ and $\{P'_{n_i}\}$, we arrive at a contradiction to (6.3), by Proposition 2.1. Therefore $\{\mathscr{L}(T_n|P'_n)\}$ is tight.

That tightness of $\{\mathscr{L}(T_n|P'_n)\}$ implies tightness of $\{\mathscr{L}(T_n|P_n)\}$ is proved in a symmetric way.

In order to establish the other direction of the proposition, assume that $\{\mathscr{L}(T_n|P_n)\}$ is tight if and only if $\{\mathscr{L}(T_n|P'_n)\}$ is tight, and suppose that $V_n \to 0$ in P_n-probability; here it suffices for $\{V_n\}$ to be a sequence of r.v.s such that V_n is $\mathscr{A}_n$-measurable. We would like to show that $V_n \to 0$ in P'_n-probability. Suppose that $V_n \nrightarrow 0$ in P'_n-probability. Then there exists $\{n'\} \subseteq \{n\}$ and $\epsilon > 0$ such that
$$P'_{n'}(|V_{n'}| > \epsilon) \geqslant \epsilon \quad \text{for all } n'. \tag{6.4}$$

Consider $\{n'\}$. Then, by the fact that $V_{n'} \to 0$ in $P_{n'}$-probability, it follows that there exists $\{m\} \subseteq \{n'\}$ and $a_m \downarrow 0$ such that
$$P_m(|V_m| > a_m) \to 0. \tag{6.5}$$

Set $\tilde{V}_m = V_m/a_m$. Then $\{\mathscr{L}(\tilde{V}_m|P_m)\}$ is, clearly, tight by virtue of (6.5). But $\epsilon \leqslant P'_m(|V_m| > \epsilon) = P'_m(|\tilde{V}_m| > \epsilon/a_m)$ on account of (6.4), and therefore $\{\mathscr{L}(\tilde{V}_m|P'_m)\}$ cannot be tight by Proposition 1.1. Setting $\tilde{\tilde{V}}_n = \tilde{V}_m$ for $n = m$ and equal to zero otherwise, we have that $\{\mathscr{L}(\tilde{\tilde{V}}_n|P_n)\}$ is tight while $\{\mathscr{L}(\tilde{\tilde{V}}_n|P'_n)\}$ is not. This, however, is a contradiction to our assumption. Therefore $V_n \to 0$ in P'_n-probability. That $V_n \to 0$ in P'_n-probability implies $V_n \to 0$ in P_n-probability, follows by symmetry. Thus $\{P_n\}$ and $\{P'_n\}$ are contiguous and the proof of the proposition is completed. ▮

REMARK 6.1 According to the proposition just proved, contiguity of $\{P_n\}$ and $\{P'_n\}$ expresses nearness of these sequences in the sense that they produce distributions, which either simultaneously do not escape to infinity or they simultaneously do.

This section is closed with a theorem, where the various characterizations of contiguity are gathered together for easy reference.

THEOREM 6.1 Let $\mathscr{L}_n, \mathscr{L}'_n$ be defined by (3.4), and let $\{T_n\}$ be any sequence of m-dimensional random vectors such that T_n is $\mathscr{A}_n$-measurable. Then the following statements are equivalent.

(i) $\{P_n\}$ and $\{P'_n\}$ are contiguous.

(ii) For $A_n \in \mathscr{A}_n$, $P_n(A_n) \to 0$ if and only if $P'_n(A_n) \to 0$.

(iii) $\{\mathscr{L}_n\}$ and $\{\mathscr{L}'_n\}$ are relatively compact (equivalently, tight).

(iv) $\{\mathscr{L}_n\}$ is a relatively compact (equivalently, tight) and if $\{m\} \subseteq \{n\}$ is such that $\mathscr{L}_m \Rightarrow \mathscr{L}$, a probability measure, then $\int \exp \lambda \, d\mathscr{L} = 1$.

(v) $\{\mathscr{L}(T_n | P_n)\}$ is tight if and only if $\{\mathscr{L}(T_n | P'_n)\}$ is so.

Proof It follows from Proposition 3.1 and Proposition 6.1. ▮

All four characterizations of contiguity summarized in Theorem 6.1 have been mentioned by LeCam in [4], where there is also some discussion regarding their equivalence. In connection with this, the reader is also referred to Roussas and Soms [8], Roussas [6] and LeCam [8].

7 Some results following from contiguity

In this section, we discuss some important consequences of the concept of contiguity with a view of using them in statistical applications. Although their statistical significance may not be apparent in the present context, it will become so in the subsequent chapters. In some of the following derivations, we make use of the convenient but unnecessary assumption $P_n \approx P'_n$ for all sufficiently large n. Under this assumption $\Lambda_n = \log (dP'_n/dP_n)$ is well-defined a.s. $[P_n]$, $[P'_n]$. Recall that $\mathscr{L}_n = \mathscr{L}(\Lambda_n | P_n)$, $\mathscr{L}'_n = \mathscr{L}(\Lambda_n | P'_n)$. Next, let $\{T_n\}$ be a sequence of k-dimensional random vectors such that T_n is $\mathscr{A}_n$-measurable and set

$$\tilde{\mathscr{L}}_n = \mathscr{L}[(\Lambda_n, T_n) | P_n], \quad \tilde{\mathscr{L}}'_n = \mathscr{L}[(\Lambda_n, T_n) | P'_n]. \qquad (7.1)$$

(Here and in the sequel, by (Λ_n, T_n) we mean the $(k+1)$-dimensional random vector $(\Lambda_n, T'_n)'$.) Then the first main result of this section relates to the determination of the (weak) limit of the sequence $\{\tilde{\mathscr{L}}'_n\}$ in terms of the (weak) limit of $\{\tilde{\mathscr{L}}_n\}$. More precisely, we have the following theorem.

THEOREM 7.1 Suppose $\{P_n\}$ and $\{P'_n\}$ are contiguous and let $\tilde{\mathscr{L}}_n$ and $\tilde{\mathscr{L}}'_n$ be defined by (7.1). It is further assumed that

$\mathscr{L}_n \Rightarrow \mathscr{L}$, a probability measure. Then $\mathscr{L}'_n \Rightarrow \mathscr{L}'$, where $\mathscr{L}' = \exp \lambda \mathscr{L}$ in the sense that $d\mathscr{L}'/d\mathscr{L} = \exp \lambda$.

Proof In this proof we work with sufficiently large n, so that the assumption $P_n \approx P'_n$ is satisfied. The contiguity assumption of $\{P_n\}$ and $\{P'_n\}$ implies that $\{\mathscr{L}_n\}$ is relatively compact. This is so by Theorem 6.1. Therefore there exists a subsequence $\{m\} \subseteq \{n\}$ such that $\mathscr{L}_m \Rightarrow \mathscr{L}$, a probability measure, and $\int \exp \lambda \, d\mathscr{L} = 1$ by the theorem just cited. On the other hand, by setting $\mathscr{L} = \tilde{\mathscr{L}}(\lambda, t)$, we have that $\mathscr{L}_n \Rightarrow \tilde{\mathscr{L}}$ implies $\mathscr{L}_n \Rightarrow \tilde{\mathscr{L}}(\lambda)$, where $\tilde{\mathscr{L}}(\lambda)$ is a marginal of $\tilde{\mathscr{L}}(\lambda, t)$. Hence $\tilde{\mathscr{L}}(\lambda) = \mathscr{L}(\lambda) = \mathscr{L}$, by Theorem 1.5 A, and therefore

$$\int \exp \lambda \, d\mathscr{L} = 1. \tag{7.2}$$

Next, let f be real-valued, continuous, defined on R^{k+1} and vanishing outside a compact set. Then we have

$$\int f \, d\mathscr{L}'_n = \int f(\lambda, t) \, d\mathscr{L}[(\Lambda_n, T_n)|P'_n] = \int f(\Lambda_n, T_n) \, dP'_n$$

$$= \int f(\Lambda_n, T_n) \exp \Lambda_n \, dP_n = \int f(\lambda, t) \exp \lambda \, d\mathscr{L}_n$$

$$\to \int f(\lambda, t) \exp \lambda \, d\mathscr{L} = \int f(\lambda, t) \, d\mathscr{L}',$$

where $\mathscr{L}' = \exp \lambda \mathscr{L}$.

By (7.2), it follows that $\mathscr{L}'$ is a probability measure. Since $\int f \, d\mathscr{L}'_n \to \int f \, d\mathscr{L}'$ for all real-valued, continuous functions defined on R^{k+1} each of which vanishes outside a compact set, one has that $\mathscr{L}'_n \Rightarrow \mathscr{L}'$, as was to be shown.$\rrbracket$

To this theorem we have the following two important corollaries.

COROLLARY 7.1 If $\mathscr{L}(\Lambda_n|P_n) \Rightarrow N(\mu, \sigma^2)$, then $\mu = -\frac{1}{2}\sigma^2$.

Proof By taking $T_n = 1$ in the theorem, we have that $\mathscr{L}(\Lambda_n|P'_n) \Rightarrow \exp \lambda N(\mu, \sigma^2)$ and hence

$$1 = \int \exp \lambda \, dN(\mu, \sigma^2) = \frac{1}{\sqrt{(2\pi)}\,\sigma} \int \exp \left[-\frac{(\lambda - \mu)^2}{2\sigma^2} + \lambda \right] d\lambda$$

$$= \exp \left(\frac{\sigma^2 + 2\mu}{2} \right) \frac{1}{\sqrt{(2\pi)}\,\sigma} \int \exp \left[-\frac{[\lambda - (\mu + \sigma^2)]^2}{2\sigma^2} \right] d\lambda = \exp \left(\frac{\sigma^2 + 2\mu}{2} \right).$$

From this we obtain $\exp\left(\dfrac{\sigma^2 + 2\mu}{2}\right) = 1$, so that $\mu = -\tfrac{1}{2}\sigma^2$, as was to be seen.]

COROLLARY 7.2   If $\mathscr{L}(\Lambda_n | P_n) \Rightarrow N(-\tfrac{1}{2}\sigma^2, \sigma^2)$, then

$$\mathscr{L}(\Lambda_n | P_n') \Rightarrow N(\tfrac{1}{2}\sigma^2, \sigma^2).$$

Proof   By taking again $T_n = 1$ in the theorem, we have that $\mathscr{L}(\Lambda_n | P_n') \Rightarrow \mathscr{L}'$, where $\mathscr{L}' = \tilde{\mathscr{L}}'(\lambda)$ is a marginal of

$$\tilde{\mathscr{L}}'(\lambda, t) = \tilde{\mathscr{L}}'$$

and, by Corollary 7.1, its d.f. F' is given by

$$F'(x) = \int_{-\infty}^{x} \exp \lambda \, dN(-\tfrac{1}{2}\sigma^2, \sigma^2)$$

$$= \frac{1}{\sqrt{(2\pi)}\,\sigma} \int_{-\infty}^{x} \exp\left[-\frac{(\lambda - \sigma^2/2)^2}{2\sigma^2}\right] d\lambda.$$

Thus $\mathscr{L}'$ is $N(\tfrac{1}{2}\sigma^2, \sigma^2)$, as was to be seen.]

REMARK 7.1   In the proof of Theorem 7.1, we made use of the assumption $P_n \approx P_n'$ in passing from the integral $\int f(\Lambda_n, T_n) \, dP_n'$ to the integral $\int f(\Lambda_n, T_n) \exp \Lambda_n \, dP_n$. However, the conclusion of the theorem remains true without this assumption by Theorem 5.1 and Proposition 1.2.

In terms of the notation introduced so far, we now give a description of a typical problem which arises in statistical applications. The situation is the following: it is known that $\{P_n\}$ and $\{P_n'\}$ are contiguous and that P_n' depends on $h \in R^k$ in a certain specified way. Also $\mathscr{L}(T_n | P_n) \Rightarrow N(0, \Gamma)$, where Γ is a non-singular $k \times k$ covariance matrix, while

$$\mathscr{L}(\Lambda_n | P_n) \Rightarrow N(-\tfrac{1}{2}\sigma^2, \sigma^2) \quad \text{and} \quad \Lambda_n - h'T_n \to -\tfrac{1}{2}\sigma^2$$

in P_n-probability; here $\sigma^2 = \sigma^2(h) = h'\Gamma h$. Our problem then is that of determining the asymptotic distribution of T_n under P_n'.

An outline of the solution of this problem is this: one derives the asymptotic distribution of (Λ_n, T_n) under P_n first, and then the asymptotic distribution of the same random vector (Λ_n, T_n), under P_n'. From this the desired distribution will follow.

To facilitate the derivations, first we state and prove the following result.

LEMMA 7.1   Suppose that P'_n depends on $h \in R^k$ in a specified way. Let $\{T_n\}$ be a sequence of k-dimensional random vectors such that T_n is $\mathscr{A}_n$-measurable, and assume that the following convergences hold true:

$$\mathscr{L}(T_n | P_n) \Rightarrow N(0, \Gamma), \tag{7.3}$$

where Γ is a $k \times k$ non-singular covariance matrix, and

$$\Lambda_n - h'T_n \to -\tfrac{1}{2}\sigma^2 \quad \text{in } P_n\text{-probability}, \tag{7.4}$$

where $\sigma^2 = \sigma^2(h) = h'\Gamma h$.

Then $\mathscr{L}[(\Lambda_n, T_n) | P_n]$ converges (weakly) to a $((k+1)$-dimensional) normal law with mean and covariance given by

$$(-\tfrac{1}{2}\sigma^2, 0, \ldots, 0)' \quad \text{and} \quad \begin{pmatrix} \sigma^2 & h'\Gamma \\ \Gamma h & \Gamma \end{pmatrix}, \quad \text{respectively.}$$

Proof   In order to somewhat simplify the notation, we write c for $-\tfrac{1}{2}\sigma^2$. For $u \in R$ and $v \in R^k$, we have

$$iu\Lambda_n + iv'T_n = iu(\Lambda_n - h'T_n - c) + i(v + uh)'T_n + iuc.$$

Then

$$I_n = \left| \int \exp(iu\Lambda_n + iv'T_n)\, dP_n - \int \exp[i(v + uh)'T_n + iuc]\, dP_n \right|$$

$$= \left| \int \{\exp[iu(\Lambda_n - h'T_n - c) + i(v + uh)'T_n + iuc] \right.$$

$$\left. - \exp[i(v + uh)'T_n + iuc]\} \, dP_n \right|$$

$$= \left| \int \exp[i(v + uh)'T_n + iuc]\{\exp[iu(\Lambda_n - h'T_n - c) - 1]\}\, dP_n \right|$$

$$\leqslant \int |\exp(iuZ_n) - 1|\, dP_n,$$

where we set $Z_n = \Lambda_n - h'T_n - c. \tag{7.5}$

But, for $\epsilon > 0$,

$$\int |\exp(iuZ_n) - 1|\, dP_n = \int_{(|Z_n| \leqslant \epsilon)} |\exp(iuZ_n) - 1|\, dP_n$$

$$+ \int_{(|Z_n| > \epsilon)} |\exp(iuZ_n) - 1|\, dP_n$$

$$\leqslant |u|\, \epsilon + P_n(|Z_n| > \epsilon).$$

Since this is true for every $\epsilon > 0$, equation (7.5) and assumption (7.4) show that
$$I_n \to 0. \tag{7.6}$$

Next relation (7.3) (see also Exercises 4 and 5) implies that
$$\mathscr{L}[(v+uh)'\, T_n + uc\,|\,P_n] \Rightarrow N[uc, (v+uh)'\, \Gamma(v+uh)]$$
from which we obtain
$$\int \exp\left[i(v+uh)'\, T_n + iuc\right] dP_n \to \exp\left[iuc - \tfrac{1}{2}(v+uh)'\, \Gamma(v+uh)\right]. \tag{7.7}$$

Relations (7.6) and (7.7) taken together imply that the characteristic function of (Λ_n, T_n) converges to the function
$$\exp\left[iuc - \tfrac{1}{2}(v+uh)'\, \Gamma(v+uh)\right]. \tag{7.8}$$
But
$$uc = \begin{pmatrix} u \\ v \end{pmatrix}' (c, 0, ..., 0)' \quad\text{and}\quad (v+uh)'\, \Gamma(v+uh)$$
$$= \begin{pmatrix} u \\ v \end{pmatrix}' \begin{pmatrix} \sigma^2 & h'\Gamma \\ \Gamma h & \Gamma \end{pmatrix} \begin{pmatrix} u \\ v \end{pmatrix},$$
as is easily seen by utilizing the fact that $\sigma^2 = h'\Gamma h$.

Therefore the function in (7.8) is the characteristic function of the $(k+1)$-dimensional normal distribution with mean $(c, 0, ..., 0)'$ and covariance
$$\begin{pmatrix} \sigma^2 & h'\Gamma \\ \Gamma h & \Gamma \end{pmatrix},$$
and this completes the proof of the lemma. ∎

For later use, we denote by $\mathscr{L}$ the $(k+1)$-dimensional normal distribution with mean and covariance as above, that is,
$$\mathscr{L} = N\left[(-\tfrac{1}{2}\sigma^2, 0, ..., 0)', \begin{pmatrix} \sigma^2 & h'\Gamma \\ \Gamma h & \Gamma \end{pmatrix}\right]. \tag{7.9}$$

By an application of Theorem 7.1, one has the following corollary to the lemma just proved.

COROLLARY 7.3 Let $\mathscr{L}$ be defined by relation (7.9). Then, if $\{P_n\}$, $\{P'_n\}$ are contiguous,
$$\mathscr{L}[(\Lambda_n, T_n)\,|\,P'_n] \Rightarrow \tilde{\mathscr{L}}',$$
where $\tilde{\mathscr{L}}' = \exp\lambda\tilde{\mathscr{L}}$.

We are now in a position to prove the second main result of this section.

THEOREM 7.2   Under the same assumptions as those in Lemma 7.1, and the condition that $\{P_n\}$ and $\{P'_n\}$ are contiguous, we have

$$\mathscr{L}(T_n|P'_n) \Rightarrow N(\Gamma h, \Gamma) \quad (h \in R^k).$$

Proof It is convenient to use also the notation $\mathscr{L}(\lambda, t)$ for the normal distribution defined by (7.9), $\mathscr{L}(\lambda|t)$ for the conditional distribution, given t, and $\mathscr{L}(t)$ for the normal distribution with mean 0 and covariance Γ. It is well known that $\mathscr{L}(\lambda|t)$ is normal with mean given by

$$-\tfrac{1}{2}\sigma^2 + \Sigma_{12}\Sigma_{22}^{-1}(t-0) = -\tfrac{1}{2}h'\Gamma h + h'\Gamma\Gamma^{-1}t = -\tfrac{1}{2}h'\Gamma h + h't$$

and variance given by

$$\Sigma_{11} - \Sigma_{12}\Sigma_{22}^{-1}\Sigma_{21} = \sigma^2 - h'\Gamma\Gamma^{-1}\Gamma h = -h'\Gamma h + h'\Gamma h = 0.$$

(For this, see, e.g. Rao [3], p. 441 (v).)

We have then

$$\int \exp \lambda \, \mathrm{d}\mathscr{L} = \int \exp \lambda \, \mathscr{L}(\mathrm{d}\lambda, \mathrm{d}t) = \int \exp \lambda \, \mathscr{L}(\mathrm{d}\lambda|t)\, \mathscr{L}(\mathrm{d}t).$$

$$(7.10)$$

Taking into consideration the form of $\mathscr{L}(\lambda|t)$, the first integration in (7.10) with respect to λ over the interval $(-\infty, \lambda']$, gives

$$\int_{-\infty}^{\lambda'} \exp \lambda \, \mathscr{L}(\mathrm{d}\lambda|t) = \begin{cases} \exp\left(-\tfrac{1}{2}h'\Gamma h + h't\right) & \text{if} \quad \lambda' \geqslant -\tfrac{1}{2}h'\Gamma h + h't \\ 0 & \text{otherwise.} \end{cases}$$

Next $P'_n(\Lambda_n \leqslant \lambda', T_n \leqslant t') \to \displaystyle\int_{-\infty}^{\lambda'}\int_{-\infty}^{t'} \exp \lambda \, \mathscr{L}(\mathrm{d}\lambda, \mathrm{d}t),$

by Corollary 7.3, and this implies that

$$P'_n(T_n \leqslant t') \to \int_{-\infty}^{t'}\int_{-\infty}^{\infty} \exp \lambda \, \mathscr{L}(\mathrm{d}\lambda, \mathrm{d}t)$$

$$= \int_{-\infty}^{t'}\left[\int_{-\infty}^{\infty} \exp \lambda \, \mathscr{L}(\mathrm{d}\lambda|t)\right]\mathscr{L}(\mathrm{d}t)$$

$$= \int_{-\infty}^{t'} \exp\left(-\tfrac{1}{2}h'\Gamma h + h't\right)\mathscr{L}(\mathrm{d}t)$$

$$= \int_{-\infty}^{t'} (2\pi)^{-\frac{1}{2}k}|\Gamma|^{-\frac{1}{2}}\exp\left[-\tfrac{1}{2}(t-\Gamma h)'\,\Gamma^{-1}(t-\Gamma h)\right]\mathrm{d}t,$$

where $|\Gamma|$ denotes the determinant of Γ. Therefore

$$\mathscr{L}(T_n|P_n') \Rightarrow N(\Gamma h, \Gamma),$$

as was to be seen.]

REMARK 7.2 In [1], p. 202, Hájek and Šidák define contiguity as follows: $\{P_n'\}$ is said to be contiguous to $\{P_n\}$ if, for $A_n \in \mathscr{A}_n$, $P_n(A_n) \to 0$ implies $P_n'(A_n) \to 0$. However, in this work we have adopted and used throughout the more symmetric definition of contiguity as was introduced by LeCam [4], p. 41.

Exercises

1. Consider the measurable space $(\Omega, \mathscr{F})$ and let P, Q be two probability measures on $\mathscr{F}$ dominated by a σ-finite measure μ (e.g. $\mu = P + Q$). Set $f = \mathrm{d}P/\mathrm{d}\mu$, $g = \mathrm{d}Q/\mathrm{d}\mu$ and define the quantities d and ρ as follows

$$d = \|P - Q\| = \int |f - g|\, \mathrm{d}\mu, \quad \rho = \int (fg)^{\frac{1}{2}}\, \mathrm{d}\mu.$$

Then show that $\quad 2(1 - \rho^2)^{\frac{1}{2}} \geqslant d \geqslant 2(1 - \rho).$

(See, e.g. Lemma 1 in Kraft [1].)

2. For each integer $n \geqslant 1$, consider the measurable space $(\mathscr{X}, \mathscr{A}_n)$. Also consider the probability density

$$f(x; \theta) = \frac{\Gamma(\theta + 2)}{\Gamma(\theta + 1)} x^\theta \quad (0 < x < 1, \theta > -1)$$

and for $\theta_n \to 2$, $\theta_n' \to 0$, let P_{θ_n}, P_{θ_n}' be the probability measures on $\mathscr{A}_n$ determined by $f(\cdot; \theta_n), f(\cdot; \theta_n')$, respectively. Determine whether or not the sequences $\{P_n\}$ and $\{P_n'\}$ are contiguous.

3. Consider the probability space $(\Omega, \mathscr{F}, P)$ and suppose that $(A_i, i \in I)$ is a collection of events such that $A_i \cap A_j = \varnothing$ for $i \neq j$; I is an uncountable index set. Then show that $P(A_i)$ is positive for only countably many events.

4. For each integer $n \geqslant 1$, let X_n, Y_n be r.v.s defined on the probability space $(\Omega_n, \mathscr{F}_n, P_n)$ and let X be an r.v. defined on the probability space $(\Omega, \mathscr{F}, P)$. Suppose that $\mathscr{L}(X_n|P_n) \Rightarrow \mathscr{L}(X|P)$. Then:

(i) If $Y_n - X_n \to c$ in P_n-probability, it follows that

$$\mathscr{L}(Y_n|P_n) \Rightarrow \mathscr{L}(X+c|P).$$

(ii) If $Y_n \to c \ (> 0)$ in P_n-probability, it follows that

$$\mathscr{L}(X_n Y_n|P_n) \Rightarrow \mathscr{L}(cX|P).$$

5. For each integer $n \geqslant 1$, consider the probability space $(\Omega_n, \mathscr{F}_n, P_n)$ and let T_n be an $\mathscr{A}_n$-measurable k-dimensional random vector such that $\mathscr{L}(T_n|P_n) \Rightarrow N(\mu, \Gamma)$. Then for any (constant) k-dimensional vector c, one has that

$$\mathscr{L}(c'T_n|P_n) \Rightarrow N(c'\mu, c'\Gamma c).$$

2. *Asymptotic expansion and asymptotic distribution of likelihood functions*

Summary

In this chapter, some fundamental results regarding the asymptotic expansion, in the probability sense, and also the asymptotic distribution of certain likelihood functions are derived. These results constitute the backbone of the remaining chapters in this monograph and their derivation rests heavily on the material discussed in Chapter 1.

The underlying probability model involved in our discussions is that of a Markov process satisfying certain reasonable regularity conditions. This model includes, of course, as a special but important case the model consisting of independent identically distribution (i.i.d.) random variables (r.v.s) which is assumed more often in statistical literature.

We now proceed to present a brief outline of what is done in this chapter, since the various derivations are rather involved and the reader might lose sight of the essence of the results. In Section 2, we gather together the various assumptions which are used in the present chapter and which also are basic for what is discussed in the subsequent chapters. The new element here is the assumption of differentiability in quadratic mean of the square root of the probability density function. It replaces the assumption usually made in statistical literature about the existence of two or three pointwise derivatives of the logarithm of the density. As is shown by LeCam [6], the classical Cramér-type assumptions imply the one made here. The underlying conditions are then verified in a number of examples which are used throughout this monograph for illustrative purposes. Additional examples where these conditions were found to hold true may be found in Lind and Roussas [1] and Philippou [1].

In anticipation of the statistical applications discussed in Chapters 4, 5 and 6, it should be mentioned here that the covariance function $\Gamma(\theta)$ defined by (2.1) plays the role of Fisher's information matrix $I(\theta)$ under the standard assumptions. Also the random vector $\Delta_n(\theta)$ defined by (4.2) replaces the maximum likelihood estimate when θ is replaced by a known parameter point θ_0.

Section 4 is concerned with the asymptotic behaviour of the log-likelihood function corresponding to an arbitrary but fixed parameter point θ and a 'neighbouring' point. It is shown in Theorem 4.1 (see also Remark 4.1) that asymptotically and in the neighbourhood of any parameter point the likelihood function behaves as if it were an exponential family. Actually, this is an involved assertion and its precise formulation and rigorous justification is the main objective of Chapter 3. The statistical implications of this fact are apparent: in the neighbourhood of each parameter point and for certain statistical purposes, the given family of probability measures may be used as if it were an exponential family.

From a statistical point of view, one is, clearly, interested in the asymptotic distribution of the log-likelihood function and the random vector $\Delta_n(\theta)$ mentioned above. These distributions are provided by Theorems 4.2–4.6. The proofs of these theorems are given in Section 6 and they are based on a series of lemmas established in Section 5.

1 Preliminaries

In this section, we introduce the basic notation and definitions required for the formulation of the assumptions made below. To start with, let $(\mathscr{X}, \mathscr{A})$ be a measurable space and suppose Θ is a k-dimensional open subset of the Euclidean space R^k, k a positive integer. The underlying topology in R^k is the usual one. The pair $(R, \mathscr{B})$ stands for the Borel real line and for each $\theta \in \Theta$, one assumes the existence of a probability measure, denoted by $p_\theta(\cdot)$ and defined on $\mathscr{B}$ and also of a transition probability measure $p_\theta(\cdot, \cdot)$ defined on $R \times \mathscr{B}$. Then by the Kolmogorov consistency theorem (see, e.g. Loève [1], p. 93),

$p_\theta(\cdot)$ and $p_\theta(\cdot, \cdot)$ determine a probability measure on the product σ-field $\prod_{j=1}^{\infty} \mathscr{B}_j$ of subsets of $\prod_{j=1}^{\infty} R_j$, where $R_j = R$ and $\mathscr{B}_j = \mathscr{B}$ for all j. Let P_θ denote the so-defined probability measure. Without loss of generality (see, e.g. Doob [1], p. 14, and other references cited there), we may and shall assume from now on that $(\mathscr{X}, \mathscr{A}) = \left(\prod_{j=1}^{\infty} R_j, \prod_{j=1}^{\infty} \mathscr{B}_j \right)$. Thus for each $\theta \in \Theta$, one has the probability space $(\mathscr{X}, \mathscr{A}, P_\theta)$.

Next let $\{X_n\}$, $n \geqslant 0$, n an integer, be the coordinate process defined on $(\mathscr{X}, \mathscr{A})$; i.e. $X_n(\omega) = x_n$, where $\omega = (x_1, x_2, \ldots) \in \mathscr{X}$. Then it is well known that for each $\theta \in \Theta$, this process is a Markov process with initial distribution $p_\theta(\cdot)$ and transition measure $p_\theta(\cdot, \cdot)$. This is the stochastic process we are going to deal with from now on.

Suppressing the random element ω, let $g(\theta)$ be an r.v. defined on $(\mathscr{X}, \mathscr{A}, P)$ for each $\theta \in \Theta$, where P is some probability measure on $\mathscr{A}$. Then we can give the following definition which will be needed in the formulation of our assumptions.

DEFINITION 1.1   Suppose $\int g^2(\theta) \, dP < \infty$. The random function (r.f.) g is said to be *differentiable in quadratic mean* (q.m.) at θ when the probability measure P is employed if

$$\frac{1}{\|h\|} [g(\theta + h) - g(\theta) - h' \dot{g}(\theta)] \to 0 \quad \text{in q.m. } [P] \text{ as} \quad (0 \neq) \|h\| \to 0,$$

where $\|\cdot\|$ is the usual Euclidean norm and $\dot{g}(\theta)$ is the derivative in q.m. (a $k \times 1$ r. vector) of $g(\theta)$ at θ. Here and in similar situations, ''' denotes transpose and $h' \dot{g}(\theta)$ is, of course, the inner product of the vectors indicated.

REMARK 1.1   It is clear that Definition 1.1 is equivalent to the following one, which is used more conveniently in actually checking differentiability in q.m. Namely,

DEFINITION 1.1'   Suppose $\int g^2(\theta) \, dP < \infty$. Then the r.f. g is said to be *differentiable in q.m.* at θ when P is employed if

$$\frac{1}{\lambda} [g(\theta + \lambda h) - g(\theta)] \to h' \dot{g}(\theta) \quad \text{in q.m. } [P] \text{ as} \quad (0 <) \lambda \to 0$$

uniformly on bounded sets of values of h; again $\dot{g}(\theta)$ is the derivative in q.m. of $g(\theta)$ at θ (see also Exercise 1).

Now, in the case that Θ is a one-dimensional set, $\dot{g}(\theta)$ is simply an r.v. Then it is clear that Definition 1.1 is equivalent to the following one.

DEFINITION 1.1″   Suppose $\int g^2(\theta)\,\mathrm{d}P < \infty$. Then the r.f. g is said to be *differentiable in q.m.* at θ when P is employed if

$$\frac{1}{h}[g(\theta+h) - g(\theta)] \to \dot{g}(\theta) \quad \text{in q.m. } [P] \text{ as} \quad (0 \neq)\,h \to 0,$$

where $\dot{g}(\theta)$ is the derivative in q.m. of $g(\theta)$ at θ.

Let now $\mathscr{A}_n$ be the σ-field induced by the r.v.s $\{X_0, X_1, \ldots, X_n\}$ and denote by $P_{n,\theta}$ the restriction of P_θ to $\mathscr{A}_n$. It will be assumed below that for any two $\theta, \theta^* \in \Theta$, $P_{n,\theta}$ and P_{n,θ^*} are mutually absolutely continuous for all n. Therefore we may set

$$\left.\begin{aligned} [\mathrm{d}P_{0,\theta^*}/\mathrm{d}P_{0,\theta}] &= q(X_0; \theta, \theta^*), \\ [\mathrm{d}P_{1,\theta^*}/\mathrm{d}P_{1,\theta}] &= q(X_0, X_1; \theta, \theta^*), \end{aligned}\right\} \tag{1.1}$$

where $q(X_0; \theta, \theta^*)$ and $q(X_0, X_1; \theta, \theta^*)$ are well-defined except perhaps on a null set. We also set

$$q(X_j | X_{j-1}; \theta, \theta^*) = q(X_{j-1}, X_j; \theta, \theta^*)/q(X_{j-1}; \theta, \theta^*) \tag{1.2}$$

and $\qquad\qquad \phi_j(\theta, \theta^*) = [q(X_j | X_{j-1}; \theta, \theta^*)]^{\frac{1}{2}}. \tag{1.3}$

It follows that

$$[\mathrm{d}P_{n,\theta^*}/\mathrm{d}P_{n,\theta}] = q(X_0; \theta, \theta^*) \prod_{j=1}^{n} \phi_j^2(\theta, \theta^*).$$

Of course, the quantities defined by (1.1) and (1.2) are simply the initial and transition density, respectively, of the process, both considered as r.v.s (which are obtained by replacing the arguments x_{j-1} and x_j by the r.v.s X_{j-1} and X_j, respectively). Then the quantity defined by (1.4) is the $(n+1)$-dimensional density of the process, again considered as an r.v.

2 Assumptions

In this section all of the assumptions, under which various results are going to be derived, are gathered together for easy reference.

(A1) For each $\theta \in \Theta$, the Markov process $\{X_n\}$, $n \geqslant 0$ is (strictly) stationary and metrically transitive (ergodic).

(A2) The probability measures $\{P_{n,\theta}; \theta \in \Theta\}$ are mutually absolutely continuous for all $n \geqslant 0$.

(A3) (i) For each $\theta \in \Theta$, the r.f. $\phi_1(\theta, \theta^*)$ is differentiable in q.m. with respect to θ^* at the point (θ, θ) when P_θ is employed.

Let $\dot\phi_1(\theta)$ be the derivative in q.m. of $\phi_1(\theta, \theta^*)$ with respect to θ^* at (θ, θ). Then

(ii) $\dot\phi_1(\theta)$ is $\mathscr{A}_1 \times \mathscr{C}$-measurable, where $\mathscr{C}$ is the σ-field of Borel subsets of Θ.

Let $\Gamma(\theta)$ be the covariance function defined by

$$\Gamma(\theta) = 4\mathscr{E}_\theta[\dot\phi_1(\theta)\,\dot\phi_1'(\theta)]. \tag{2.1}$$

Then

(iii) $\Gamma(\theta)$ is positive definite for every $\theta \in \Theta$.

(A4) (i) For each $\theta \in \Theta$, $q(X_0, X_1; \theta, \theta^*) \to 1$ in $P_{1,\theta}$-probability as $\theta^* \to \theta$.

(ii) For each fixed $\theta \in \Theta$, $q(X_0; \theta, \theta^*)$ is $\mathscr{A}_0 \times \mathscr{C}$-measurable and $q(X_0, X_1; \theta, \theta^*)$ is $\mathscr{A}_1 \times \mathscr{C}$-measurable.

COMMENTS ON THE ASSUMPTIONS For the definition of strict stationarity and ergodicity, the reader is referred to, e.g. Doob [1], p. 191 and p. 460.

Although (A1) is a basic assumption, (A2) is made to simplify the derivations. One could dispense with it.

From the definition of $\phi_1(\theta, \theta^*)$ by (1.3), one notices that $\int \phi_1^2(\theta, \theta^*)\,\mathrm{d}P_\theta$ is finite and indeed equal to 1; therefore it makes sense to talk about differentiability in q.m. (see Definition 1.1). Actually this is the reason for considering the square root of the transition density rather than the density itself. Furthermore it is felt that differentiability in q.m. is more probabilistic an assumption than the classical assumption of pointwise differentiability. The remaining parts of (A3), are in the nature of regularity conditions which are likely to be satisfied in most

cases of practical importance. Finally notice that in forming the
quotient $\frac{1}{\|h\|}[\phi_1(\theta, \theta + h) - \phi_1(\theta, \theta)]$ (or $\frac{1}{\lambda}[\phi_1(\theta, \theta + \lambda h) - \phi_1(\theta, \theta)]$)
the term $\phi_1(\theta, \theta) = 1$.

As for (A4)(i), this is evidently an exceedingly mild restriction
on the two-dimensional density of the process. Assumption
(A4)(ii) is of the same nature as (A3)(ii) and is, actually used
only in the proof of Lemma 7.1 in Chapter 5.

Assumptions (A1–4) take a simpler form if the underlying
process consists of i.i.d. r.v.s. They are listed below for the sake
of record.

In the first place, (A1) is automatically satisfied. For this, the
reader is referred to Doob [1], Example 2 and Theorem 1.2,
p. 460.

Next, in the present case, clearly,

$$\phi_j(\theta, \theta^*) = [q(X_j; \theta, \theta^*)]^{\frac{1}{2}} \tag{2.2}$$

and (A4)(i) follows from (A3)(i), since differentiability in q.m.
obviously implies continuity in probability. Thus the simplified
assumptions are as follows.

(A1') The probability measures $\{P_{n,\theta}; \theta \in \Theta\}$ are mutually
absolutely continuous for all $n \geq 1$.

(A2')(i) For each $\theta \in \Theta$, the r.f. $\phi_1(\theta, \theta^*)$ defined by (2.2) is
differentiable in q.m. with respect to θ^* at the point (θ, θ) when
P_θ is employed. Let $\dot{\phi}_1(\theta)$ be the derivative in q.m. of $\phi_1(\theta, \theta^*)$
with respect to θ^* at (θ, θ). Then

(ii) $\dot{\phi}_1(\theta)$ is $X_1^{-1}(\mathscr{B}) \times \mathscr{C}$-measurable.

Let $\Gamma(\theta)$ be the covariance function defined by (2.1), where
$\dot{\phi}_1(\theta)$ is given by (2.2). Then

(iii) $\Gamma(\theta)$ is positive definite for every $\theta \in \Theta$.

(iv) For each fixed $\theta \in \Theta$, $q(X_0; \theta, \theta^*)$ is $\mathscr{A}_0 \times \mathscr{C}$-measurable.

In order to avoid trivial repetition, in what follows limits are
always taken as $\{n\}$, or subsequences thereof, converges to ∞
unless otherwise specified.

All results in this monograph will be derived under assumptions (A1–4) (or (A1'–2') for the i.i.d. case) and this will not be
mentioned explicitly again. Whenever additional assumptions
are required, they will be spelled out explicitly.

3 Some examples

In this section, a number of examples are presented which are shown to satisfy assumptions (A 1–4), or (A 1′–2′) for the i.i.d. case. Some versions of these examples are also discussed in Roussas [2] and Johnson and Roussas [1]. In showing differentiability in q.m., Theorem 2.1 A (Vitali's theorem) is instrumental.

3.1 Let $\{X_n\}$, $n \geqslant 1$ be i.i.d. from $N(\theta, \sigma^2)$, $\theta \in R$, where σ is assumed to be known.

Then (A 1′) is evidently true. Next it is easily seen that

$$\phi_1(\theta, \theta^*) = \exp\left\{\frac{1}{2\sigma^2}[(\theta^* - \theta)X_1 - \tfrac{1}{2}(\theta^{*2} - \theta^2)]\right\}.$$

The (pointwise) derivative of $\phi_1(\theta, \theta^*)$ with respect to θ^* at (θ, θ) is equal to $\dfrac{1}{2\sigma^2}(X_1 - \theta)$, while

$$\mathscr{E}_\theta\left[\frac{1}{2\sigma^2}(X_1 - \theta)\right]^2 = \frac{1}{4\sigma^2}. \tag{3.1}$$

Also $\mathscr{E}_\theta\left\{\dfrac{1}{h}[\phi_1(\theta, \theta + h) - 1]\right\}^2 = \dfrac{2}{h^2}[1 - \mathscr{E}_\theta \phi_1(\theta, \theta + h)],$

where

$$\phi_1(\theta, \theta + h) = \exp\left[-\frac{1}{2\sigma^2}\left(\frac{h^2}{2} + \theta h\right)\right]\exp\left(\frac{1}{2\sigma^2}hX_1\right).$$

But $\mathscr{E}_\theta \exp\left(\dfrac{1}{2\sigma^2}hX_1\right) = \exp\left[\dfrac{1}{2\sigma^2}\left(\dfrac{h^2}{4} + \theta h\right)\right],$

so that $\mathscr{E}_\theta \phi_1(\theta, \theta + h) = \exp\left(-\dfrac{h^2}{8\sigma^2}\right)$

and therefore

$$\mathscr{E}_\theta\left\{\frac{1}{h}[\phi_1(\theta, \theta + h) - 1]\right\}^2 = \frac{2}{h^2}\left[1 - \exp\left(-\frac{h^2}{8\sigma^2}\right)\right]$$

$$= \frac{2}{h^2} \cdot \frac{h^2}{8\sigma^2}[1 - \mathrm{o}(1)] = \frac{1}{4\sigma^2}[1 - \mathrm{o}(1)]$$

$$\to \frac{1}{4\sigma^2} \quad \text{as} \quad h \to 0.$$

This result together with (3.1) and Theorem 2.1 A, implies that

$$\dot{\phi}_1(\theta) = \frac{1}{2\sigma^2}(X_1 - \theta). \tag{3.2}$$

Furthermore by (2.1) and (3.1),

$$\Gamma(\theta) = 4\mathscr{E}_\theta[\dot{\phi}_1(\theta)]^2 = \frac{1}{\sigma^2}, \tag{3.3}$$

so that (A 2′) (i)–(iii) are fulfilled.

3.2 Let $\{X_n\}$, $n \geqslant 1$ be i.i.d. from $N(\mu, \theta)$, $\theta > 0$, where μ is assumed to be known.

Since (A 1′) obviously holds true, it suffices to check (A 2′). We have

$$\phi_1(\theta, \theta^*) = \left(\frac{\theta}{\theta^*}\right)^{\frac{1}{4}} \exp\left[\frac{(X_1 - \mu)^2}{4}\left(\frac{1}{\theta} - \frac{1}{\theta^*}\right)\right]$$

and the (pointwise) derivative of $\phi_1(\theta, \theta^*)$ with respect to θ^* at (θ, θ) is equal to

$$-\frac{1}{4\theta} + \frac{(X_1 - \mu)^2}{4\theta^2}. \tag{3.4}$$

Furthermore $\quad \mathscr{E}_\theta\left[-\frac{1}{4\theta} + \frac{(X_1 - \mu)^2}{4\theta^2}\right]^2 = \frac{1}{8\theta^2}, \tag{3.5}$

as is easily seen. Next

$$\phi_1(\theta, \theta + h) = \left(\frac{\theta}{\theta + h}\right)^{\frac{1}{4}} \exp\left[\frac{(X_1 - \mu)^2}{4}\left(\frac{1}{\theta} - \frac{1}{\theta + h}\right)\right],$$

so that $\quad \mathscr{E}_\theta\phi_1(\theta, \theta + h) = \left[\frac{4\theta(\theta + h)}{(2\theta + h)^2}\right]^{\frac{1}{4}}.$

Therefore

$$\mathscr{E}_\theta\left\{\frac{1}{h}[\phi_1(\theta, \theta + h) - \phi_1(\theta, \theta)]\right\}^2 = \frac{2}{h^2}[1 - \mathscr{E}_\theta\phi_1(\theta, \theta + h)]$$

$$= \frac{2}{h^2}\left\{1 - \left[\frac{4\theta(\theta + h)}{(2\theta + h)^2}\right]^{\frac{1}{4}}\right\}$$

$$\rightarrow \frac{1}{8\theta^2} \quad \text{as} \quad h \rightarrow 0, \tag{3.6}$$

as is easily checked.

Relations (3.4)–(3.6) and Theorem 2.1 A show that

$$\dot{\phi}_1(\theta) = -\frac{1}{4\theta} + \frac{(X_1 - \mu)^2}{4\theta^2}. \tag{3.7}$$

Since also by means of (2.1) and (3.5),

$$\Gamma(\theta) = 4\mathscr{E}_\theta[\dot{\phi}_1(\theta)]^2 = \frac{1}{2\theta^2}, \tag{3.8}$$

it follows that (A 2′) (i)–(iii) are fulfilled.

3.3 Let $\{X_n\}$, $n \geqslant 1$ be i.i.d. with the double exponential density

$$p(x; \theta) = \tfrac{1}{2}\exp[-|x-\theta|] \quad (\theta \in R).$$

Once more (A 1′) is trivially true and therefore we occupy ourselves with the task of verifying (A 2′). Clearly one has

$$\phi_1(\theta, \theta^*) = \exp[\tfrac{1}{2}|X_1 - \theta| - \tfrac{1}{2}|X_1 - \theta^*|].$$

However, the pointwise derivative of $\phi_1(\theta, \theta^*)$ with respect to θ^* does not exist at (θ, θ) when $\theta = X_1$. We will show that the derivative in q.m. does exist. In fact, for each $\theta \in \Theta$, define the r.v. $Z_j(\theta)$ as follows

$$Z_j(\theta) = \begin{cases} -\tfrac{1}{2} & \text{if} \quad X_j < \theta \\ 0 & \text{if} \quad X_j = \theta \\ \tfrac{1}{2} & \text{otherwise} \quad (j = 1, ..., n). \end{cases} \tag{3.9}$$

Then $\mathscr{E}_\theta Z_j(\theta) = \mathscr{E}_\theta Z_1(\theta) = 0$ and $\mathscr{E}_\theta Z_j^2(\theta) = \mathscr{E}_\theta Z_1^2(\theta) = \tfrac{1}{4}$, since $P_\theta(X_1 < \theta) = P_\theta(X_1 > \theta) = \tfrac{1}{2}$. Furthermore

$$\frac{1}{h}[\phi_1(\theta, \theta+h) - 1] = \frac{1}{h}\{\exp[\tfrac{1}{2}|X_1 - \theta| - \tfrac{1}{2}|X_1 - (\theta+h)|] - 1\},$$

so that for each $\theta \in \Theta$, one has that $\dfrac{1}{h}[\phi_1(\theta, \theta+h) - 1] \to Z_1(\theta)$ with P_θ-probability 1, as $h \to 0$ which implies that

$$\frac{1}{h}[\phi_1(\theta, \theta+h) - 1] \to Z_1(\theta) \quad \text{in } P_\theta\text{-probability, as } h \to 0. \tag{3.10}$$

Next $\quad \mathscr{E}_\theta\left\{\dfrac{1}{h}[\phi_1(\theta, \theta+h) - 1]\right\}^2 = \dfrac{2}{h^2}[1 - \mathscr{E}_\theta\phi_1(\theta, \theta+h)],$

where
$$\mathscr{E}_\theta \phi_1(\theta, \theta+h) = \begin{cases} \frac{1}{2}(2-h)\exp\frac{1}{2}h & \text{if } h < 0 \\ \frac{1}{2}(2+h)\exp(-\frac{1}{2}h) & \text{if } h > 0, \end{cases}$$

as is easily seen by integration.

Considering the case $h < 0$ first, one has

$$\frac{2}{h^2}[1 - \mathscr{E}_\theta \phi_1(\theta, \theta+h)] = \frac{2}{h^2}\left[1 - \tfrac{1}{2}(2-h)\exp\frac{h}{2}\right]$$

$$= \frac{1}{2t^2}(1 - \exp t + t \exp t)$$

by setting $\frac{1}{2}h = t$, and this is equal to $\frac{1}{4} + \frac{1}{2}o(1) \to \frac{1}{4}$ as $t \to 0$. That is,

$$\mathscr{E}_\theta \left\{\frac{1}{h}[\phi_1(\theta, \theta+h) - 1]\right\}^2 \to \frac{1}{4} \quad \text{as } h \to 0, \qquad (3.11)$$

provided $h < 0$, and the same is seen to be true when $h > 0$.

The fact that $\mathscr{E}_\theta Z_1^2(\theta) = \frac{1}{4}$ and relations (3.10) and (3.11) together with Theorem 2.1 A imply then that

$$\phi_1(\theta) = Z_1(\theta). \qquad (3.12)$$

Furthermore
$$\Gamma(\theta) = 4\mathscr{E}_\theta [\phi_1(\theta)]^2 = 1, \qquad (3.13)$$

so that (A 2′) (i)–(iii) are also fulfilled.

3.4 Let $\{X_n\}$, $n \geq 0$ be centred at expectation and suppose they constitute a Gaussian process with covariance given by $\mathscr{E}_\theta(X_m X_n) = \exp(-\theta|m-n|)$, $\theta > 0$.

Then the process is a stationary Markov process, as follows from Doob [1], Example 4, p. 218 and pp. 233–4. Furthermore the process is metrically transitive according to statements on p. 476 and p. 637, section 7 of the reference just cited. Therefore (A1) is satisfied. Assumption (A2) is clearly true. As for (A3), from the fact that the Xs have expectation 0 and variance 1, it follows that the stationary initial distribution is $N(0, 1)$. By setting $p(\theta) = \exp(-\theta)$, we have that the transition density here is given by

$$\frac{1}{\{2\pi[1 - p^2(\theta)]\}^{\frac{1}{2}}} \exp\left\{-\frac{[x_1 - p(\theta)x_0]^2}{2[1 - p^2(\theta)]}\right\}. \qquad (3.14)$$

Therefore it follows that

$$\phi_1(\theta, \theta^*) = \phi_1^*[p(\theta), p(\theta^*)]$$

$$= \left[\frac{1-p^2(\theta)}{1-p^2(\theta^*)}\right]^{\frac{1}{4}} \exp\left\{-\frac{X_0^2 - 2p(\theta^*)X_0X_1 + X_1^2}{4[1-p^2(\theta^*)]}\right.$$

$$\left. + \frac{X_0^2 - 2p(\theta)X_0X_1 + X_1^2}{4[1-p^2(\theta)]}\right\}. \tag{3.15}$$

From (3.15), one has for the (pointwise) derivatives indicated below

$$\frac{\partial}{\partial\theta^*}\phi_1(\theta,\theta^*)\bigg|_{\theta^*=\theta}$$

$$= \frac{\partial}{\partial p(\theta^*)}\phi_1^*[p(\theta),p(\theta^*)]\frac{dp(\theta^*)}{d\theta^*}\bigg|_{\theta^*=\theta}$$

$$= \frac{p^2(\theta)[X_0^2 - 2p(\theta)X_0X_1 + X_1^2]}{2[1-p^2(\theta)]^2} - \frac{p^2(\theta)+p(\theta)X_0X_1}{2[1-p^2(\theta)]}. \tag{3.16}$$

Since the Xs are distributed as $N(0,1)$ under P_θ, we have

$$\mathscr{E}_\theta X_0^2 = \mathscr{E}_\theta X_1^2 = 1 \quad \text{and} \quad \mathscr{E}_\theta X_0^4 = \mathscr{E}_\theta X_1^4 = 3, \tag{3.17}$$

while by our assumption about the covariance of successive Xs, it follows that

$$\mathscr{E}_\theta(X_0X_1) = p(\theta). \tag{3.18}$$

Next by means of (3.14) and (3.17), one has

$$\mathscr{E}_\theta(X_0^2X_1^2) = \mathscr{E}_\theta[\mathscr{E}_\theta(X_0^2X_1^2)|X_0] = \mathscr{E}_\theta[X_0^2\mathscr{E}_\theta(X_1^2|X_0)]$$

$$= \mathscr{E}_\theta\{X_0^2[1-p^2(\theta)+p^2(\theta)X_0^2]\} = 1+2p^2(\theta);$$

i.e. $$\mathscr{E}_\theta(X_0^2X_1^2) = 1+2p^2(\theta), \tag{3.19}$$

and in a similar fashion

$$\mathscr{E}_\theta(X_0^3X_1) = \mathscr{E}_\theta(X_0X_1^3) = 3p(\theta). \tag{3.20}$$

Then by means of (3.16)–(3.20), one finds that

$$\mathscr{E}_\theta\left[\frac{\partial}{\partial\theta^*}\phi_1(\theta,\theta^*)\bigg|_{\theta^*=\theta}\right]^2 = \frac{p^2(\theta)[1+p^2(\theta)]}{4[1-p^2(\theta)]^2}. \tag{3.21}$$

Next

$$\mathscr{E}_\theta \left\{ \frac{1}{h} [\phi_1(\theta, \theta+h) - 1] \right\}^2$$

$$= \frac{2}{h^2} [1 - \mathscr{E}_\theta \phi_1(\theta, \theta+h)]$$

$$= \frac{2}{h^2} \left\{ 1 - \frac{2[1 - p^2(\theta)]^{\frac{3}{4}} [1 - p^2(\theta+h)]^{\frac{3}{4}}}{[\{2 - [p^2(\theta) + p^2(\theta+h)]\}^2 - [p(\theta) + p(\theta+h)]^2 [1 - p(\theta)p(\theta+h)]^2]^{\frac{1}{2}}} \right\}$$

$$(3.22)$$

as follows from (3.15) by replacing $\theta*$ by $\theta+h$ and carrying out the integration.

Finally, letting $h \to 0$ in (3.22), we see that the limit on the right-hand side is equal to

$$\frac{p^2(\theta) [1 + p^2(\theta)]}{4[1 - p^2(\theta)]^2}.$$

This result together with (3.21) and Theorem 2.1 A implies that

$$\phi_1(\theta) = \frac{p^2(\theta) [X_0^2 - 2p(\theta) X_0 X_1 + X_1^2]}{2[1 - p^2(\theta)]^2} - \frac{p^2(\theta) + p(\theta) X_0 X_1}{2[1 - p^2(\theta)]}. \quad (3.23)$$

From (3.21), it follows that

$$\Gamma(\theta) = \frac{p^2(\theta) [1 + p^2(\theta)]}{[1 - p^2(\theta)]^2}, \quad (3.24)$$

so that (A 3 (i)–(iii)) are satisfied. Finally it is clear that

$$q(X_0, X_1; \theta, \theta*) = \phi_1^2(\theta, \theta*) = \phi_1^{*2}[p(\theta), p(\theta*)],$$

so that, by (3.15), (A 4) is also satisfied.

4   Asymptotic expansion and asymptotic normality of likelihood functions

In this section we formulate the basic results of the present chapter which relate to the asymptotic expansion in the probability sense, and also asymptotic normality of certain likelihood functions.

In what follows, $X_0, X_1, ..., X_n$ are $n+1$ observations from the underlying Markov process and θ is an arbitrarily chosen but thereafter kept fixed point of Θ. Let $\{h_n\}$ be a sequence of points of R^k such that $h_n \to h \in R^k$, and set $\theta_n = \theta + h_n n^{-\frac{1}{2}}$, so that $\theta_n \in \Theta$ for sufficiently large n, since Θ is open and $\theta \in \Theta$. Then on account of (A 2), the log-likelihood of $[\mathrm{d}P_{n,\theta_n}/\mathrm{d}P_{n,\theta}]$ is well-defined except perhaps on a P_{θ^*}-null set for all $\theta^* \in \Theta$. In the sequel, we will work outside this exceptional set. Set

$$\Lambda(\theta, \theta_n) = \log [\mathrm{d}P_{n,\theta_n}/\mathrm{d}P_{n,\theta}]$$

$$= \log q(X_0; \theta, \theta_n) + 2 \sum_{j=1}^{n} \log \phi_j(\theta, \theta_n), \qquad (4.1)$$

and also define the quantities $\Delta_n(\theta)$ and $A(h, \theta)$ as follows

$$\Delta_n(\theta) = 2n^{-\frac{1}{2}} \sum_{j=1}^{n} \dot{\phi}_j(\theta), \quad A(h, \theta) = \tfrac{1}{2} h' \Gamma(\theta) h, \qquad (4.2)$$

where, as it follows from (2.1),

$$h' \Gamma(\theta) h = 4 \mathscr{E}_\theta [h' \dot{\phi}_1(\theta)]^2. \qquad (4.3)$$

Then one has the following result.

THEOREM 4.1 Let $h_n, h \in R^k$ be such that $h_n \to h$ and set $\theta_n = \theta + h_n n^{-\frac{1}{2}}$. Let $\Lambda(\theta, \theta_n)$, $\Delta_n(\theta)$ and $A(h, \theta)$ be defined by (4.1) and (4.2), respectively. Then

$$\Lambda(\theta, \theta_n) - h' \Delta_n(\theta) \to -A(h, \theta) \quad \text{in } P_{n,\theta}\text{-probability}.$$

REMARK 4.1 The following is a heuristic interpretation of the theorem. It says that one has

$$[\mathrm{d}P_{n,\theta_n}/\mathrm{d}P_{n,\theta}] = \exp \Lambda(\theta, \theta_n) \approx \exp [h' \Delta_n(\theta) - A(h, \theta)];$$

i.e. the likelihood behaves approximately as if it were an exponential family. This is made precise and also proved in Chapter 3, Theorem 3.1.

From the loose interpretation of Theorem 4.1 just presented, it is evident that the r. vector $\Delta_n(\theta)$, being the r. vector appearing in the exponent of an exponential family, is bound to play a very important role in the sequel. Therefore it is eminently important to derive its asymptotic distribution. This also is done in the present chapter and the relevant result goes as follows.

THEOREM 4.2 Let $\Delta_n(\theta)$ and $\Gamma(\theta)$ be defined by (4.2) and (2.1), respectively. Then

$$\mathscr{L}[\Delta_n(\theta)|P_{n,\theta}] \Rightarrow N(0, \Gamma(\theta)).$$

Then as a consequence of Theorems 4.1 and 4.2, one has

THEOREM 4.3 Let h_n, $h \in R^k$ be such that $h_n \to h$ and set $\theta_n = \theta + h_n n^{-\frac{1}{2}}$. Let $\Lambda(\theta, \theta_n)$ and $h'\Gamma(\theta)h$ be defined by (4.1) and (4.3), respectively. Then

$$\mathscr{L}[\Lambda(\theta, \theta_n)|P_{n,\theta}] \Rightarrow N(-\tfrac{1}{2}h'\Gamma(\theta)h, h'\Gamma(\theta)h).$$

Now for the purpose of statistical applications to be discussed in other chapters, the asymptotic distribution of $\Lambda(\theta, \theta_n)$ as well as $\Delta_n(\theta)$, under P_{n,θ_n}, is also needed. These distributions are provided by Theorems 4.5 and 4.6 below. Also Theorem 4.1 remains true when $P_{n,\theta}$ is replaced by P_{n,θ_n}. This result is stated as Theorem 4.4 below.

THEOREM 4.4 With the same notation as that of Theorem 4.1, one has

$$\Lambda(\theta, \theta_n) - h'\Delta_n(\theta) \to -A(h, \theta) \quad \text{in } P_{n,\theta_n}\text{-probability.}$$

THEOREM 4.5 With the same notation as that of Theorem 4.3, one has

$$\mathscr{L}[\Lambda(\theta, \theta_n)|P_{n,\theta_n}] \Rightarrow N(\tfrac{1}{2}h'\Gamma(\theta)h, h'\Gamma(\theta)h).$$

THEOREM 4.6 With the same notation as that of Theorem 4.2, one has

$$\mathscr{L}[\Delta_n(\theta)|P_{n,\theta_n}] \Rightarrow N(\Gamma(\theta)h, \Gamma(\theta)).$$

5 Some lemmas

Since the proofs of the fundamental results stated in Section 4, and, in particular, that of Theorem 4.1, are quite long, it would be suggested from an organizational viewpoint to present the intermediate steps as lemmas. Actually some of these lemmas are also of independent interest.

Unless otherwise specified, in what follows

$$h_n, h \in R^k \quad \text{with} \quad h_n \to h \quad \text{and} \quad \theta_n = \theta + h_n n^{-\frac{1}{2}}. \quad (5.1)$$

We also set
$$\Lambda_n(\theta) = \Lambda(\theta, \theta_n) \quad \text{and} \quad \phi_{nj}(\theta) = \phi_j(\theta, \theta_n), \qquad (5.2)$$
where $\Lambda(\theta, \theta_n)$ and $\phi_j(\theta, \theta_n)$ are given by (4.1) and (1.3), respectively, so that
$$\Lambda_n(\theta) = \log q(X_0; \theta, \theta_n) + \sum_{j=1}^{n} \log \phi_{nj}^2(\theta). \qquad (5.3)$$

Occasionally we shall also use the simpler notation Λ_n, ϕ_{nj} and ϕ_j instead of $\Lambda_n(\theta)$, $\phi_{nj}(\theta)$ and $\phi_j(\theta)$, respectively, when there is no danger of confusion. We finally mention that in the various derivations we shall use either one of the notations P_{n,θ^*} or P_{θ^*} ($\theta^* \in \Theta$).

The first lemma is an auxiliary result to be used in the proof of the next lemma.

LEMMA 5.1   With $\phi_{nj}(\theta)$ defined by (5.2), one has
$$\sum_{j=1}^{n} [\phi_{nj}(\theta) - 1]^2 \to \mathcal{E}_\theta[h'\dot{\phi}_1(\theta)]^2 \quad \text{in } P_\theta\text{-probability.}$$

Proof   Since, clearly, $\phi_j(\theta, \theta) = 1$, assumption (A 3)(i) implies by virtue of Definition 1.1′,
$$\frac{1}{\lambda}[\phi_j(\theta, \theta + \lambda h) - 1] \to h'\dot{\phi}_j(\theta) \quad \text{in q.m. } [P_\theta] \text{ as } \lambda \to 0 \qquad (5.4)$$

uniformly on bounded sets of values of h. This is true for all $j = 1, \dots, n$. By taking λ to be $n^{-\frac{1}{2}}$, and replacing h by a sequence $\{h_n\}$ which converges to h and also using the notation introduced in (5.1) and (5.2), relation (5.4) implies
$$n^{\frac{1}{2}}(\phi_{nj} - 1) \to h'\dot{\phi}_j \quad \text{in q.m. } [P_\theta], \text{ for each } j. \qquad (5.5)$$
From (5.5) it follows then that
$$n(\phi_{n1} - 1)^2 \to (h'\dot{\phi}_1)^2 \quad \text{in the first mean } [P_\theta]. \qquad (5.6)$$
See (5.9), (5.10), (5.11) below.

Next $\qquad \dfrac{1}{n}\sum_{j=1}^{n} (h'\dot{\phi}_j)^2 \to \mathcal{E}_\theta(h'\dot{\phi}_1)^2 \quad \text{a.s. } [P_\theta] \qquad (5.7)$

by the ergodic theorem and the stationarity of the process, and
$$\sum_{j=1}^{n} (\phi_{nj} - 1)^2 - \frac{1}{n}\sum_{j=1}^{n} (h'\dot{\phi}_j)^2 = \frac{1}{n}\sum_{j=1}^{n} [n(\phi_{nj} - 1)^2 - (h'\dot{\phi}_j)^2],$$

from which it follows, by the Markov inequality that, for every $\epsilon > 0$,

$$P_\theta\left[\left|\sum_{j=1}^{n}(\phi_{nj}-1)^2-\frac{1}{n}\sum_{j=1}^{n}(h'\dot{\phi}_j)^2\right| > \epsilon\right]$$

$$= P_\theta\left\{\left|\frac{1}{n}\sum_{j=1}^{n}[n(\phi_{nj}-1)^2-(h'\dot{\phi}_j)^2]\right| > \epsilon\right\}$$

$$\leqslant \frac{1}{n\epsilon}n\mathscr{E}_\theta|n(\phi_{n1}-1)^2-(h'\dot{\phi}_1)^2| = \frac{1}{\epsilon}\mathscr{E}_\theta|n(\phi_{n1}-1)^2-(h'\dot{\phi}_1)^2|.$$

However, this last expression converges to zero on account of (5.6). Thus

$$\sum_{j=1}^{n}(\phi_{nj}-1)^2-\frac{1}{n}\sum_{j=1}^{n}(h'\dot{\phi}_j)^2 \to 0 \quad \text{in } P_\theta\text{-probability}.$$

This relation together with (5.7) implies the desired result. ▮

From (5.5) it follows that $\phi_{nj}-1 \to 0$ in P_θ-probability. The following lemma strengthens this result as follows.

LEMMA 5.2   With $\phi_{nj}(\theta)$ as defined by (5.2), one has

$$\max[|\phi_{nj}(\theta)-1|; 1 \leqslant j \leqslant n] \to 0 \quad \text{in } P_\theta\text{-probability}.$$

Proof   In fact,

$$\phi_{nj}-1 = n^{-\frac{1}{2}}h'\dot{\phi}_j+n^{-\frac{1}{2}}R_{nj}, \tag{5.8}$$

where $R_{nj} = R_{nj}(\theta,h) = n^{\frac{1}{2}}(\phi_{nj}-1)-h'\dot{\phi}_j$ and hence

$$\mathscr{E}_\theta|R_{nj}|^2 = \mathscr{E}_\theta|R_{n1}|^2 = \mathscr{E}_\theta|n^{\frac{1}{2}}(\phi_{n1}-1)-h'\dot{\phi}_1|^2 \to 0 \tag{5.9}$$

by (5.5). From (5.8), one has

$$P_\theta[\max(|\phi_{nj}-1|; 1 \leqslant j \leqslant n) > \epsilon]$$

$$= P_\theta[\max(|n^{-\frac{1}{2}}h'\dot{\phi}_j+n^{-\frac{1}{2}}R_{nj}|; 1 \leqslant j \leqslant n) > \epsilon]$$

$$\leqslant P_\theta[\max(|h'\dot{\phi}_j|; 1 \leqslant j \leqslant n) > \tfrac{1}{2}\epsilon n^{\frac{1}{2}}]$$

$$+ P_\theta[\max(|R_{nj}|; 1 \leqslant j \leqslant n) > \tfrac{1}{2}\epsilon n^{\frac{1}{2}}]$$

$$\leqslant nP_\theta(|h'\dot{\phi}_1| > \tfrac{1}{2}\epsilon n^{\frac{1}{2}})+nP_\theta(|R_{n1}| > \tfrac{1}{2}\epsilon n^{\frac{1}{2}}). \tag{5.10}$$

But

$$nP_\theta(|R_{n1}| > \tfrac{1}{2}\epsilon n^{\frac{1}{2}}) \leqslant \frac{4n}{\epsilon^2 n}\mathscr{E}_\theta|R_{n1}|^2 = \frac{4}{\epsilon^2}\mathscr{E}_\theta|R_{n1}|^2 \to 0 \tag{5.11}$$

by means of (5.9), and if F denotes the distribution of the r.v. $|h'\phi_1|$ under P_θ, one has

$$\int_0^\infty x^2 \mathrm{d}F = \mathscr{E}_\theta |h'\phi_1|^2 < \infty.$$

This implies that
$$\int_{\frac{1}{2}\epsilon n^{\frac{1}{2}}}^\infty x^2 \mathrm{d}F \to 0. \tag{5.12}$$

Next

$$nP_\theta(|h'\phi_1| > \tfrac{1}{2}\epsilon n^{\frac{1}{2}}) = n\int_{\frac{1}{2}\epsilon n^{\frac{1}{2}}}^\infty \mathrm{d}F = \frac{4}{\epsilon^2}\int_{\frac{1}{2}\epsilon n^{\frac{1}{2}}}^\infty (\tfrac{1}{2}\epsilon n^{\frac{1}{2}})^2 \mathrm{d}F$$

$$\leqslant \frac{4}{\epsilon^2}\int_{\frac{1}{2}\epsilon n^{\frac{1}{2}}}^\infty x^2 \mathrm{d}F \to 0 \tag{5.13}$$

by means of (5.12).

Thus relations (5.10), (5.11) and (5.13) complete the proof of the lemma.]

LEMMA 5.3 With $\Lambda_n(\theta)$ and $\phi_{nj}(\theta)$ as defined by (5.2) and (5.3), one has

$$\Lambda_n(\theta) - 2\left\{\sum_{j=1}^n [\phi_{nj}(\theta) - 1] - \frac{1}{2}\sum_{j=1}^n [\phi_{nj}(\theta) - 1]^2\right\}$$
$$\to 0 \text{ in } P_\theta\text{-probability.}$$

Proof This lemma states, in effect, that $\Lambda_n(\theta)$ may be replaced asymptotically – and in the probability sense – by certain sums involving the r.f.s $\phi_{nj}(\theta)$, $j = 1, ..., n$ which might be easier to deal with.

Now for a given $0 < \epsilon (\leqslant \frac{1}{2})$, set

$$A_n = A_n(\theta) = [\max(|\phi_{nj} - 1|; 1 \leqslant j \leqslant n) > \epsilon]. \tag{5.14}$$

Then Lemma 5.2 implies that

$$P_\theta(A_n^c) > 1 - \epsilon \quad \text{for all sufficiently large } n, n \geqslant n_0, \quad \text{say.}$$

Consider the expansion

$$\log x = \log[1 + (x - 1)]$$
$$= (x - 1) - \tfrac{1}{2}(x - 1)^2 + c(x - 1)^3, \quad |c| \leqslant 3 \text{ (e.g.).} \tag{5.15}$$

Then (5.14) implies that the expansion in (5.15) holds uniformly in $j = 1, ..., n, n \geq n_0$ on the set A_n^c of probability $> 1 - \epsilon$ if x is replaced by ϕ_{nj}. That is, one has

$$\log \phi_{nj} = \log[1 + (\phi_{nj} - 1)]$$
$$= (\phi_{nj} - 1) - \tfrac{1}{2}(\phi_{nj} - 1)^2 + c_{nj}(\phi_{nj} - 1)^3, \ |c_{nj}| \leq 3$$
$$(j = 1, ..., n),$$

on A_n^c with $\qquad P_\theta(A_n^c) > 1 - \epsilon \quad (n \geq n_0)$.

Therefore on A_n^c and with $n \geq n_0$, we have

$$\sum_{j=1}^n \log \phi_{nj} = \sum_{j=1}^n (\phi_{nj} - 1) - \frac{1}{2} \sum_{j=1}^n (\phi_{nj} - 1)^2 + \sum_{j=1}^n c_{nj}(\phi_{nj} - 1)^3.$$

But

$$\left| \sum_{j=1}^n c_{nj}(\phi_{nj} - 1)^3 \right| \leq 3[\max (|\phi_{nj} - 1|; 1 \leq j \leq n)] \sum_{j=1}^n (\phi_{nj} - 1)^2 \to 0$$

$$\text{in } P_\theta\text{-probability}$$

by Lemmas 5.1 and 5.2. Therefore

$$\sum_{j=1}^n \log \phi_{nj} - \left[\sum_{j=1}^n (\phi_{nj} - 1) - \frac{1}{2} \sum_{j=1}^n (\phi_{nj} - 1)^2 \right] \to 0$$

$$\text{in } P_\theta\text{-probability},$$

or equivalently,

$$\sum_{j=1}^n \log \phi_{nj}^2 - 2 \left[\sum_{j=1}^n (\phi_{nj} - 1) - \frac{1}{2} \sum_{j=1}^n (\phi_{nj} - 1)^2 \right] \to 0$$

$$\text{in } P_\theta\text{-probability.} \quad (5.16)$$

Now by (5.3) and (A4)

$$\Lambda_n - \sum_{j=1}^n \log \phi_{nj}^2 = \log q(X_0; \theta, \theta_n) \to 0 \quad \text{in } P_\theta\text{-probability,} \quad (5.17)$$

so that relations (5.16) and (5.17) imply

$$\Lambda_n - 2 \left[\sum_{j=1}^n (\phi_{nj} - 1) - \frac{1}{2} \sum_{j=1}^n (\phi_{nj} - 1)^2 \right] \to 0 \quad \text{in } P_\theta\text{-probability,}$$

which is what the lemma asserts. ∎

The following lemma will also be needed.

LEMMA 5.4 With $\phi_{nj}(\theta)$ as defined by (5.3), one has that, for each j,

$$n^{\frac{1}{2}}[\phi_{nj}^2(\theta) - 1] \to 2h'\dot{\phi}_j(\theta) \quad \text{in the first mean } [P_\theta].$$

Proof Consider the following identity

$$n^{\frac{1}{2}}(\phi_{nj}^2 - 1) - 2h'\dot{\phi}_j = \{\phi_{nj}[n^{\frac{1}{2}}(\phi_{nj} - 1) - h'\dot{\phi}_j]\} + [h'\dot{\phi}_j(\phi_{nj} - 1)]$$
$$+ [n^{\frac{1}{2}}(\phi_{nj} - 1) - h'\dot{\phi}_j]. \quad (5.18)$$

Now by taking absolute values on both sides of (5.18) and the expectations under P_θ, we obtain

$$\mathscr{E}_\theta |n^{\frac{1}{2}}(\phi_{nj}^2 - 1) - 2h'\dot{\phi}_j| \leq \mathscr{E}_\theta |\phi_{nj}[n^{\frac{1}{2}}(\phi_{nj} - 1) - h'\dot{\phi}_j]|$$
$$+ \mathscr{E}_\theta |h'\dot{\phi}_j(\phi_{nj} - 1)|$$
$$+ \mathscr{E}_\theta |n^{\frac{1}{2}}(\phi_{nj} - 1) - h'\dot{\phi}_j|.$$

Next by means of Hölder inequality, applied to the first two terms on the right-hand side of the last expression above, and the fact that $\mathscr{E}_\theta \phi_{nj}^2 = 1$, we have

$$\mathscr{E}_\theta |n^{\frac{1}{2}}(\phi_{nj}^2 - 1) - 2h'\dot{\phi}_j| \leq \mathscr{E}_\theta^{\frac{1}{2}}[n^{\frac{1}{2}}(\phi_{nj} - 1) - h'\dot{\phi}_j]^2$$
$$+ \mathscr{E}_\theta^{\frac{1}{2}}(h'\dot{\phi}_j)^2 \mathscr{E}_\theta^{\frac{1}{2}}(\phi_{nj} - 1)^2 + \mathscr{E}_\theta |n^{\frac{1}{2}}(\phi_{nj} - 1) - h'\dot{\phi}_j|. \quad (5.19)$$

But $\quad \mathscr{E}_\theta |n^{\frac{1}{2}}(\phi_{nj} - 1) - h'\dot{\phi}_j| \leq \mathscr{E}_\theta^{\frac{1}{2}}[n^{\frac{1}{2}}(\phi_{nj} - 1) - h'\dot{\phi}_j]^2,$

and hence (5.19) becomes

$$\mathscr{E}_\theta |n^{\frac{1}{2}}(\phi_{nj}^2 - 1) - 2h'\dot{\phi}_j|$$
$$\leq 2\mathscr{E}_\theta^{\frac{1}{2}}[n^{\frac{1}{2}}(\phi_{nj} - 1) - h'\dot{\phi}_j]^2 + \mathscr{E}_\theta^{\frac{1}{2}}(h'\dot{\phi}_j)^2 \mathscr{E}_\theta^{\frac{1}{2}}(\phi_{nj} - 1)^2. \quad (5.20)$$

Now by (5.5) $\quad \mathscr{E}_\theta[n^{\frac{1}{2}}(\phi_{nj} - 1) - h'\dot{\phi}_j]^2 \to 0, \quad (5.21)$

which by Theorem 2.1 A implies $\mathscr{E}_\theta[n^{\frac{1}{2}}(\phi_{nj} - 1)]^2 \to \mathscr{E}_\theta(h'\dot{\phi}_j)^2$, and hence
$$\mathscr{E}_\theta(\phi_{nj} - 1)^2 \to 0. \quad (5.22)$$

By (5.21), (5.22) and the fact that $\mathscr{E}_\theta(h'\dot{\phi}_j)^2 < \infty$ (which follows from (A3)(i)), relation (5.20) gives

$$\mathscr{E}_\theta |n^{\frac{1}{2}}(\phi_{nj}^2 - 1) - 2h'\dot{\phi}_j| \to 0,$$

as was to be shown. ∎

We recall that $\mathscr{A}_j$ denotes the σ-field induced by the r.v.s $X_0, X_1, \ldots, X_j$. Then by taking into consideration the definition of ϕ_{nj} and utilizing the Markov property, it is an easy matter to see that

$$\mathscr{E}_\theta(\phi_{nj}^2 | \mathscr{A}_{j-1}) = 1 \quad \text{a.s.} [P_\theta]. \tag{5.23}$$

Now we formulate and prove a lemma which together with the previous ones will help to establish parts of the main results stated in Section 4. To this end, we set

$$\psi_{nj}(\theta) = \mathscr{E}_\theta[\phi_{nj}(\theta) | \mathscr{A}_{j-1}] \quad (j = 1, \ldots, n), \tag{5.24}$$

and then one has the following lemma, keeping in mind that $\dot{\phi}_j$ is a k-dimensional r. vector and that equality of vectors means coordinatewise equality.

LEMMA 5.5   With $\phi_{nj}(\theta)$ and $\psi_{nj}(\theta)$ as defined by (5.3) and (5.24), respectively, one has

(i) $\mathscr{E}_\theta[\dot{\phi}_j(\theta) | \mathscr{A}_{j-1}] = 0 \quad \text{a.s.} [P_\theta] \quad (j = 1, \ldots, n).$

(ii) $\sum_{j=1}^{n} [\psi_{nj}(\theta) - 1] \to -\frac{1}{2}\mathscr{E}_\theta[h'\dot{\phi}_j(\theta)]^2 \quad \text{in } P_{n,\theta}\text{-probability.}$

(iii) $\sum_{j=1}^{n} \{[\phi_{nj}(\theta) - 1] - n^{-\frac{1}{2}}h'\dot{\phi}_j(\theta)\} - \sum_{j=1}^{n} [\psi_{nj}(\theta) - 1] \to 0$

$$\text{in } P_{n,\theta}\text{-probability.}$$

Proof   (i) By Lemma 5.4, we have that for each j,

$$n^{\frac{1}{2}}(\phi_{nj}^2 - 1) \to 2h'\dot{\phi}_j$$

in the first mean $[P_\theta]$. Then the same convergence is true by conditioning both sides, given $\mathscr{A}_{j-1}$; i.e.

$$\mathscr{E}_\theta[n^{\frac{1}{2}}(\phi_{nj}^2 - 1) | \mathscr{A}_{j-1}] \to \mathscr{E}_\theta(2h'\dot{\phi}_j | \mathscr{A}_{j-1}) \quad \text{in the first mean } [P_\theta]. \tag{5.25}$$

This follows from well-known properties of conditioning (see, e.g. Loève [1], p. 348). But the left-hand side of (5.25) is equal to 0 a.s. $[P_\theta]$ by virtue of (5.23). Therefore

$$\mathscr{E}_\theta(h'\dot{\phi}_j | \mathscr{A}_{j-1}) = 0 \quad \text{a.s.} [P_\theta].$$

Since this is true for any $h \in R^k$, it follows that

$$\mathscr{E}_\theta(\dot{\phi}_j | \mathscr{A}_{j-1}) = 0 \quad \text{a.s.} [P_\theta],$$

as was to be seen. (Of course, it is understood that this latter 0 is the $k \times 1$ zero vector.)

(ii) In the first place, the ergodic theorem implies that

$$\frac{1}{n} \sum_{j=1}^{n} \mathscr{E}_{\theta}[(h'\dot{\phi}_1)^2|\mathscr{A}_{j-1}] \to \mathscr{E}_{\theta}(h'\dot{\phi}_j)^2 \quad \text{a.s. } [P_\theta]. \quad (5.26)$$

Next

$$P_\theta\left\{\left|\sum_{j=1}^{n} \mathscr{E}_{\theta}[(\phi_{nj}-1)^2|\mathscr{A}_{j-1}] - \frac{1}{n}\sum_{j=1}^{n}\mathscr{E}_{\theta}[(h'\dot{\phi}_j)^2|\mathscr{A}_{j-1}]\right| > \epsilon\right\}$$

$$= P_\theta\left\{\left|\sum_{j=1}^{n}\mathscr{E}_{\theta}[n(\phi_{nj}-1)^2-(h'\dot{\phi}_j)^2|\mathscr{A}_{j-1}]\right| > \epsilon n\right\}$$

$$\leqslant \frac{n}{\epsilon n}\mathscr{E}_{\theta}|\mathscr{E}_{\theta}[n(\phi_{n1}-1)^2-(h'\dot{\phi}_1)^2|\mathscr{A}_0]|$$

$$\leqslant \frac{1}{\epsilon}\mathscr{E}_{\theta}|n(\phi_{n1}-1)^2-(h'\dot{\phi}_1)^2|. \quad (5.27)$$

But by (5.5), $n^{\frac{1}{2}}(\phi_{n1}-1) \to h'\dot{\phi}_1$ in q.m. $[P_\theta]$ which by means of Theorem 2.1 A implies that

$$n(\phi_{n1}-1)^2 \to (h'\dot{\phi}_1)^2 \quad \text{in the first mean } [P_\theta].$$

Thus (5.27) yields

$$\sum_{j=1}^{n}\mathscr{E}_{\theta}[(\phi_{nj}-1)^2|\mathscr{A}_{j-1}]$$

$$-\frac{1}{n}\sum_{j=1}^{n}\mathscr{E}_{\theta}[(h'\dot{\phi}_j)^2|\mathscr{A}_{j-1}] \to 0 \quad \text{in } P_\theta\text{-probability.}$$

$$(5.28)$$

Relations (5.26) and (5.28) taken together give then

$$\sum_{j=1}^{n}\mathscr{E}_{\theta}[(\phi_{nj}-1)^2|\mathscr{A}_{j-1}] \to \mathscr{E}_{\theta}(h'\dot{\phi}_1)^2 \quad \text{in } P_\theta\text{-probability.}$$

$$(5.29)$$

Now consider the identity

$$\phi_{nj}^2 - 1 = (\phi_{nj}-1)^2 + 2(\phi_{nj}-1)$$

and condition both sides, given $\mathscr{A}_{j-1}$. By taking into consideration (5.23), one obtains then

$$0 = \mathscr{E}_{\theta}[(\phi_{nj}-1)^2|\mathscr{A}_{j-1}] + 2\mathscr{E}_{\theta}[(\phi_{nj}-1)|\mathscr{A}_{j-1}] \quad \text{a.s. } [P_\theta],$$

or, by virtue of (5.24),

$$2(\psi_{nj}-1)+\mathscr{E}_\theta[(\phi_{nj}-1)^2|\mathscr{A}_{j-1}] = 0 \quad \text{a.s.} [P_\theta].$$

Hence

$$2\sum_{j=1}^n (\psi_{nj}-1)+ \sum_{j=1}^n \mathscr{E}_\theta[(\phi_{nj}-1)^2|\mathscr{A}_{j-1}] = 0 \quad \text{a.s.} [P_\theta].$$

Letting $n \to \infty$ and taking into consideration (5.29), one has

$$2\sum_{j=1}^n (\psi_{nj}-1) \to -\mathscr{E}_\theta(h'\phi_1)^2 \quad \text{in } P_\theta\text{-probability,}$$

as was to be shown.

(iii) Consider the r.v.s

$$Y_j = (\phi_{nj}-1)-n^{-\frac{1}{2}}h'\phi_j-(\psi_{nj}-1) \quad (j=1,...,n). \tag{5.30}$$

Then

$$\begin{aligned}
\mathscr{E}_\theta(Y_j|Y_1,...,Y_{j-1}) &= \mathscr{E}_\theta[\mathscr{E}_\theta(Y_j|\mathscr{A}_{j-1})|Y_1,...,Y_{j-1}] \\
&= \mathscr{E}_\theta[(\psi_{nj}-1)-0-(\psi_{nj}-1)|Y_1,...,Y_{j-1}] \\
&= 0 \quad \text{a.s.} [P_\theta],
\end{aligned}$$

the first equality being true on the basis of well-known properties of conditioning (see, e.g. Loève [1], p. 350) and the second equality holding by (5.24) and the first part of the lemma.

So

$$\mathscr{E}_\theta(Y_j|Y_1,...,Y_{j-1}) = 0 \quad \text{a.s.} [P_\theta] \quad (j=1,...,n). \tag{5.31}$$

Then the extended Kolmogorov inequality (see, e.g. Loève [1], p. 386) applies and gives

$$\begin{aligned}
P_\theta\left(\left|\sum_{j=1}^n Y_j\right| > \epsilon\right) &\leqslant \frac{1}{\epsilon^2}\sum_{j=1}^n \sigma_\theta^2(Y_j) = \frac{n}{\epsilon^2}\sigma_\theta^2(Y_1) \\
&= \frac{n}{\epsilon^2}\mathscr{E}_\theta Y_1^2 = \frac{n}{\epsilon^2}\mathscr{E}_\theta[(\phi_{n1}-1)-n^{-\frac{1}{2}}h'\phi_1-(\psi_{n1}-1)]^2 \\
&= \frac{1}{\epsilon^2}\mathscr{E}_\theta\{[n^{\frac{1}{2}}(\phi_{n1}-1)-h'\phi_1]-n^{\frac{1}{2}}(\psi_{n1}-1)\}^2 \\
&\leqslant \frac{1}{\epsilon^2}\{\mathscr{E}_\theta^{\frac{1}{2}}[n^{\frac{1}{2}}(\phi_{n1}-1)-h'\phi_1]^2+\mathscr{E}_\theta^{\frac{1}{2}}[n^{\frac{1}{2}}(\psi_{n1}-1)]^2\}.
\end{aligned}$$

$$\tag{5.32}$$

At the right-hand side above, we also used the stationarity of the Ys which follows from that of the Xs in the first equality sign, the fact that $\mathscr{E}_\theta Y_1 = 0$ which follows from (5.31), and also the Minkowski inequality in order to obtain the last inequality. Now the convergence in q.m. $[P_\theta]$ in (5.5) implies that

$$\mathscr{E}_\theta[n^{\frac{1}{2}}(\phi_{n1}-1)-h'\dot{\phi}_1]^2 \to 0. \tag{5.33}$$

By also taking into consideration (5.24) and well-known conditioning properties also cited before (see Loève [1], p. 348), (5.5) implies, in conjunction with the first part of the lemma, that

$$n^{\frac{1}{2}}(\psi_{n1}-1) \to 0 \quad \text{in q.m. } [P_\theta],$$

so that $$\mathscr{E}_\theta[n^{\frac{1}{2}}(\psi_{n1}-1)]^2 \to 0. \tag{5.34}$$

Thus relation (5.32) becomes by means of (5.33) and (5.34)

$$\sum_{j=1}^{n} Y_j \to 0 \quad \text{in } P_\theta\text{-probability}.$$

The definition of $Y_j, j = 1, ..., n$ by (5.30) completes the proof of (iii) and hence the proof of the lemma itself. **]**

6   Proof of theorems of Section 4

The lemmas established in Section 5 will suffice for the proof of some of the main theorems stated in Section 4. In order to complete their proof, however, some additional results will be needed. These auxiliary facts will be established in the present section and some of them will also be employed on other occasions as well.

Proof of Theorem 4.1   We want to show that

$$\Lambda_n - 2n^{-\frac{1}{2}}\sum_{j=1}^{n} h'\dot{\phi}_j \to -2\mathscr{E}_\theta(h'\dot{\phi}_1)^2 \quad \text{in } P_\theta\text{-probability}. \tag{6.1}$$

By Lemma 5.3, we have

$$\Lambda_n - 2\sum_{j=1}^{n}(\phi_{nj}-1) + \sum_{j=1}^{n}(\phi_{nj}-1)^2 \to 0 \quad \text{in } P_\theta\text{-probability}, \tag{6.2}$$

while by Lemma 5.1,

$$\sum_{j=1}^{n} (\phi_{nj} - 1)^2 \to \mathcal{E}_\theta (h' \dot\phi_1)^2 \quad \text{in } P_\theta\text{-probability.} \qquad (6.3)$$

Subtracting (6.3) from (6.2), one obtains

$$\Lambda_n - 2 \sum_{j=1}^{n} (\phi_{nj} - 1) \to -\mathcal{E}_\theta (h' \dot\phi_1)^2 \quad \text{in } P_\theta\text{-probability.} \qquad (6.4)$$

Next adding up the relationships in (ii) and (iii) of Lemma 5.5, one has

$$\sum_{j=1}^{n} (\phi_{nj} - 1) - n^{-\frac{1}{2}} \sum_{j=1}^{n} h' \dot\phi_j \to -\tfrac{1}{2}\mathcal{E}_\theta (h' \dot\phi_1)^2 \quad \text{in } P_\theta\text{-probability,}$$

so that

$$2 \sum_{j=1}^{n} (\phi_{nj} - 1) - 2n^{-\frac{1}{2}} \sum_{j=1}^{n} h' \dot\phi_j \to -\mathcal{E}_\theta (h' \dot\phi_1)^2 \quad \text{in } P_\theta\text{-probability.}$$

$$(6.5)$$

From (6.4) and (6.5), it follows that

$$\Lambda_n - 2n^{-\frac{1}{2}} \sum_{j=1}^{n} h' \dot\phi_j \to -2\mathcal{E}_\theta (h' \dot\phi_1)^2 \quad \text{in } P_\theta\text{-probability}$$

which is (6.1). ▮

Proof of Theorem 4.2 By writing Δ_n and Γ instead of $\Delta_n(\theta)$ and $\Gamma(\theta)$, we wish to show that

$$\mathcal{L}(\Delta_n | P_\theta) \Rightarrow N(0, \Gamma). \qquad (6.6)$$

This convergence is equivalent to the following one

$$\mathcal{L}(t' \Delta_n | P_\theta) \Rightarrow N(0, t'\Gamma t) \quad \text{for every } t \in R^k,$$

or equivalently,

$$\mathcal{L}\left(n^{-\frac{1}{2}} \sum_{j=1}^{n} 2t' \dot\phi_j \Big| P_\theta\right) \Rightarrow N(0, t'\Gamma t) \quad \text{for every } t \in R^k. \qquad (6.7)$$

Set $$Z_j = 2t' \dot\phi_j \quad (j = 1, ..., n).$$

Then by (A 1), the process $\{Z_n\}$, $n \geqslant 0$ is stationary and ergodic, while $\mathcal{E}_\theta Z_1 = 0$, as it follows from Lemma 5.5 (i). Also $\mathcal{E}_\theta Z_1^2 < \infty$

by (A3)(i). Finally, utilizing well-known properties of conditioning (see Loève [1], p. 350) and Lemma 5.5(i), one obtains

$$\mathscr{E}_\theta(Z_j|Z_1,...,Z_{j-1}) = \mathscr{E}_\theta[\mathscr{E}_\theta(Z_j|\mathscr{A}_{j-1})|Z_1,...,Z_{j-1}]$$
$$= \mathscr{E}_\theta(0|Z_1,...,Z_{j-1}) = 0 \quad \text{a.s. } [P_\theta].$$

Therefore the assumptions of Theorem 2.2A are fulfilled and hence its conclusion implies that

$$\mathscr{L}\left(n^{-\frac{1}{2}}\sum_{j=1}^n Z_j|P_\theta\right) \Rightarrow N(0,\mathscr{E}_\theta Z_1^2).$$

However, this convergence is the same as (6.7) which in turn is equivalent to (6.6). The proof of Theorem 4.2 is completed. ∎

Proof of Theorem 4.3   This is an immediate consequence of Theorems 4.1, 4.2 and the standard Slutsky theorems (see, e.g. Rao [3], p. 102 (x, b)). In fact, by (6.7), we have

$$\mathscr{L}\left(n^{-\frac{1}{2}}\sum_{j=1}^n 2h'\dot{\phi}_j|P_\theta\right) \Rightarrow N(0,h'\Gamma h), \tag{6.8}$$

while by (6.1)

$$\Lambda_n - n^{-\frac{1}{2}}\sum_{j=1}^n 2h'\dot{\phi}_j \to -2\mathscr{E}_\theta(h'\dot{\phi}_1)^2 = -\tfrac{1}{2}h'\Gamma h \quad \text{in } P_\theta\text{-probability}. \tag{6.9}$$

Thus, (6.8) and (6.9) imply

$$\mathscr{L}(\Lambda_n|P_\theta) \Rightarrow N(-\tfrac{1}{2}h'\Gamma h, h'\Gamma h),$$

as the theorem asserts. ∎

For the proof of Theorem 4.4, we need the following proposition which will also be employed on other occasions.

PROPOSITION 6.1   Let $\{h_n^*\}$ be a bounded sequence in R^k and set $\theta_n^* = \theta + h_n^* n^{-\frac{1}{2}}$. Then the sequences $\{P_{n,\theta}\}$ and $\{P_{n,\theta_n^*}\}$ are contiguous.

Proof   Because of the boundedness of $\{h_n^*\}$, for any subsequence $\{n'\} \subset \{n\}$ there exists a further subsequence $\{m\} \subset \{n'\}$ such that $h_m^* \to h$, some $h \in R^k$. Then Theorem 4.3 applies and gives that

$$\mathscr{L}[\Lambda(\theta,\theta_m^*)|P_\theta] \Rightarrow N(-\tfrac{1}{2}h'\Gamma h, h'\Gamma h).$$

It follows that $\{\mathscr{L}[\Lambda(\theta, \theta_n^*)|P_{n,\theta}]\}$ is relatively compact. Furthermore, it is easily seen as in the proof of Corollary 3.1 of Chapter 1, that

$$\int \exp \lambda \, d\mathscr{L} = 1, \quad \text{where } \mathscr{L} = N(-\tfrac{1}{2}h'\Gamma h, h'\Gamma h).$$

Then Corollary 2.2 in Chapter 1 applies and gives the desired result.]

Proof of Theorem 4.4   By Proposition 6.1, $\{P_{n,\theta}\}$ and $\{P_{n,\theta_n}\}$ are contiguous. Then Theorem 4.1 together with the definition of contiguity (Definition 2.1, Chaper 1) implies the assertion of the theorem.

Proof of Theorem 4.5   It is an immediate consequence of Theorem 4.3, Proposition 6.1 and Corollary 3.2 in Chapter 1.]

Finally we have all we need in order to establish the last theorem of Section 4.

Proof of Theorem 4.6   Assumptions (3.3)–(3.5) in Lemma 3.1 of Chapter 1 are satisfied with $P_n = P_{n,\theta}$, $P'_n = P_{n,\theta_n}$, $T_n = \Delta_n(\theta)$ and $\Gamma = \Gamma(\theta)$. This is so because of Theorems 4.2, 4.3 and 4.1, respectively. Furthermore, $\{P_{n,\theta}\}$ and $\{P_{n,\theta_n}\}$ are contiguous by Proposition 6.1. Therefore, the assumptions of Theorem 3.2 in Chapter 1 are fulfilled and then its conclusion gives

$$\mathscr{L}[\Delta_n(\theta)|P_{n,\theta_n}] \Rightarrow N(\Gamma(\theta)\,h, \Gamma(\theta)),$$

as was to be shown.]

Exercise

Justify the equivalence of Definitions 1.1 and 1.1′ asserted in Remark 1.1.

3. *Approximation of a given family of probability measures by an exponential family – asymptotic sufficiency*

Summary

It was mentioned in Chapter 2 that, under the basic assumptions (A 1–4), one may approximate a given family of probability measures by an exponential family of probability measures in the neighbourhood of each parameter point. By this, we mean the following. For $\theta \in \Theta$, set $\theta_n = \theta + hn^{-\frac{1}{2}}$, so that $\theta_n \in \Theta$ for all sufficiently large n. Then there is a probability measure $R_{n,h}$, where the parameter now involved is $h \in R^k$, such that $R_{n,h}$ is close to P_{n,θ_n} in the L_1-norm sense. More precisely, one has

$$\sup [\|P_{n,\theta_n} - R_{n,h}\| ; h \in B, \theta_n = \theta + hn^{-\frac{1}{2}}] \to 0,$$

where B is any bounded set of hs. The statistical implications of this approximation were also mentioned in Chapter 2.

In the present chapter, we give an explicit way of constructing the measures $R_{n,h}$ (see (1.7)) and also establish the approximation mentioned above (see Corollary 3.1). In the process of doing so, one has to introduce a certain truncated version, $\Delta_n^*(\theta)$, of the random vector $\Delta_n(\theta)$, defined by (4.2) in Chapter 2 and also establish some auxiliary results. This is done in Sections 1 and 2. The proof of the main result is presented in Section 3.

In Section 4, we introduce the concept of differential (asymptotic) sufficiency of a sequence of statistics at a given parameter point θ_0, say, for the given family of probability measures. According to this concept, the sequence of statistics $\{V_n\}$, say, is differentially (asymptotically) sufficient at θ_0 for the given family of probability measures if for each n, V_n is sufficient for a family of probability measures indexed by θ and in the neighbourhood of θ_0 they lie close to the given family in the L_1-norm sense (see

Definition 4.2). It is then shown that the random vector $\Delta_n(\theta_0)$ mentioned above is actually differentially (asymptotically) sufficient for the given family at the point θ_0 (see Corollary 4.2).

In the final section, we consider some of the statistical implications of the results obtained in the previous two sections. It is shown that for certain testing hypotheses problems and from an asymptotic power point of view, one may base his tests on the random vector $\Delta_n(\theta_0)$ above, where θ_0 is known (see Theorem 5.1). It is also shown that tests may actually be based on $\Delta_n^*(\theta_0)$ rather than $\Delta_n(\theta_0)$, when again asymptotic power is the criterion of optimality of a test (see Theorem 5.2). This result is important in that it provides a way of setting up tests, since $\Delta_n^*(\theta_0)$ happens to be the all-important statistic which appears in the exponent of the approximating exponential family. However, the actual random vector used is $\Delta_n(\theta_0)$ rather than $\Delta_n^*(\theta_0)$.

The results of this chapter are based on the paper [4] by LeCam cited previously. The result regarding the exponential approximation also has been established by Hájek in a recent paper [2] (see Lemma 1) from which we also borrow some ideas. A brief and somewhat different account of the derivations can also be found in Johnson and Roussas [2].

1 Formulation of the problem and some preliminary results

In Chapter 2, Remark 4.1, it was heuristically stated that, under suitable conditions

$$[\mathrm{d}P_{n,\,\theta_n}/\mathrm{d}P_{n,\,\theta}] = \exp \Lambda(\theta, \theta_n) \approx \exp\left[h'\Delta_n(\theta) - A(h, \theta)\right],$$

where $\theta_n = \theta + h_n n^{-\frac{1}{2}}$ and $h_n \to h \in R^k$. That is, the likelihood behaves approximately as if it were an exponential family with h playing the role of the parameter and θ being kept fixed. One of the main purposes of the present chapter is to make this statement precise and also prove it.

To start with, in order for $\exp\left[h'\Delta_n(\theta) - A(h, \theta)\right]$ to be a candidate for an exponential family, it has to have a finite integral with respect to $P_{n,\,\theta}$, so that when properly normalized it will, indeed, be an exponential family. There is no guarantee, however, that $\int \exp h' \Delta_n(\theta) \, dP_{n,\theta}$ is finite. Then the appropriate thing to do would be to replace $\Delta_n(\theta)$ by something else, $\Delta_n^*(\theta)$, say, so that $\int \exp h' \Delta_n^*(\theta) \, dP_{n,\theta}$ is finite. This is done in (1.5).

Next, a probability measure R_{n,h_n} on $\mathscr{A}_n$ is defined in terms of $\Delta_n^*(\theta)$ (see relation (1.7)), which has a density of the exponential form with respect to $P_{n,\theta}$. It is then shown in Theorem 3.1 that

$$\|P_{n,\theta_n} - R_{n,h_n}\| \to 0.$$

It is in this sense that the likelihood mentioned above is approximately exponential. Namely, the original family of probability measures is approximated by another family of probability measures in the L_1-norm sense and in a neighbourhood of θ. Furthermore, the members of the approximating family are absolutely continuous with respect to $P_{n,\theta}$ and their densities are of an exponential form with parameter h and θ being kept fixed.

In what follows, an arbitrary point θ_0 in Θ will be chosen and kept fixed thereafter, and we set Δ_n for $\Delta_n(\theta_0)$. We shall also write P_θ rather than $P_{n,\theta}$. Next define the real-valued function ρ on R^k by

$$\rho(z) = \|z\|, \tag{1.1}$$

where $\|\cdot\|$ is the usual Euclidean norm in R^k, and ρ is obviously continuous.

Set
$$Q = N(0, \Gamma), \quad \Delta = \text{identity mapping on } R^k, \tag{1.2}$$

so that $\mathscr{L}(\Delta|Q) = Q$.

Then we have the following proposition.

PROPOSITION 1.1 Let ρ and Q, Δ be defined by (1.1) and (1.2), respectively. Then

$$\mathscr{L}[\rho(\Delta_n)|P_{\theta_0}] \Rightarrow \mathscr{L}[\rho(\Delta)|Q].$$

Proof It follows from Theorem 4.2, Chapter 2, and the continuity of ρ (see, e.g. Rao [3], p. 104 (xii)). ∎

The following result is also true.

PROPOSITION 1.2 For every $\lambda > 0$, we have

$$\mathscr{E}_Q[\exp \lambda \rho(\Delta)] < \infty.$$

Proof Let $\Delta = (\Delta_1, \ldots, \Delta_k)'$. Then $\|\Delta\| \leqslant \sum_{j=1}^{k} |\Delta_j|$ and therefore

$$\mathscr{E}_Q[\exp \lambda \rho(\Delta)] = \mathscr{E}_Q[\exp \lambda \|\Delta\|] \leqslant \mathscr{E}_Q\left[\exp \lambda \sum_{j=1}^{k} |\Delta_j|\right]$$

$$= \mathscr{E}_Q\left[\prod_{j=1}^{k} \exp \lambda |\Delta_j|\right] \leqslant \prod_{j=1}^{k} \mathscr{E}_Q^{1/k}[\exp k\lambda |\Delta_j|]$$

by a generalized version of Hölder inequality (see Exercise 1). But $\mathscr{E}_Q[\exp k\lambda |\Delta_j|] < \infty$ since Δ_j is normal under $Q, j = 1, \ldots, k$. This completes the proof of the proposition. ∎

Consider the random vectors Δ_n, Δ and for each $\alpha > 0$, define the following truncated versions of them denoted by Δ_n^α and Δ^α.

$$\Delta_n^\alpha = \begin{cases} \Delta_n & \text{if } \rho(\Delta_n) \leqslant \alpha \\ 0 & \text{otherwise} \end{cases}; \quad \Delta^\alpha = \begin{cases} \Delta & \text{if } \rho(\Delta) \leqslant \alpha \\ 0 & \text{otherwise} \end{cases} \right\} \quad (1.3)$$

Then the following proposition is true.

PROPOSITION 1.3 For every $\alpha > 0$, we have

(i) $$\mathscr{L}(\Delta_n^\alpha | P_{\theta_0}) \Rightarrow \mathscr{L}(\Delta^\alpha | Q),$$

and

(ii) $$\mathscr{L}[\rho(\Delta_n^\alpha) | P_{\theta_0}] \Rightarrow \mathscr{L}[\rho(\Delta^\alpha) | Q].$$

Proof (i) For every real-valued, bounded and continuous function f on R^k, one has

$$\int f(z) \, \mathrm{d}\mathscr{L}(\Delta_n^\alpha | P_{\theta_0}) = \int f(\Delta_n^\alpha) \, \mathrm{d}P_{\theta_0}$$

$$= \int f\{\Delta_n I_{[\rho(\Delta_n) \leqslant \alpha]}(\Delta_n)\} \, \mathrm{d}P_{\theta_0}$$

$$= \int f\{z I_{[\rho(z) \leqslant \alpha]}(z)\} \, \mathrm{d}\mathscr{L}(\Delta_n | P_{\theta_0})$$

and this last integral converges to

$$\int f\{z I_{[\rho(z) \leqslant \alpha]}(z)\} \, \mathrm{d}Q = \int f\{z I_{[\rho(z) \leqslant \alpha]}(z)\} \, \mathrm{d}\mathscr{L}(\Delta | Q)$$

$$= \int f\{\Delta I_{[\rho(\Delta) \leqslant \alpha]}(\Delta)\} \, \mathrm{d}Q$$

$$= \int f(\Delta^\alpha) \, \mathrm{d}Q$$

$$= \int f(z) \, \mathrm{d}\mathscr{L}(\Delta^\alpha | Q),$$

since $\mathscr{L}(\Delta_n | P_{\theta_0}) \Rightarrow Q$ and the set of discontinuities of

$$f\{z I_{[\rho(z) \leqslant \alpha]}(z)\}$$

is a subset of the surface of the sphere $S(0, \alpha)$ which is assigned (k-dimensional) Lebesgue measure zero and hence Q-measure zero. (See Corollary 1.1, Chapter 1.)

(ii) This follows from (i) and the continuity of ρ. (See also proof of Proposition 1.1.)∎

By the definition of ρ and Δ^α, Δ_n^α, it follows that $\rho(\Delta^\alpha) \leqslant \alpha$, $\rho(\Delta_n^\alpha) \leqslant \alpha$ for every $\alpha > 0$. Thus, whereas $\mathscr{E}_{\theta_0}[\exp \lambda \rho(\Delta_n)]$ need not be finite, the expectations $\mathscr{E}_Q[\exp \lambda \rho(\Delta^\alpha)]$ and $\mathscr{E}_{\theta_0}[\exp \lambda \rho(\Delta_n^\alpha)]$ are finite for every $\alpha, \lambda > 0$ and all n. What is more, the following proposition holds true.

PROPOSITION 1.4   For every $\alpha, \lambda > 0$, we have

$$\mathscr{E}_{\theta_0}[\exp \lambda \rho(\Delta_n^\alpha)] \to \mathscr{E}_Q[\exp \lambda \rho(\Delta^\alpha)].$$

Proof   Since the distributions $\mathscr{L}[\rho(\Delta_n^\alpha) | P_{\theta_0}]$ and $\mathscr{L}[\rho(\Delta^\alpha) | Q]$ have supports confined to the interval $[0, \alpha]$, it suffices to restrict the integration to this interval. Then we obtain

$$\mathscr{E}_{\theta_0}[\exp \lambda \rho(\Delta_n^\alpha)] = \int_{[0, \alpha]} \exp \lambda x \, \mathrm{d}\mathscr{L}[\rho(\Delta_n^\alpha) | P_{\theta_0}]$$

$$\to \int_{[0, \alpha]} \exp \lambda x \, \mathrm{d}\mathscr{L}[\rho(\Delta^\alpha) | Q]$$

by Proposition 1.3 (ii) and the fact that the integrand is bounded and continuous over the interval $[0, \alpha]$.∎

Now let $\{\alpha_\nu\}$ and $\{\epsilon_\nu\}$ be two sequences such that $0 < \alpha_\nu \uparrow \infty$ and $0 < \epsilon_\nu \downarrow 0$ as $\nu \to \infty$. Consider $\lambda_0 > 0$ to be specified later on (see relation (3.10)). For every α_ν, ϵ_ν and λ_0 as above, there exists a positive integer $N_\nu = N(\alpha_\nu, \epsilon_\nu, \lambda_0)$ such that

$$|\mathscr{E}_{\theta_0}[\exp \lambda_0 \rho(\Delta_n^{\alpha_\nu})] - \mathscr{E}_Q[\exp \lambda_0 \rho(\Delta^{\alpha_\nu})]| \leqslant \epsilon_\nu$$
$$(n \geqslant N_\nu, \nu = 1, 2, \ldots). \quad (1.4)$$

This is so by Proposition 1.4. Clearly, $\{N_\nu\}$ can be chosen so that $N_\nu \uparrow \infty$. Then we define a subsequence of k-dimensional random vectors Δ_n^* by

$$\Delta_n^* = \Delta_n^{\alpha_\nu} \quad \text{for } n \text{ such that } N_\nu \leqslant n < N_{\nu+1} \quad (\nu = 1, 2, \ldots). \quad (1.5)$$

Set $\Delta_n^* = 0$ for $n < N_1$.

Since for any $h \in R^k$,

$$|h'\Delta_n^\alpha| \leqslant \|h\| \, \|\Delta_n^\alpha\| \leqslant \alpha\|h\|,$$

it follows that $\mathscr{E}_{\theta_0}(\exp h'\Delta_n^\alpha) < \infty$. In particular,

$$\mathscr{E}_{\theta_0}(\exp h'\Delta_n^*) < \infty$$

by virtue of (1.5). Thus we may define the following quantity

$$\exp B_n(h) = \mathscr{E}_{\theta_0}(\exp h'\Delta_n^*) \quad (h \in R^k) \quad (1.6)$$

and also the probability measure $R_{n,h}$ on $\mathscr{A}_n$ by

$$R_{n,h}(A) = \exp\left[-B_n(h)\right] \int_A \exp\left(h'\Delta_n^*\right) dP_{\theta_0}. \quad (1.7)$$

Then we can formulate the following fundamental result.

THEOREM 1.1   Let $R_{n,h}$ be defined by (1.7) and for a bounded sequence $\{h_n\}$ in R^k, set $\theta_n = \theta_0 + h_n n^{-\frac{1}{2}}$. Then we have

$$\|P_{n,\theta_n} - R_{n,h_n}\| \to 0. \quad (1.8)$$

The proof of the theorem is presented in Section 3 after some auxiliary results have been established in the following section.

2   Some auxiliary results

In this section, four propositions and two lemmas are formulated and proved. Their purpose is to facilitate the proof of Theorem 1.1. Some of them, however, such as Propositions 2.3 and 2.4, will be employed on other occasions as well.

PROPOSITION 2.1   For every $\lambda > 0$, we have

$$\mathscr{E}_Q[\exp \lambda \rho(\Delta^\alpha)] \to \mathscr{E}_Q[\exp \lambda \rho(\Delta)] \quad \text{as} \quad \alpha \to \infty.$$

Proof   One has

$$\mathscr{E}_Q[\exp \lambda \rho(\Delta^\alpha)] = \mathscr{E}_Q\{\exp \lambda \rho(\Delta) \, I_{[\rho(\Delta) \leqslant \alpha]}(\Delta)\}.$$

Now $\exp \lambda \rho(\Delta) I_{[\rho(\Delta) \leqslant \alpha]}(\Delta)$ is bounded by $\exp \lambda \rho(\Delta)$ which is independent of α and Q-integrable by Proposition 1.2. Also since

$$\exp \lambda \rho(\Delta) I_{[\rho(\Delta) \leqslant \alpha]}(\Delta) \to \exp \lambda \rho(\Delta) \quad \text{as} \quad \alpha \to \infty,$$

the dominated convergence theorem applies and gives the result.]

PROPOSITION 2.2   Let Δ_n^* be defined by (1.5) and let λ_0 be as in (1.4). Then

$$\mathscr{E}_{\theta_0}[\exp \lambda_0 \rho(\Delta_n^*)] \to \mathscr{E}_Q[\exp \lambda_0 \rho(\Delta)].$$

Proof   Clearly, on account of (1.4), one has

$$\left| \mathscr{E}_{\theta_0}[\exp \lambda_0 \rho(\Delta_n^*)] - \mathscr{E}_Q[\exp \lambda_0 \rho(\Delta^{\alpha_\nu})] \right| \leqslant \epsilon_\nu$$
$$(N_\nu \leqslant n < N_{\nu+1}, \nu = 1, 2, \ldots).$$

Thus

$$\mathscr{E}_{\theta_0}[\exp \lambda_0 \rho(\Delta_n^*)] - \mathscr{E}_Q[\exp \lambda_0 \rho(\Delta^{\alpha_\nu})] \to 0 \quad \text{as} \quad \nu \to \infty \quad (2.1)$$

(which implies $n \to \infty$ because of $N_\nu \to \infty$ and (1.5)). But by Proposition 2.1,

$$\mathscr{E}_Q[\exp \lambda_0 \rho(\Delta^{\alpha_\nu})] \to \mathscr{E}_Q[\exp \lambda_0 \rho(\Delta)]. \quad (2.2)$$

Relations (2.1) and (2.2) complete the proof of the proposition.]

Now it is, of course, important to know what is the relationship of Δ_n^* to Δ_n. According to the following fact, they differ only on a set whose probability tends to zero. More precisely, we have

PROPOSITION 2.3   Let Δ_n^* be defined by (1.5) and let θ_n be as in Theorem 1.1. Then

(i) $P_{\theta_0}(\Delta_n^* \neq \Delta_n) \to 0$,

and

(ii) $P_{\theta_n}(\Delta_n^* \neq \Delta_n) \to 0$.

Proof   From the definition of Δ_n^*, one has

$$\Delta_n^* = \Delta_n^{\alpha_\nu} = \Delta_n I_{[\rho(\Delta_n) \leqslant \alpha_\nu]}(\Delta_n) \quad (N_\nu \leqslant n < N_{\nu+1}, \nu = 1, 2, \ldots).$$

Thus $(\Delta_n^* \neq \Delta_n) = [\rho(\Delta_n) > \alpha_\nu]$

and therefore $P_{\theta_0}(\Delta_n^* \neq \Delta_n) = P_{\theta_0}[\rho(\Delta_n) > \alpha_\nu]. \quad (2.3)$

But $\mathscr{L}[\rho(\Delta_n) | P_{\theta_0}] \Rightarrow \mathscr{L}[\rho(\Delta) | Q]$ by Proposition 1.1 and the convergence of the corresponding distribution functions (d.f.s) is uniform. This is so by Polya's lemma (see, e.g. Rao [3], p. 100, (vi)), since Q is an absolutely continuous measure. Since $\alpha_\nu \to \infty$ as $\nu \to \infty$ (which imply $n \to \infty$), by taking the limits in (2.3), we obtain the first of the proposition. The second conclusion follows from the first part and contiguity of $\{P_{\theta_0}\}$ and $\{P_{\theta_n}\}$ by Proposition 6.1, Chapter 2. ∎

By Proposition 2.3 and Theorems 4.2 and 4.6, Chapter 2, we obtain

PROPOSITION 2.4 Let Δ_n^* be defined by (1.5). Then one has
  (i) $\mathscr{L}(\Delta_n^*|P_{\theta_0}) \Rightarrow N(0, \Gamma)$,
and
  (ii) $\mathscr{L}(\Delta_n^*|P_{\theta_0 + h_n n^{-\frac{1}{2}}}) \Rightarrow N(\Gamma h, \Gamma)$, provided $h_n \to h$.

This section is closed with two lemmas and one corollary which will be used in the following section.

For $n \geqslant 1$, let X_n be r.v.s defined on the probability space $(\Omega, \mathscr{F}, P)$. Then we recall that the r.v.s $|X_n|$, $n \geqslant 1$, are said to be *uniformly integrable* if $\displaystyle\int_{(|X_n| \geqslant a)} |X_n| \, dP \to 0$ uniformly in n as $a \to \infty$ (see, e.g. Loève [1], p. 162). The following lemma provides sufficient conditions for uniform integrability. Namely,

LEMMA 2.1 For $n \geqslant 1$, let X_n be r.v.s defined on the probability space $(\Omega, \mathscr{F}, P)$ and let X be an r.v. defined on the probability space $(\Omega', \mathscr{F}', P')$. Furthermore, suppose that $\mathscr{E}|X_n| \to \mathscr{E}|X|$ finite, $\mathscr{L}(X_n|P) \Rightarrow \mathscr{L}(X|P')$. Then the r.v.s $|X_n|$, $n \geqslant 1$, are uniformly integrable.

Proof For simplicity, set F_n, F for the distribution functions corresponding to $\mathscr{L}(X_n|P)$, $\mathscr{L}(X|P')$, respectively. Then we have

$$\int_{(|X_n| \geqslant a)} |X_n| \, dP = \int_{(|x| \geqslant a)} |x| \, dF_n \leqslant \int |x| \, dF_n - \int |x| \, dF$$

$$+ \left| \int_{(|x| < a)} |x| \, dF_n - \int_{(|x| < a)} |x| \, dF \right| + \int_{(|x| \geqslant a)} |x| \, dF$$

$$= |\mathscr{E}|X_n| - \mathscr{E}|X|| + \left| \int_{(|x| < a)} |x| \, dF_n - \int_{(|x| < a)} |x| \, dF \right|$$

$$+ \int_{(|x| \geqslant a)} |x| \, dF.$$

Now $|\mathscr{E}|X_n| - \mathscr{E}|X|| < \frac{1}{3}\epsilon$ for all $n \geqslant n_1$, say, by assumption; $\int_{(|x| \geqslant a)} |x| \, dF < \frac{1}{3}\epsilon$ for all sufficiently large a, by the assumption that $\mathscr{E}|X| < \infty$. By choosing a as just stated and also such that both a and $-a$ are continuity points of F, one has

$$\left| \int_{(|x| < a)} |x| \, dF_n - \int_{(|x| < a)} |x| \, dF \right| < \frac{1}{3}\epsilon$$

for all $n \geqslant n_2$, say, by the Helly–Bray lemma (see Loève [1], p. 180). Thus with a as before and $n \geqslant n_3 = \max(n_1, n_2)$, one has that $\int_{(|X_n| \geqslant a)} |X_n| \, dP < \epsilon$. Increasing a so that to make this integral $< \epsilon$ for $n = 1, \ldots, n_3 - 1$, we obtain the desired result.]

LEMMA 2.2 For $n \geqslant 1$, let X_n, Y_n be r.v.s defined on the probability space $(\Omega, \mathscr{F}, P)$ and suppose that the r.v.s $|X_n|$, $n \geqslant 1$, and $|Y_n|$, $n \geqslant 1$, are uniformly integrable and that $X_n - Y_n \xrightarrow{P} 0$. Then the r.v.s $|X_n - Y_n|$, $n \geqslant 1$, are uniformly integrable.

Proof We have

$$\int_{(|X_n - Y_n| \geqslant a)} |X_n - Y_n| \, dP \leqslant \int_{(|X_n - Y_n| \geqslant a)} |X_n| \, dP$$

$$+ \int_{(|X_n - Y_n| \geqslant a)} |Y_n| \, dP \qquad (2.4)$$

and

$$\int_{(|X_n - Y_n| \geqslant a)} |X_n| \, dP$$

$$= \int_{(|X_n - Y_n| \geqslant a) \cap (|X_n| \geqslant c)} |X_n| \, dP$$

$$+ \int_{(|X_n - Y_n| \geqslant a) \cap (|X_n| < c)} |X_n| \, dP$$

$$\leqslant \int_{(|X_n| \geqslant c)} |X_n| \, dP + cP(|X_n - Y_n| \geqslant a). \qquad (2.5)$$

Thus for all sufficiently large c, we have $\int_{(|X_n| \geqslant c)} |X_n| \, dP < \frac{1}{2}\epsilon$ for all n and for all sufficiently large a, we have

$$P(|X_n - Y_n| \geqslant a) < \epsilon/2c$$

for all $n \geqslant n_0$, say. Therefore for c, a as just stated and for all $n \geqslant n_0$ relation (2.4) gives $\int_{(|X_n - Y_n| \geqslant a)} |X_n| \, dP < \epsilon$. Increasing a further so that this integral becomes $< \epsilon$ for $n = 1, \ldots, n_0 - 1$, we obtain the desired result. ∎

The lemma just proved has the following corollary.

COROLLARY 2.1   Under the assumptions of Lemma 2.2, one has that $\mathscr{E} |X_n - Y_n| \to 0$.

Proof   It follows from the assumption that $X_n - Y_n \xrightarrow{P} 0$, the uniform integrability of the r.v.s $|X_n - Y_n|$, $n \geqslant 1$, as concluded in the lemma, and the L_r-convergence theorem (see Loève [1], p. 163). ∎

3   The proof of the theorem

On the basis of the results of the first two sections, we may now present the proof of Theorem 1.1.

PROOF OF THEOREM 1.1   The proof is by contradiction. Assume that (1.8) is not true. Then there exists a subsequence $\{m\} \subseteq \{n\}$ and $\{h_m\} \subseteq \{h_n\}$ with $h_m \to h$ such that

$$\|P_{\theta_m} - R_{m, h_m}\| \nrightarrow 0. \tag{3.1}$$

From Theorem 4.1 in Chapter 2, one, clearly, has that

$$[dP_{\theta_m}/dP_{\theta_0}] = \exp \Lambda(\theta_0, \theta_m) = L_m(h_m)$$
$$= \exp [h_m' \Delta_m - A(h_m) + Z_m(h_m)], \tag{3.2}$$

where

$$A(h_m) = \tfrac{1}{2} h_m' \Gamma h_m \quad \text{and} \quad Z_m(h_m) \to 0 \quad \text{in } P_{\theta_0}\text{-probability.} \tag{3.3}$$

From (1.6) and (1.7), one also has that

$$[dR_{m, h_m}/dP_{\theta_0}] = L_m^*(h_m) = \exp [-B_m(h_m) + h_m' \Delta_m^*]. \tag{3.4}$$

Then from (3.2) and (3.4), one has that

$$|L_m(h_m) - L_m^*(h_m)|$$
$$\leqslant [|h_m'(\Delta_m - \Delta_m^*)| + |Z_m(h_m)| + |B_m(h_m) - A(h_m)|] \exp T_m,$$
$$\tag{3.5}$$

where T_m is an r.v. lying between the r.v.s $h'_m \Delta_m - A(h_m) + Z_m(h_m)$ and $-B_m(h_m) + h'_m \Delta_m^*$.

The first and the second term inside the brackets on the right-hand side of (3.5) converge to 0 in P_{θ_0}-probability by Proposition 2.3 and relation (3.3), respectively. In the following, we shall also show that

$$B_m(h_m) - A(h_m) \to 0. \tag{3.6}$$

These last three conclusions imply that $\{T_m\}$ is bounded in P_{θ_0}-probability and therefore relation (3.5) implies that

$$L_m(h_m) - L_m^*(h_m) \to 0 \quad \text{in } P_{\theta_0}\text{-probability.} \tag{3.7}$$

Next Theorem 4.3 in Chapter 2 implies that

$$\mathscr{L}[L_m(h_m)|P_{\theta_0}] \Rightarrow \mathscr{L}(\exp \Delta^*|Q^*),$$

where Δ^* is the identity mapping on R and

$$Q^* = N(-\tfrac{1}{2}h'\Gamma h, \, h'\Gamma h),$$

whereas $\mathscr{E}_{Q^*} \exp \Delta^* = 1$ by Corollary 7.1, Chapter 1. Therefore Lemma 2.1 applies with $\{n\}$ replaced by $\{m\}$,

$$(\Omega, \mathscr{F}, P) = (\mathscr{X}, \mathscr{A}, P_{\theta_0}),$$

$$(\Omega', \mathscr{F}', P') = (R, \mathscr{B}, Q^*)$$

and $\qquad X_m = L_m(h_m)\,(L_m^*(h_m)), \, X = \exp \Delta^*$

and gives that the r.v.s

$$|L_m(h_m)| = L_m(h_m), \quad |L_m^*(h_m)| = L_m^*(h_m)$$

corresponding to the sequence $\{m\}$ are uniformly integrable. This result, together with relation (3.7) and Corollary 2.1, gives that $\int |L_m(h_m) - L_m^*(h_m)|\, dP_{\theta_0} \to 0$. By Theorem 1.4 A and relations (3.2) and (3.4), this is equivalent to a contradiction of (3.1). Thus the proof of the theorem will be completed by establishing (3.6).

Now $\int \exp h'\Delta\, dQ$ is the moment generating function of $h'\Delta$ evaluated at $t = 1$. Since this is equal to $\exp A(h)$ and, clearly, $A(h_m) \to A(h)$, in order to prove (3.6), it suffices to show that

$$\exp B_m(h_m) \to \int \exp h'\Delta\, dQ \left(= \int \exp h'z\, dQ \right). \tag{3.8}$$

For simplicity, set $\mathscr{L}_n = \mathscr{L}(\Delta_n^* | P_{\theta_0})$, so that $\mathscr{L}_n \Rightarrow Q$ by Proposition 2.4 (i). Next let S_r be the (closed) sphere centred at the origin and having radius r, and let M_r be a constant such that $|\exp h'_m z - \exp h'z| \leqslant M_r \|h_m - h\|$, provided $z \in S_r$. The fact that $Q(\partial S_r) = 0$, $\mathscr{L}_m \Rightarrow Q$ and Theorem 1.1, Chapter 1, imply that $\int_{S_r} \exp h'z \, \mathrm{d}\mathscr{L}_m \to \int_{S_r} \exp h'z \, \mathrm{d}Q$. Therefore

$$\left| \int_{S_r} \exp h'_m z \, \mathrm{d}\mathscr{L}_m - \int_{S_r} \exp h'z \, \mathrm{d}Q \right|$$

$$\leqslant \int_{S_r} |\exp h'_m z - \exp h'z| \, \mathrm{d}\mathscr{L}_m$$

$$+ \left| \int_{S_r} \exp h'z \, \mathrm{d}\mathscr{L}_m - \int_{S_r} \exp h'z \, \mathrm{d}Q \right|$$

$$\leqslant M_r \|h_m - h\| + \left| \int_{S_r} \exp h'z \, \mathrm{d}\mathscr{L}_m - \int_{S_r} \exp h'z \, \mathrm{d}Q \right| \to 0,$$

so that
$$\int_{S_r} \exp h'_m z \, \mathrm{d}\mathscr{L}_m \to \int_{S_r} \exp h'z \, \mathrm{d}Q. \qquad (3.9)$$

Let λ_0 be defined by

$$\|h_n\| \leqslant \lambda_0, \quad \|h\| < \lambda_0. \qquad (3.10)$$

Next $\left| \int_{S_r^c} \exp h'_m z \, \mathrm{d}\mathscr{L}_m - \int_{S_r^c} \exp h'z \, \mathrm{d}Q \right|$

$$\leqslant \int_{S_r^c} \exp |h'_m z| \, \mathrm{d}\mathscr{L}_m + \int_{S_r^c} \exp |h'z| \, \mathrm{d}Q$$

$$\leqslant \int_{S_r^c} \exp \lambda_0 \rho(z) \, \mathrm{d}\mathscr{L}_m + \int_{S_r^c} \exp \lambda_0 \rho(z) \, \mathrm{d}Q. \quad (3.10')$$

Now since $\int \exp \lambda_0 \rho(z) \, \mathrm{d}Q < \infty$ by Proposition 1.2, we may choose r sufficiently large, so that

$$\int_{S_r^c} \exp \lambda_0 \rho(z) \, \mathrm{d}Q < \epsilon. \qquad (3.11)$$

Since $\int_{S_r} \exp \lambda_0 \rho(z) \, \mathrm{d}\mathscr{L}_m \to \int_{S_r} \exp \lambda_0 \rho(z) \, \mathrm{d}Q$,

t follows from (1.4) and (1.5) that

$$\int_{S_r^c} \exp \lambda_0 \rho(z) \, \mathrm{d}\mathscr{L}_m \to \int_{S_r^c} \exp \lambda_0 \rho(z) \, \mathrm{d}Q,$$

so that for all sufficiently large m, one has that

$$\int_{S_r^c} \exp \lambda_0 \rho(z)\, \mathrm{d}\mathscr{L}_m < 2\epsilon. \qquad (3.12)$$

By (3.10'), (3.11), and (3.12), it follows that

$$\int_{S_{rj}^c} \exp h_m' z\, \mathrm{d}\mathscr{L}_m \to \int_{S_r^c} \exp h' z\, \mathrm{d}Q$$

which, together with (3.9), implies that

$$\int \exp h_m' z\, \mathrm{d}\mathscr{L}_m \to \int \exp h' z\, \mathrm{d}Q. \qquad (3.13)$$

Finally,

$$\exp B_m(h_m) = \int \exp h_m' \Delta_m^* \,\mathrm{d}P_{\theta_0} = \int \exp h_m' z\, \mathrm{d}\mathscr{L}_m \to \int \exp h' z\, \mathrm{d}Q$$

by (3.13), and then (3.8) completes the proof of the theorem.∎
The theorem just established has the following corollary.

COROLLARY 3.1   Let B be any bounded set in R^k. Then one has that
$$\sup \left[\|P_{\theta_n} - R_{n,h}\|;\ h \in B,\ \theta_n = \theta_0 + hn^{-\frac{1}{2}}\right] \to 0. \qquad (3.14)$$

Proof   The proof is by contradiction. Suppose that (3.14) is not true. Then there exists a subsequence $\{m\} \subseteq \{n\}$ and $\{h_m\}$ in B such that $\|P_{\theta_m^*} - R_{m,h_m}\| \to \delta$, some $\delta > 0$, where $\theta_m^* = \theta_0 + h_m m^{-\frac{1}{2}}$. But this contradicts Theorem 1.1. Thus the corollary holds true.∎

REMARK 3.1   A contradiction argument as the one used to establish (3.17) will be used on many occasions.

4   Differential equivalence of sequences of probability measures and differential sufficiency

Two sequences of probability measures which converge in L_1-norm are indistinguishable asymptotically. Such sequences ought to be equivalent asymptotically. For two sequences of probability measures which depend on a parameter, we have the following definition of asymptotic equivalence.

DEFINITION 4.1   For $\theta \in \Theta$, let $\{Q_{n,\theta}^{(1)}\}$ and $\{Q_{n,\theta}^{(2)}\}$ be two sequences of probability measures on $\mathscr{A}_n$. Then the two sequences are said to be *differentially (asymptotically) equivalent* at the point $\theta_0 \in \Theta$ if for every bounded set B in R^k, the following convergence holds true.

$$\sup [\|Q_{n,\theta}^{(1)} - Q_{n,\theta}^{(2)}\|; (\theta - \theta_0) n^{\frac{1}{2}} \in B] \to 0. \qquad (4.1)$$

On account of (4.1), if $\theta_n \in \Theta$ and $\theta_n \to \theta_0$ at the appropriate rate, then

$$\|Q_{n,\theta_n}^{(1)} - Q_{n,\theta_n}^{(2)}\| \to 0.$$

In other words, in a neighbourhood of θ_0, either one of the sequences $\{Q_{n,\theta_n}^{(1)}\}$, $\{Q_{n,\theta_n}^{(2)}\}$ is as good as the other.

By setting $(\theta - \theta_0) n^{\frac{1}{2}} = h$ and writing θ_n instead of θ, one has $\theta_n = \theta_0 + h n^{-\frac{1}{2}}$. Therefore the defining relation (4.1) becomes as follows:

$$\sup [\|Q_{n,\theta_n}^{(1)} - Q_{n,\theta_n}^{(2)}\|; h \in B, \theta_n = \theta_0 + h n^{-\frac{1}{2}}] \to 0. \qquad (4.2)$$

At this point it will be convenient to introduce the following notation. For each n and all $\theta \in \Theta$, define the probability measure $R_{n,\theta}^*$ as follows:

$$R_{n,\theta}^* = R_{n,(\theta - \theta_0) n^{\frac{1}{2}}}. \qquad (4.3)$$

Then we can present the following important corollary to Theorem 1.1.

COROLLARY 4.1   Let $R_{n,\theta}^*$ be given by (4.3). Then the sequences of probability measures $\{P_{n,\theta}\}$ and $\{R_{n,\theta}^*\}$ are differentially equivalent at θ_0.

Proof   According to (4.2), one has to show that

$$\sup [\|P_{\theta_n} - R_{n,\theta_n}^*\|; h \in B, \theta_n = \theta_0 + h n^{-\frac{1}{2}}] \to 0$$

for every bounded set B in R^k.

By virtue of (4.3), this is equivalent to proving that

$$\sup [\|P_{\theta_n} - R_{n,h}\|; h \in B, \theta_n = \theta_0 + h n^{-\frac{1}{2}}] \to 0. \qquad (4.4)$$

However, (4.4) is true because of Corollary 3.1.∎

The following definition introduces the concept of asymptotic sufficiency of a sequence of statistics, or σ-fields, for a given family of probability measures. Specifically, one has

DEFINITION 4.2   For $\theta \in \Theta$, let $\{Q_{n,\theta}^{(1)}\}$ be a sequence of probability measures on $\mathscr{A}_n$ and let $\{V_n\}$ be a sequence of k-dimensional, $\mathscr{A}_n$-measurable random vectors. Denote by $\mathscr{B}_n$ the σ-field induced by V_n. Then the sequence $\{V_n\}$ or $\{\mathscr{B}_n\}$ is said to be *differentially (asymptotically) sufficient* at θ_0 for the family $\{Q_{n,\theta}^{(1)}; \theta \in \Theta\}$, if there exists a family of probability measures $\{Q_{n,\theta}^{(2)}; \theta \in \Theta\}$ such that, for each n, V_n or $\mathscr{B}_n$ is sufficient for the family $\{Q_{n,\theta}^{(2)}; \theta \in \Theta\}$ and $\{Q_{n,\theta}^{(1)}\}, \{Q_{n,\theta}^{(2)}\}$ are differentially equivalent at θ_0.

The interpretation of asymptotic sufficiency is obvious on account of Definition 4.1 and the comments following it. Namely, for sufficiently large n, in a neighbourhood of θ_0, V_n, or $\mathscr{B}_n$, may be used as if it were sufficient for $\{Q_{n,\theta}^{(1)}; \theta \text{ in a neighbourhood of } \theta_0\}$.

Now according to the following result, the random vector Δ_n^* possesses the property of being differentially sufficient at θ_0 for the family $\{P_{n,\theta}; \theta \in \Theta\}$. More precisely

COROLLARY 4.2 Let Δ_n^* be defined by (1.5). Then the sequence $\{\Delta_n^*\}$ is differentially sufficient at θ_0 for the family $\{P_{n,\theta}; \theta \in \Theta\}$.

Proof By (4.3) and the definition of $R_{n,h}$ in (1.7), one has

$$[\mathrm{d}R_{n,\theta}^*/\mathrm{d}P_{\theta_0}]$$
$$= [\mathrm{d}R_{n,(\theta-\theta_0)n^{\frac{1}{2}}}/\mathrm{d}P_{\theta_0}]$$
$$= \exp\{-B_n[(\theta-\theta_0)n^{\frac{1}{2}}]\}\exp(n^{\frac{1}{2}}\theta'\Delta_n^*)\exp(-n^{\frac{1}{2}}\theta_0'\Delta_n^*).$$

From this it follows that, for each n, Δ_n^* is sufficient for the family $\{R_{n,\theta}^*; \theta \in \Theta\}$. Since $\{P_{n,\theta}\}$ and $\{R_{n,\theta}^*\}$ are differentially equivalent at θ_0 by Corollary 4.1, the desired result follows. ∎

5   Some statistical implications of Theorem 1.1

Some of the statistical implications of Theorem 1.1 are summarized in the following two theorems. The first of these results states that, for testing the hypothesis $H_0: \theta = \theta_0$ and from asymptotic power viewpoint, we may confine ourselves to tests which depend on Δ_n alone. That is,

THEOREM 5.1 Let $\{Z_n\}$ be a sequence of r.v.s such that $|Z_n| \leqslant 1$ for all n and set $\bar{Z}_n = \mathscr{E}_{\theta_0}(Z_n|\Delta_n)$. Then

$$\sup|\mathscr{E}_{\theta_n}Z_n - \mathscr{E}_{\theta_n}\bar{Z}_n| \to 0,$$

where $\theta_n = \theta_0 + hn^{-\frac{1}{2}}$ and, for each n, the sup is taken over all r.v.s Z_n bounded by 1 in absolute value and over all hs in any bounded subset B of R^k.

Proof   We have

$$\mathscr{E}_{\theta_n} Z_n - \mathscr{E}_{\theta_n} \bar{Z}_n = I_1(n,h) + I_2(n,h) + I_3(n,h),$$

where
$$I_1(n,h) = \mathscr{E}(Z_n | P_{\theta_n}) - \mathscr{E}(Z_n | R_{n,h})$$

$$I_2(n,h) = \mathscr{E}(Z_n | R_{n,h}) - \mathscr{E}(\bar{Z}_n | R_{n,h})$$

$$I_3(n,h) = \mathscr{E}(\bar{Z}_n | R_{n,h}) - \mathscr{E}(\bar{Z}_n | P_{\theta_n}).$$

From the assumption that $|Z_n| \leqslant 1$, the definition of $\bar{Z}_n$ and the fact that $R_{n,h} \ll P_{\theta_0}$, $P_{\theta_n} \ll P_{\theta_0}$, one has that $\bar{Z}_n$ is also bounded by 1 a.s. $[R_{n,h}]$ and $[P_{\theta_n}]$. Then with the sup as above, both $\sup |I_1(n,h)|$ and $\sup |I_3(n,h)|$ are bounded by

$$\sup [\|P_{\theta_n} - R_{n,h}\| ; h \in B]$$

which tends to 0 by (4.4). We next show that $I_2(n,h) = 0$, which will complete the proof of the theorem. In fact,

$$\mathscr{E}(\bar{Z}_n | R_{n,h}) = \mathscr{E}[\mathscr{E}_{\theta_0}(Z_n | \Delta_n) | R_{n,h}] = \int [\mathscr{E}_{\theta_0}(Z_n | \Delta_n)] \, \mathrm{d}R_{n,h}$$

$$= \int [\mathscr{E}_{\theta_0}(Z_n | \Delta_n)] \exp[-B_n(h)] \exp(h' \Delta_n^*) \, \mathrm{d}P_{\theta_0}$$

$$= \int [\mathscr{E}_{\theta_0}\{Z_n \exp[-B_n(h)] \exp(h' \Delta_n^*) | \Delta_n\}] \, \mathrm{d}P_{\theta_0},$$

the last equality holding true because Δ_n^* is a function of Δ_n.

By rewriting the last expression in terms of expectations and using the definition of $R_{n,h}$ once more, one obtains

$$\mathscr{E}_{\theta_0}[\mathscr{E}_{\theta_0}\{Z_n \exp[-B_n(h)] \exp(h' \Delta_n^*) | \Delta_n\}]$$

$$= \mathscr{E}_{\theta_0}\{Z_n \exp[-B_n(h)] \exp(h' \Delta_n^*)\}$$

$$= \mathscr{E}(Z_n | R_{n,h}).$$

That is,
$$\mathscr{E}(\bar{Z}_n | R_{n,h}) = \mathscr{E}(Z_n | R_{n,h}),$$

as was to be seen. ▮

Again for testing the hypothesis H_0 and when our interest lies in the asymptotic behaviour of the power of tests, the theorem

below is quite relevant. It states that one can actually base all
tests on Δ_n^* alone. The significance of this statement is apparent
by taking into consideration that Δ_n^* plays the all-important role
of the statistic appearing in the exponent of an exponential
family. The exact statement of the theorem is as follows.

THEOREM 5.2　Let $\{Z_n\}$ be a sequence of tests defined on
R^k and set $\theta_n = \theta_0 + hn^{-\frac{1}{2}}$, $h \in R^k$. Then for any bounded subset B
of R^k, we have

$$\sup\left[\left|\mathscr{E}_{\theta_n}Z_n(\Delta_n) - \mathscr{E}_{\theta_n}Z_n(\Delta_n^*)\right|; h \in B\right] \to 0.$$

Proof　We have

$$\mathscr{E}_\theta Z_n(\Delta_n) - \mathscr{E}_{\theta_n}Z_n(\Delta_n^*) = \int Z_n(\Delta_n)\,\mathrm{d}P_{\theta_n} - \int Z_n(\Delta_n^*)\,\mathrm{d}P_{\theta_n}$$

$$= \int [Z_n(\Delta_n) - Z_n(\Delta_n^*)]\,\mathrm{d}P_{\theta_n}$$

$$= \int_{(\Delta_n = \Delta_n^*)} [Z_n(\Delta_n) - Z_n(\Delta_n^*)]\,\mathrm{d}P_{\theta_n}$$

$$+ \int_{(\Delta_n \neq \Delta_n^*)} [Z_n(\Delta_n) - Z_n(\Delta_n^*)]\,\mathrm{d}P_{\theta_n}$$

$$= \int_{(\Delta_n \neq \Delta_n^*)} [Z_n(\Delta_n) - Z_n(\Delta_n^*)]\,\mathrm{d}P_{\theta_n}.$$

Hence

$$\left|\mathscr{E}_{\theta_n}Z_n(\Delta_n) - \mathscr{E}_{\theta_n}Z_n(\Delta_n^*)\right| \leqslant 2P_{\theta_n}(\Delta_n \neq \Delta_n^*)$$

and therefore

$$\sup\left[\left|\mathscr{E}_{\theta_n}Z_n(\Delta_n) - \mathscr{E}_{\theta_n}Z_n(\Delta_n^*)\right|; h \in B\right] \leqslant 2\sup\left[P_{\theta_n}(\Delta_n \neq \Delta_n^*); h \in B\right].$$
$$(5.1)$$

Now let $\{h_n\}$ be a sequence in B and set $\theta_n^* = \theta_0 + h_n n^{-\frac{1}{2}}$. Then, by
Proposition 2.3 (ii), we obtain

$$P_{\theta_n^*}(\Delta_n \neq \Delta_n^*) \to 0. \qquad (5.2)$$

From (5.2), it easily follows by a contradiction argument, that

$$\sup\left[P_{\theta_n}(\Delta_n \neq \Delta_n^*); h \in B\right] \to 0. \qquad (5.3)$$

Relations (5.1) and (5.3) give the desired result.]

REMARK 5.1 The various tests shall be actually expressed in terms of Δ_n. The introduction of Δ_n^* is a mathematical device by which the original family was approximated by an exponential family defined in terms of Δ_n^*. Exploiting the distinguished role that Δ_n^* plays in connection with this exponential family, one is able to set up tests based on Δ_n^*. Then Theorem 5.2 allows one to pass on to tests based on Δ_n. Furthermore, one need not be concerned with tests not based on Δ_n. This is so by Theorem 5.1. Concrete statistical applications will illustrate the point even further.

Exercises

1. Let $X_1, ..., X_n$ be r.v.s defined on the probability space $(\Omega, \mathscr{F}, P)$. Then show that

$$\mathscr{E} |X_1 ... X_n| \leqslant \mathscr{E}^{1/q_1} |X_1|^{q_1} ... \mathscr{E}^{1/q_n} |X_n|^{q_n},$$

where $q_1, ..., q_n > 0$ and $1/q_1 + ... + 1/q_n = 1$.

2. (i) For each $n \geqslant 1$, let X_n, Y_n be two r.v.s defined on the probability space $(\Omega, \mathscr{F}_n, P_n)$ and let Z be an r.v. defined on the probability space $(\Omega', \mathscr{F}', P')$. Suppose that

$$\mathscr{L}(X_n|P_n) \Rightarrow \mathscr{L}(Z|P') \quad \text{and} \quad Y_n - X_n \to c \quad \text{in } P_n\text{-probability,}$$

where c is a constant. Then show that

$$\mathscr{L}(Y_n|P_n) \Rightarrow \mathscr{L}(Z+c|P').$$

(ii) Let $\{X_n\}$, $\{Y_n\}$ be as above and let X, Y be r.v.s defined on the probability space $(\Omega', \mathscr{F}', P')$. Then show that

$$\mathscr{L}[(X_n, Y_n)|P_n] \Rightarrow \mathscr{L}[(X, Y)|P']$$

if and only if

$$\mathscr{L}(\lambda_1 X_n + \lambda_2 Y_n | P_n) \Rightarrow \mathscr{L}(\lambda_1 X + \lambda_2 Y | P') \quad \text{for every } \lambda, \lambda_2 \text{ in } R.$$

4. Some statistical applications: AUMP and AUMPU tests for certain testing hypotheses problems

Summary

In the present chapter, we apply results obtained in the last two chapters to certain testing hypotheses problems and for the case that Θ is a subset of R.

All results derived so far in Chapters 2 and 3 were obtained under the basic assumptions (A 1–4). According to these results, the asymptotic expansion of the log-likelihood given in Theorem 4.1, Chapter 2, as well as the asymptotic distribution of the log-likelihood and also of the fundamental random vector $\Delta_n(\theta)$ (for its definition, see (1.1) below) given by Theorems 4.2–4.6, Chapter 2, were valid for moving parameter points θ_n essentially of the following form: $\theta_n = \theta + hn^{-\frac{1}{2}}$, $h \in R$. Thus as $n \to \infty$, θ_n approaches the fixed parameter point θ and at a prescribed rate. If θ_n either does not approach θ at this rate or it stays away from θ, equivalently, if $n^{\frac{1}{2}}(\theta_n - \theta) \to \infty(-\infty)$, the above mentioned theorems need not be true. However, in the problems to be considered in this chapter and also Chapter 6, one has to consider θ_ns for which $n^{\frac{1}{2}}(\theta_n - \theta) \to \infty(-\infty)$. This suggests the need for additional assumptions to accommodate this case. These assumptions, labelled as (A 5) and (A 5'), are introduced in the first section of the present chapter and require that the random vector $\Delta_n(\theta_0)$ converges to $\infty(-\infty)$ in P_{θ_n}-probability whenever

$$n^{\frac{1}{2}}(\theta_n - \theta_0) \to \infty(-\infty).$$

Assumptions (A 5), (A 5') are rather mild assumptions to make and for the sake of illustrations, they have been checked and found to be true for the examples of Chapter 2. The relevant discussion is presented in Examples 1.1–1.4 below.

The testing hypotheses problems referred to above and discussed in this chapter are essentially the following ones:

$$H_0: \theta = \theta_0 (\in \Theta) \quad \text{against} \quad A: \theta > \theta_0 (\theta \in \Theta)$$
$$H_0: \theta = \theta_0 \quad \text{against} \quad A'': \theta \neq \theta_0$$
$$H_1: \theta \leqslant \theta_0 \quad \text{against} \quad A: \theta > \theta_0.$$

On the basis of results obtained in Chapter 2, it is shown in Theorem 3.1 below that the test defined by (3.2) and (3.3) and based on the random vector $\Delta_n(\theta_0)$ is asymptotically uniformly most powerful (AUMP; see Definition 3.1) for testing H_0 against A above, provided assumptions (A1–4), (A5) hold. Under assumptions (A1–4) alone, the test in question is asymptotically locally most powerful (see Corollary 3.1).

In the following section, Section 4, we derive the AUMP tests for the illustrative Examples 1.1–1.4 and also compare them with existing tests for the same examples.

For testing H_0 against A'', it is shown in Theorem 5.1 that the test defined by (5.17) and (5.18) is AUMP unbiased (AUMPU; Definition 5.1) under assumptions (A1–4), (A5), (A5'). Again under (A1–4) alone, the test is asymptotically locally most powerful unbiased, as Corollary 5.4 asserts. For the proof of Theorem 5.1, one utilizes results regarding the exponential approximation discussed in Chapter 3. At this point, one may refer again to the examples of Section 4 in order to see what the AUMPU tests look like for the concrete examples discussed there.

Finally, in Section 6, we consider the third testing hypothesis problem described above and show (see Theorem 6.1) that, under (A1–4), (A5), (A5'), the test defined by (3.2) and (3.3) is AUMP among all tests which satisfy (6.1). As before, under (A1–4), the test in question is asymptotically locally most powerful (see Corollary 6.1). Again for the proof of Theorem 6.1, one utilizes results of Chapter 3. Also tests for concrete cases are provided by the tests for Examples 1.1–1.4.

1 Additional assumptions – Examples

In the framework assumed in Chapter 2, let $X_0, X_1, ..., X_n$ be $n+1$ r.v.s from the underlying Markov process and let

$\phi_j(\theta)$, $j = 1, \ldots, n$ be the derivatives in q.m. introduced in (A 3). For each $\theta \in \Theta$, define the k-dimensional random vector $\Delta_n(\theta)$ as in (4.2); i.e.

$$\Delta_n(\theta) = 2n^{-\frac{1}{2}} \sum_{j=1}^{n} \phi_j(\theta). \tag{1.1}$$

In all of this chapter except for Section 9, it will be assumed that Θ is an open subset of R, so that $\Delta_n(\theta)$ is simply an r.v.

Now let $\theta_0 \in \Theta$ and define

$$\omega = \{\theta \in \Theta; \, \theta > \theta_0\}, \quad \omega' = \{\theta \in \Theta; \, \theta < \theta_0\}. \tag{1.2}$$

We can then formulate the following assumptions.

ASSUMPTIONS (A5) Consider a sequence $\{\theta_n\}$ with $\theta_n \in \omega$ for all n. Then $\Delta_n(\theta_0) \to \infty$ in P_{n, θ_n}-probability whenever

$$n^{\frac{1}{2}}(\theta_n - \theta_0) \to \infty.$$

(A5′) Consider a sequence $\{\theta_n\}$ with $\theta_n \in \omega'$ for all n. Then $\Delta_n(\theta_0) \to -\infty$ in P_{n, θ_n}-probability whenever $n^{\frac{1}{2}}(\theta_n - \theta_0) \to -\infty$.

Of course, ω and ω' above are defined by (1.2) and $\Delta_n(\theta_0)$ is taken from (1.1) for $\theta = \theta_0$.

Before we go any further, it would be appropriate to check and see whether (A5) and (A5′) are satisfied in the Examples 3.1–3.4 discussed in Chapter 2.

EXAMPLES

1.1 Refer to Example 3.1 in Chapter 2. Then, by means of (3.2), one has

$$\Delta_n(\theta_0) = \frac{1}{\sigma^2} n^{-\frac{1}{2}} \sum_{j=1}^{n} (X_j - \theta_0). \tag{1.3}$$

Clearly $\quad \sigma^2 \Delta_n(\theta_0) = n^{-\frac{1}{2}} \sum_{j=1}^{n} (X_j - \theta_n) + n^{\frac{1}{2}}(\theta_n - \theta_0)$

and, under P_{θ_n}, $n^{-\frac{1}{2}} \sum_{j=1}^{n} (X_j - \theta_n)$ is distributed as $N(0, \sigma^2)$. Therefore, for any $M > 0$,

$$P_{\theta_n}[\sigma^2 \Delta_n(\theta_0) > M] = P_{\theta_n}\left[n^{-\frac{1}{2}} \sum_{j=1}^{n} (X_j - \theta_n) > M - n^{\frac{1}{2}}(\theta_n - \theta_0) \right] \to 1$$

as $n^{\frac{1}{2}}(\theta_n - \theta_0) \to \infty$. Thus (A5) is fulfilled. Assumption (A5′) is also satisfied because

$$P_{\theta_n}[\sigma^2 \Delta_n(\theta_0) < -M]$$
$$= P_{\theta_n}\left[n^{-\frac{1}{2}} \sum_{j=1}^{n} (X_j - \theta_n) < -M - n^{\frac{1}{2}}(\theta_n - \theta_0) \right] \to 1$$

as above, whenever $n^{\frac{1}{2}}(\theta_n - \theta_0) \to -\infty$.

1.2 Refer to Example 3.2 in Chapter 2. Then by virtue of (3.7), one has

$$\Delta_n(\theta_0) = n^{-\frac{1}{2}} \sum_{j=1}^{n} \left[\frac{1}{2\theta_0^2}(X_j - \mu)^2 - \frac{1}{2\theta_0} \right]. \qquad (1.4)$$

As it is easily checked, relation (1.4) can be rewritten as follows

$$\Delta_n(\theta_0) = \frac{\theta_n}{\theta_0^2\sqrt{2}} n^{-\frac{1}{2}} \sum_{j=1}^{n} \frac{1}{\sqrt{2}} \left[\left(\frac{X_j - \mu}{\sqrt{\theta_n}} \right)^2 - 1 \right] + \frac{1}{2\theta_0^2} n^{\frac{1}{2}}(\theta_n - \theta_0). \qquad (1.5)$$

Now, under P_{θ_n}, the r.v. $\left(\dfrac{X_j - \mu}{\sqrt{\theta_n}} \right)^2$ is distributed as χ_1^2 and therefore its mean and variance are 1 and 2, respectively. It follows that

$$\mathscr{L} \left\{ n^{-\frac{1}{2}} \sum_{j=1}^{n} \frac{1}{\sqrt{2}} \left[\left(\frac{X_j - \mu}{\sqrt{\theta_n}} \right)^2 - 1 \right] \right\} \Rightarrow N(0, 1). \qquad (1.6)$$

This is so by the normal convergence criterion (see, e.g. Loève [1], p. 245). Suppose first that $\theta_n < \theta_0$ and $n^{\frac{1}{2}}(\theta_n - \theta_0) \to -\infty$. Then for any $M > 0$, one has

$$P_{\theta_n}[\Delta_n(\theta_0) < -M] = P_{\theta_n} \left\{ n^{-\frac{1}{2}} \sum_{j=1}^{n} \frac{1}{\sqrt{2}} \left[\left(\frac{X_j - \mu}{\sqrt{\theta_n}} \right)^2 - 1 \right] < M_n \right\}, \qquad (1.7)$$

where

$$M_n = \frac{-2\theta_0^2 M - n^{\frac{1}{2}}(\theta_n - \theta_0)}{\theta_n \sqrt{2}}.$$

Next for all $n \geqslant n_1 = n_1(M)$, $-2\theta_0^2 M > \frac{1}{2}n^{\frac{1}{2}}(\theta_n - \theta_0)$, so that

$$M_n > -\frac{n^{\frac{1}{2}}(\theta_n - \theta_0)}{\theta_n 2\sqrt{2}} \geqslant -\frac{n^{\frac{1}{2}}(\theta_n - \theta_0)}{\theta_0 2\sqrt{2}} \to \infty.$$

This result, together with (1.6), (1.7) and Polya's lemma (see, e.g. Rao [3], p. 100, (vi)) implies that $\Delta_n(\theta_0) \to -\infty$ in P_{θ_n}-probability. So (A 5′) is satisfied.

Suppose now that $\theta_n > \theta_0$ and $n^{\frac{1}{2}}(\theta_n - \theta_0) \to \infty$. Then we distinguish two cases:

(i) $\{\theta_n\}$ stays bounded and let C be a bound of it.

In this case and with M as before, one has

$$P_{\theta_n}[\Delta_n(\theta_0) > M] = P_{\theta_n} \left\{ n^{-\frac{1}{2}} \sum_{j=1}^{n} \frac{1}{\sqrt{2}} \left[\left(\frac{X_j - \mu}{\sqrt{\theta_n}} \right)^2 - 1 \right] > M_n' \right\}, \qquad (1.8)$$

where
$$M'_n = \frac{2\theta_0^2 M - n^{\frac{1}{2}}(\theta_n - \theta_0)}{\theta_n \sqrt{2}}. \qquad (1.9)$$

Next for all $n \geq n_2 = n_2(M)$, $\frac{1}{2}n^{\frac{1}{2}}(\theta_n - \theta_0) > 2\theta_0^2 M$, so that

$$M'_n < -\frac{n^{\frac{1}{2}}(\theta_n - \theta_0)}{\theta_n 2 \sqrt{2}} \leq -\frac{n^{\frac{1}{2}}(\theta_n - \theta_0)}{C 2 \sqrt{2}} \to -\infty.$$

Again this result together with (1.6), (1.8) and Polya's lemma implies that $\Delta_n(\theta_0) \to \infty$ in P_{θ_n}-probability; i.e. (A5) is fulfilled in case (i) above.

Consider now

(ii) $\{\theta_n\}$ is unbounded.

From (1.8), we have

$$P_{\theta_n}[\Delta_n(\theta_0) > M] = P_{\theta_n}\left[\sum_{j=1}^n \left(\frac{X_j - \mu}{\sqrt{\theta_n}}\right)^2 > M'_n \sqrt{(2n)} + n \right]$$

$$= \int_{M'_n \sqrt{(2n)} + n}^{\infty} f(t)\, \mathrm{d}t, \qquad (1.10)$$

where f is the density of a χ_n^2 r.v., and in a similar manner

$$P_{\theta_n^*}[\Delta_n(\theta_0) > M] = \int_{M_n^* \sqrt{(2n)} + n}^{\infty} f(t)\, \mathrm{d}t, \quad . \qquad (1.11)$$

where
$$M_n^* = \frac{2\theta_0^2 M - n^{\frac{1}{2}}(\theta_n^* - \theta_0)}{\theta_n^* \sqrt{2}}. \qquad (1.12)$$

From (1.9) and (1.12), it readily follows that

$$M_n^* \geq M'_n \quad \text{if and only if} \quad \theta_n \geq \theta_n^*.$$

Define $\{\theta_n^*\}$ as follows

$$\theta_n^* = \begin{cases} \theta_n & \text{if} \quad \theta_n \leq 2\theta_0 \\ 2\theta_0 & \text{otherwise.} \end{cases}$$

Then $\theta_n \geq \theta_n^*$. On the other hand, $\{\theta_n^*\}$ remains bounded and, clearly, $n^{\frac{1}{2}}(\theta_n^* - \theta_0) \to \infty$. But for such a choice of $\{\theta_n^*\}$, case (i) applies and gives $P_{\theta_n^*}[\Delta_n(\theta_0) > M] \to 1$, for any $M > 0$. This result, together with (1.10), (1.11), and the fact that $M_n^* \geq M_n$ for the above choice of $\{\theta_n^*\}$, implies that $P_{\theta_n}[\Delta_n(\theta_0) > M] \to 1$ for any $M > 0$. The verification of (A5) is completed. ∎

1.3 Refer to Example 3.3 in Chapter 2. Then by means of (3.9) and (3.12) one has that, with P_θ-probability equal to one,

$$\phi_j(\theta) = \tfrac{1}{2}I_{(\theta,\,\infty)}(X_j) - \tfrac{1}{2}I_{(-\infty,\,\theta)}(X_j) = I_{(\theta,\,\infty)}(X_j) - \tfrac{1}{2},$$

so that with P_θ-probability 1 for all $\theta \in R$,

$$\Delta_n(\theta_0) = 2n^{-\frac{1}{2}} \sum_{j=1}^{n} [I_{(\theta_0,\,\infty)}(X_j) - \tfrac{1}{2}] = n^{-\frac{1}{2}} \sum_{j=1}^{n} [2I_{(\theta_0,\,\infty)}(X_j) - 1]$$

$$= n^{-\frac{1}{2}} \sum_{j=1}^{n} \operatorname{sgn}(X_j - \theta_0). \tag{1.13}$$

Suppose now $\theta_n > \theta_0$ and $n^{\frac{1}{2}}(\theta_n - \theta_0) \to \infty$. Then

$$\begin{aligned}
P_{\theta_n}[\operatorname{sgn}(X_1 - \theta_0) = 1] = P_{\theta_n}(X_1 - \theta_0 > 0) &= P_{\theta_n}(X_1 > \theta_0) \\
&= P_0(X_1 > \theta_0 - \theta_n) = P_0(X_1 + \theta_n - \theta_0 > 0) \\
&= P_0[\operatorname{sgn}(X_1 + \theta_n - \theta_0) = 1],
\end{aligned}$$

the third equality being true because θ is a location parameter.

This result shows that the P_{θ_n}-distribution of $\Delta_n(\theta_0)$ is identical to the P_0-distribution of $n^{-\frac{1}{2}} \sum_{j=1}^{n} \operatorname{sgn}(X_j + \theta_n - \theta_0)$. Consequently for any $M > 0$, one has

$$P_{\theta_n}[\Delta_n(\theta_0) > M] = P_0\left[n^{-\frac{1}{2}} \sum_{j=1}^{n} \operatorname{sgn}(X_j + \theta_n - \theta_0) > M \right].$$

Next $\quad \mathcal{E}_0 \operatorname{sgn}(X_j + \theta_n - \theta_0) = 2P_0(X_1 + \theta_n - \theta_0 > 0) - 1$

$$= 1 - \exp(\theta_0 - \theta_n),$$

as is easily seen, and the second moment of $\operatorname{sgn}(X_j + \theta_n - \theta_0)$, clearly, is equal to 1. It follows that

$$\sigma_0^2 \operatorname{sgn}(X_j + \theta_n - \theta_0) = \exp(\theta_0 - \theta_n)[2 - \exp(\theta_0 - \theta_n)].$$

Set

$$Y_{nj} = \operatorname{sgn}(X_j + \theta_n - \theta_0) \quad \text{and} \quad Z_{nj} = \frac{Y_{nj} - \mathcal{E}_0 Y_{nj}}{n^{\frac{1}{2}} \sigma_0 Y_{nj}} \quad (j = 1, \dots, n).$$

Then $\quad P_{\theta_n}[\Delta_n(\theta_0) > M] = P_0\left(\sum_{j=1}^{n} Z_{nj} > M_n \right), \tag{1.14}$

where $\quad M_n = \dfrac{M - n^{\frac{1}{2}}\{1 - \exp[-(\theta_n - \theta_0)]\}}{\exp[-\tfrac{1}{2}(\theta_n - \theta_0)]\sqrt{\{2 - \exp[-(\theta_n - \theta_0)]\}}}. \tag{1.15}$

At this point we employ the following inequalities

$$1 - \exp(-x) \geqslant \tfrac{1}{2}x \quad \text{if} \quad x < \log 2$$

and
$$1 - \exp(-x) \geqslant \tfrac{1}{2} \quad \text{if} \quad x \geqslant \log 2.$$

On the basis of these inequalities,

$$n^{\frac{1}{2}}\{1 - \exp[-(\theta_n - \theta_0)]\} \geqslant \begin{cases} \tfrac{1}{2}n^{\frac{1}{2}}(\theta_n - \theta_0) & \text{if} \quad \theta_n - \theta_0 < \log 2 \\ \tfrac{1}{2}n^{\frac{1}{2}} & \text{if} \quad \theta_n - \theta_0 \geqslant \log 2. \end{cases}$$

Thus by setting d_n for the denominator in (1.15), one has

$$M_n \leqslant \begin{cases} \dfrac{2M - n^{\frac{1}{2}}(\theta_n - \theta_0)}{2d_n} & \text{if} \quad \theta_n - \theta_0 < \log 2 \\[2ex] \dfrac{2M - n^{\frac{1}{2}}}{2d_n} & \text{if} \quad \theta_n - \theta_0 \geqslant \log 2. \end{cases}$$

For all sufficiently large n, $n \geqslant n_3 = n_3(M)$, say, one has

$$\tfrac{1}{2}n^{\frac{1}{2}}(\theta_n - \theta_0) > 2M \quad \text{and} \quad \tfrac{1}{2}n^{\frac{1}{2}} > 2M.$$

Therefore

$$M_n \leqslant \begin{cases} \dfrac{-n^{\frac{1}{2}}(\theta_n - \theta_0)}{4d_n} & \text{if} \quad \theta_n - \theta_0 < \log 2 \\[2ex] -\dfrac{n^{\frac{1}{2}}}{4d_n} & \text{if} \quad \theta_n - \theta_0 \geqslant \log 2. \end{cases}$$

Finally by taking into consideration the fact that $d_n \leqslant \sqrt{2}$, one has

$$M_n \leqslant \begin{cases} -\dfrac{n^{\frac{1}{2}}(\theta_n - \theta_0)}{4\sqrt{2}} & \text{if} \quad \theta_n - \theta_0 < \log 2 \\[2ex] -\dfrac{n^{\frac{1}{2}}}{4\sqrt{2}} & \text{if} \quad \theta_n - \theta_0 \geqslant \log 2. \end{cases} \tag{1.16}$$

Letting M'_n stand for either one of the expressions at the right-hand side of (1.16), we have then

$$M'_n \to -\infty \tag{1.17}$$

and
$$P_0\left(\sum_{j=1}^{n} Z_{nj} > M_n\right) \geqslant P_0\left(\sum_{j=1}^{n} Z_{nj} > M'_n\right). \tag{1.18}$$

But for each n, $Z_{nj}, j = 1, \ldots, n$ are independent with $\mathscr{E}_0 Z_{nj} = 0$ and $\sum\limits_{j=1}^{n} \sigma_0^2 Z_{nj} = 1$. Then the normal convergence criterion (see Loève [1], p. 295) applies and gives that

$$\mathscr{L}\left(\sum_{j=1}^{n} Z_{nj} | P_0\right) \Rightarrow N(0, 1). \tag{1.19}$$

This result together with (1.17) and Polya's lemma implies that

$$P_0\left(\sum_{j=1}^{n} Z_{nj} > M_n'\right) \to 1. \tag{1.20}$$

Finally relations (1.14), (1.18) and (1.20) show that $\Delta_n(\theta_0) \to \infty$ in P_{θ_n}-probability; i.e. (A5) is fulfilled. The verification of (A5′) is done in a similar fashion and is left as an exercise (see Exercise 1).∎

1.4 Refer to Example 3.4 in Chapter 2. Then by virtue of (3.23), one has

$$\Delta_n[p(\theta_0)] = \frac{n^{-\frac{1}{2}}p^2(\theta_0)}{[1-p^2(\theta_0)]^2} \sum_{j=1}^{n} [X_{j-1}^2 - 2p(\theta_0) X_{j-1}X_j + X_j^2]$$

$$- n^{-\frac{1}{2}}\frac{1}{[1-p^2(\theta_0)]} \sum_{j=1}^{n} [p^2(\theta_0) + p(\theta_0) X_{j-1}X_j],$$

and as is easily seen, this is rewritten as follows

$$\Delta_n[p(\theta_0)] = \frac{p^2(\theta_0)}{[1-p^2(\theta_0)]^2} n^{-\frac{1}{2}} \sum_{j=1}^{n} (X_{j-1}^2 - 1)$$

$$+ \frac{p^2(\theta_0)}{[1-p^2(\theta_0)]^2} n^{-\frac{1}{2}} \sum_{j=1}^{n} (X_j^2 - 1)$$

$$- \frac{p(\theta_0)[1+p^2(\theta_0)]}{[1-p^2(\theta_0)]^2} n^{-\frac{1}{2}} \sum_{j=1}^{n} [X_{j-1}X_j - p(\theta_n)]$$

$$+ \frac{p(\theta_0)[1+p^2(\theta_0)]}{[1-p^2(\theta_0)]^2} n^{\frac{1}{2}}[p(\theta_0) - p(\theta_n)]. \tag{1.21}$$

Introduce the notation

$$c_1 = c_2 = \frac{p^2(\theta_0)}{[1-p^2(\theta_0)]^2}, \quad c_3 = -\frac{p(\theta_0)\,[1+p^2(\theta_0)]}{[1-p^2(\theta_0)]^2} = -c_4,$$

$$Y_n = n^{-\frac{1}{2}} \sum_{j=1}^n (X_{j-1}^2 - 1), \quad Z_n = n^{-\frac{1}{2}} \sum_{j=1}^n (X_j^2 - 1),$$

$$V_n = n^{-\frac{1}{2}} \sum_{j=1}^n [X_{j-1} X_j - p(\theta_n)],$$

$$\alpha_n = n^{\frac{1}{2}} [p(\theta_0) - p(\theta_n)],$$

$$\tag{1.22}$$

and also set $\quad W_n = c_1 Y_n + c_2 Z_n + c_3 V_n, \quad \beta_n = c_4 \alpha_n.$ $\qquad(1.23)$

Then we have $\qquad \Delta_n[p(\theta_0)] = W_n + \beta_n.$ $\qquad(1.24)$

For $0 < x < y$, one has $\exp(-x) - \exp(-y) = \int_x^y \exp(-t)\,\mathrm{d}t$. Since also $\exp(-y) \leqslant \exp(-t)$, we get

$$(y-x)\exp(-y) \leqslant \exp(-x) - \exp(-y).$$

Let now $\theta_n < \theta_0$ and $n^{\frac{1}{2}}(\theta_n - \theta_0) \to -\infty$. Then with $x = \theta_n$ and $y = \theta_0$, the above inequality gives

$$n^{\frac{1}{2}}[p(\theta_0) - p(\theta_n)]$$
$$= -n^{\frac{1}{2}}[\exp(-\theta_n) - \exp(-\theta_0)] \leqslant \exp(-\theta_0)\, n^{\frac{1}{2}}(\theta_n - \theta_0) \to -\infty.$$

Next let $\theta_n > \theta_0$ and $n^{\frac{1}{2}}(\theta_n - \theta_0) \to \infty$. Consider two cases:

(i) $\theta_n \leqslant c$. Then with $x = \theta_0$ and $y = \theta_n$, the above inequality gives $\quad n^{\frac{1}{2}}[p(\theta_0) - p(\theta_n)] = n^{\frac{1}{2}}[\exp(-\theta_0) - \exp(-\theta_n)]$

$$\geqslant \exp(-\theta_n)\, n^{\frac{1}{2}}(\theta_n - \theta_0)$$
$$\geqslant \exp(-c)\, n^{\frac{1}{2}}(\theta_n - \theta_0) \to \infty.$$

(ii) Let now $\{\theta_n\}$ be unbounded. Then for bounded subsequences $\{\theta_m\}$, one has by (i) that $m^{\frac{1}{2}}[p(\theta_0) - p(\theta_m)] \to \infty$, whereas for $\theta_m \to \infty$, one has

$$m^{\frac{1}{2}}[\exp(-\theta_0) - \exp(-\theta_m)] \to \infty \exp(-\theta_0) = \infty.$$

Thus with θ_n as above, we have

$$n^{\frac{1}{2}}(\theta_n - \theta_0) \to \infty(-\infty) \quad \text{implies} \quad n^{\frac{1}{2}}[p(\theta_0) - p(\theta_n)] \to \infty(-\infty).$$
$$\tag{1.25}$$

At this point we assume for a moment that $\{W_n\}$ is bounded in P_{θ_n}-probability; i.e. for every $\epsilon > 0$, there exists $M = M(\epsilon)$ sufficiently large such that

$$P_{\theta_n}(|W_n| \geq M) < \epsilon \quad \text{for all } n. \tag{1.26}$$

Let us suppose first that $n^{\frac{1}{2}}(\theta_n - \theta_0) \to \infty$. Then (1.24) in conjunction with (1.25) implies that

$$
\begin{aligned}
P_{\theta_n}\{\Delta_n[p(\theta_0)] > M'\} &= P_{\theta_n}(W_n + \beta_n > M') \\
&\geq P_{\theta_n}(|W_n| < M', \beta_n > 2M') \\
&= 1 - P_{\theta_n}[(|W_n| \geq M') \cup (\beta_n \leq 2M')] \\
&\geq 1 - P_{\theta_n}(|W_n| \geq M') - P_{\theta_n}(\beta_n \leq 2M') \\
&= 1 - P_{\theta_n}(|W_n| \geq M')
\end{aligned}
$$

for any $M' > 0$ and all sufficiently large n. By choosing $M' > M$, we get

$$1 - P_{\theta_n}(|W_n| \geq M') \geq 1 - P_{\theta_n}(|W_n| \geq M) > 1 - \epsilon \quad \text{for all } n$$

by means of (1.26). Therefore

$$P_{\theta_n}\{\Delta_n[p(\theta_0)] > M'\} > 1 - \epsilon \quad \text{for all sufficiently large } n,$$

which is equivalent to saying that

$$\Delta_n[p(\theta_0)] \to \infty \quad \text{in } P_{\theta_n}\text{-probability}.$$

If we now suppose that $n^{\frac{1}{2}}(\theta_n - \theta_0) \to -\infty$, then it is shown in a similar fashion that

$$\Delta_n[p(\theta_0)] \to -\infty \quad \text{in } P_{\theta_n}\text{-probability}.$$

Thus all that remains for us to do is to establish (1.26). We have

$$
\begin{aligned}
\sigma_{\theta_n}^2\left[\sum_{j=1}^n (X_j^2 - 1)\right] &= n\sigma_{\theta_n}^2(X_1^2 - 1) \\
&\quad + 2\sum_{1 \leq i < j \leq n} \text{Cov}_{\theta_n}[(X_i^2 - 1), (X_j^2 - 1)] \\
&\leq 2n + 2\sum_{1 \leq i < j \leq n} 2[p(\theta_n)]^{2(j-i)},
\end{aligned}
$$

as follows from Anderson [1], pp. 38–9.

Now since

$$\sum_{1 \leqslant i < j \leqslant n} [p(\theta_n)]^{2(j-i)}$$

$$= n \sum_{j=1}^{n-1} [p^2(\theta_n)]^j - \sum_{j=1}^{n-1} j[p^2(\theta_n)]^j < n \sum_{j=1}^{n-1} [p^2(\theta_n)]^j$$

$$< n \sum_{j=1}^{\infty} [p^2(\theta_n)]^j = \frac{np^2(\theta_n)}{1-p^2(\theta_n)},$$

we have
$$\sigma_{\theta_n}^2 \left[\sum_{j=1}^{n} (X_j^2 - 1) \right] < \frac{4n}{1-p^2(\theta_n)}. \tag{1.27}$$

Similarly
$$\sigma_{\theta_n}^2 \left[\sum_{j=1}^{n} (X_{j-1}^2 - 1) \right] < \frac{4n}{1-p^2(\theta_n)}. \tag{1.28}$$

Next

$$\sigma_{\theta_n}^2 \left\{ \sum_{j=1}^{n} [X_{j-1}X_j - p(\theta_n)] \right\}$$

$$= n\sigma_{\theta_n}^2 [X_0 X_1 - p(\theta_n)] + 2 \sum_{1 \leqslant i < j \leqslant n} \text{Cov}_{\theta_n} \{ [X_{i-1}X_i - p(\theta_n)],$$

$$[X_{j-1}X_j - p(\theta_n)] \}$$

$$= n[1 + p^2(\theta_n)] + 2 \sum_{1 \leqslant i < j \leqslant n} 2[p(\theta_n)]^{2(j-i)},$$

as it follows from the reference last cited.

Therefore it follows as before that

$$\sigma_{\theta_n}^2 \left\{ \sum_{j=1}^{n} [X_{j-1}X_j - p(\theta_n)] \right\} < \frac{4n}{1-p^2(\theta_n)}. \tag{1.29}$$

By means of (1.22), (1.27)–(1.29), we have then

$$P_{\theta_n}(|c_1 Y_n| \geqslant M) \leqslant \frac{4c_1^2}{1-p^2(\theta_n)} \frac{1}{M^2},$$

$$P_{\theta_n}(|c_2 Z_n| \geqslant M) \leqslant \frac{4c_2^2}{1-p^2(\theta_n)} \frac{1}{M^2}$$

and
$$P_{\theta_n}(|c_3 V_n| \geqslant M) \leqslant \frac{4c_3^2}{1-p^2(\theta_n)} \frac{1}{M^2}.$$

On the other hand,

$$P_{\theta_n}(|W_n| < 3M)$$

$$= P_{\theta_n}(|c_1 Y_n + c_2 Z_n + c_3 V_n| < 3M)$$

$$\geq P_{\theta_n}(|c_1 Y_n| < M, |c_2 Z_n| < M, |c_3 V_n| < M)$$

$$\geq 1 - P_{\theta_n}(|c_1 Y_n| \geq M) - P_{\theta_n}(|c_2 Z_n| \geq M) - P_{\theta_n}(|c_3 V_n| \geq M),$$

so that $\qquad P_{\theta_n}(|W_n| < 3M) \geq 1 - \dfrac{4(c_1^2 + c_2^2 + c_3^2)}{1 - p^2(\theta_n)} \dfrac{1}{M^2}.$ $\qquad$ (1.30)

Therefore if $\theta_n > \theta_0$, equivalently, $p^2(\theta_n) < p^2(\theta_0)$, one has by means of

$$P_{\theta_n}(|W_n| < 3M) \geq 1 - \frac{4(c_1^2 + c_2^2 + c_3^2)}{1 - p^2(\theta_0)} \frac{1}{M^2}. \qquad (1.31)$$

It follows that if for a given $\epsilon > 0$, M is chosen so that

$$M^2 > \frac{4(c_1^2 + c_2^2 + c_3^2)}{[1 - p^2(\theta_0)]\epsilon},$$

then $P_{\theta_n}(|W_n| < 3M) > 1 - \epsilon$, which is what (1.26) asserts.

If we are interested in testing the hypothesis H_0: $\theta > \theta_0$ against the alternative A': $\theta < \theta_0$, in which case $\theta_n < \theta_0$, or equivalently, $p^2(\theta_n) > p^2(\theta_0)$, then we impose the (not serious) restriction that $p(\theta)$ is bounded away from one, i.e. $p^2(\theta) \geq 1 - \delta$, $\theta \in \Theta$, for some $\delta > 0$. Then

$$P_{\theta_n}(|W_n| < 3M) \geq 1 - \frac{4(c_1^2 + c_2^2 + c_3^2)}{\delta} \frac{1}{M^2}$$

and a repetition of the previous argument establishes relation (1.26) again.

To summarize: (A5') is satisfied and so is (A5) under the restriction that $p(\theta)$ is bounded away from one.]

2   Some lemmas

In the testing hypotheses problems to be discussed in this monograph, the following auxiliary results will prove useful. They are formulated and proved here for easy reference.

LEMMA 2.1 Let $\{Y_n\}$ be a sequence of r.v.s defined on some measurable space $(\Omega, \mathscr{F})$, and for each n, let Q_n be a probability measure on $\mathscr{F}$. It is assumed that

$$\mathscr{L}(Y_n | Q_n) \Rightarrow N(\mu, \sigma^2). \tag{2.1}$$

Then

(i) For any numerical sequence $\{y_n\}$, one has

$$Q_n(Y_n = y_n) \to 0.$$

Let now the sequences of numbers $\{c_n\}$ and $\{\gamma_n\}$ with $0 \leqslant \gamma_n \leqslant 1$ for all n be defined by

$$Q_n(Y_n > c_n) + \gamma_n Q_n(Y_n = c_n) = \alpha \quad (0 < \alpha < 1).$$

Then

(ii) $c_n \to \mu + \sigma \xi_\alpha$, where ξ_α is the upper αth quantile of Φ, the d.f. of the standard normal.

Proof (i) Let F_n be the d.f. of Y_n under Q_n. Then, for any $\epsilon > 0$, one has

$$Q_n(Y_n = y_n) \leqslant F_n(y_n) - F_n(y_n - \epsilon) = \left[F_n(y_n) - \Phi\left(\frac{y_n - \mu}{\sigma}\right) \right]$$

$$- \left[F_n(y_n - \epsilon) - \Phi\left(\frac{y_n - \epsilon - \mu}{\sigma}\right) \right]$$

$$+ \left[\Phi\left(\frac{y_n - \mu}{\sigma}\right) - \Phi\left(\frac{y_n - \epsilon - \mu}{\sigma}\right) \right]. \tag{2.2}$$

The first two terms on the right-hand side of (2.2) converge to zero by (2.1) and Polya's lemma. As for the third term, we have

$$\Phi\left(\frac{y_n - \mu}{\sigma}\right) - \Phi\left(\frac{y_n - \epsilon - \mu}{\sigma}\right) = \frac{\epsilon}{\sigma} \frac{1}{\sqrt{(2\pi)}} \exp\left(-\frac{z_n^2}{2}\right)$$

for some z_n with $\dfrac{y_n - \mu}{\sigma} \geqslant z_n \geqslant \dfrac{y_n - \epsilon - \mu}{\sigma}$,

while $\dfrac{\epsilon}{\sigma} \dfrac{1}{\sqrt{(2\pi)}} \exp\left(-\dfrac{z_n^2}{2}\right) \leqslant \dfrac{\epsilon}{\sigma} \dfrac{1}{\sqrt{(2\pi)}} \to 0$ as $\epsilon \to 0$.

Therefore bounding the third term on the right-hand side of (2.2) by $\dfrac{\epsilon}{\sqrt{(2\pi)}\,\sigma}$ and then taking the limits first as $n \to \infty$ and then as $\epsilon \to 0$, it follows that $Q_n(Y_n = y_n) \to 0$.

(ii) We have

$$\alpha = \lim\left[Q_n(Y_n > c_n) + \gamma_n Q_n(Y_n = c_n)\right] = \lim Q_n(Y_n > c_n) \quad \text{by (i);}$$

or
$$1 - \lim F_n(c_n) = \alpha$$

from which it follows that

$$\lim F_n(c_n) = 1 - \alpha.$$

But as above
$$\lim\left[F_n(c_n) - \Phi\left(\frac{c_n - \mu}{\sigma}\right)\right] = 0.$$

Therefore
$$\Phi\left(\frac{c_n - \mu}{\sigma}\right) \to 1 - \alpha,$$

hence
$$\frac{c_n - \mu}{\sigma} \to \xi_\alpha \quad \text{and} \quad c_n \to \mu + \sigma\xi_\alpha,$$

as was to be shown.∎

The following result was stated in LeCam [5] without proof.

LEMMA 2.2   Consider the measurable space $(\Omega, \mathscr{F})$ and let P and Q be two probability measures on $\mathscr{F}$ such that $P \approx Q$. Let $f = \mathrm{d}P/\mathrm{d}\mu$, $g = \mathrm{d}Q/\mathrm{d}\mu$ for some dominating σ-finite measure μ (e.g. $\mu = P$ or Q or $P + Q$) and set $Z = \log(g/f)$. Then, for every $\epsilon > 0$, one has

$$\|P - Q\| \leqslant 2[1 - \exp(-\epsilon)] + 2P(|Z| > \epsilon).$$

Proof By Theorem 1.4A, we have

$$\|P - Q\| = 2\sup\left[|P(A) - Q(A)|; A \in \mathscr{F}\right] = \int |f - g|\, \mathrm{d}\mu.$$

Let now $B = (f - g > 0)$. Then

$$\int |f - g|\, \mathrm{d}\mu = \int_B |f - g|\, \mathrm{d}\mu + \int_{B^c} |f - g|\, \mathrm{d}\mu$$

$$= \int_B (f - g)\, \mathrm{d}\mu + \int_{B^c} (g - f)\, \mathrm{d}\mu$$

$$= [P(B) - Q(B)] + [Q(B^c) - P(B^c)] = 2[P(B) - Q(B)],$$

so that
$$\|P - Q\| = 2[P(B) - Q(B)]. \tag{2.3}$$

Set $C = (|Z| > \epsilon)$. Then

$$P(B) - Q(B) = P(B \cap C) + P(B \cap C^c) - Q(B \cap C) - Q(B \cap C^c)$$
$$\leqslant P(C) + P(B \cap C^c) - Q(B \cap C^c)$$
$$\leqslant P(|Z| > \epsilon) + P(B \cap C^c) - Q(B \cap C^c). \qquad (2.4)$$

But

$$Q(B \cap C^c) = \int_{B \cap C^c} g \, d\mu = \int_{B \cap C^c \cap (f > 0)} \frac{g}{f} f \, d\mu + \int_{B \cap C^c \cap (f = 0)} g \, d\mu$$
$$= \int_{B \cap C^c} \exp Z \, dP \geqslant \exp(-\epsilon) P(B \cap C^c) \qquad (2.5)$$

because on $B, f > g$ and hence $f > 0$, and on $C^c, |Z| \leqslant \epsilon$, so that $\exp(-\epsilon) \leqslant \exp(Z)$.

Relations (2.3)–(2.5) show that

$$\|P - Q\| \leqslant 2P(|Z| > \epsilon) + 2P(B \cap C^c)[1 - \exp(-\epsilon)]$$
$$\leqslant 2[1 - \exp(-\epsilon)] + 2P(|Z| > \epsilon),$$

as was to be seen. ∎

3 Testing a simple hypothesis against one-sided alternatives

In this section, we consider the problem of testing the simple hypothesis $H_0 : \theta = \theta_0 (\in \Theta)$ against the (composite) alternative $A : \theta \in \omega$, where ω is defined by (1.2). We shall restrict ourselves to sequences of tests whose level of significance is α and we shall exhibit such a sequence with the optimal property of being asymptotically uniformly most powerful (or asymptotically most powerful in Wald's [2] terminology), according to the following definition.

DEFINITION 3.1   For testing $H_0 : \theta = \theta_0$ against

$$A : \theta \in \omega = \{\theta \in \Theta ; \theta > \theta_0\},$$

the sequence of level α tests $\{\lambda_n\}$ is said to be *asymptotically uniformly most powerful* (AUMP) if for any other sequence of level α tests $\{\omega_n\}$, one has

$$\limsup [\sup(\mathcal{E}_\theta \omega_n - \mathcal{E}_\theta \lambda_n ; \theta \in \omega)] \leqslant 0. \qquad (3.1)$$

REMARK 3.1 Now objections have been raised regarding the optimality of an AUMP sequence of tests. Actually Lehmann [1] (see also Kendall and Stuart [1], p. 263, Example 25.1) constructed an example for which two sequences of tests, which are both AUMP, have the property that the quotient of the sequence of type II errors converges to infinity. This shows that the criterion for a sequence of tests of being AUMP is not a very selective one. Perhaps an additional criterion ought to be superimposed which would single out an 'efficient' sequence of AUMP tests.

Since the purpose of the present section is to illustrate how the results obtained in Chapter 2 may be used for discussing statistical problems, rather than breaking new ground (in the sense of selecting an 'efficient' AUMP test), we adhere to Definition 3.1.

I am indebted to Dr O. Krafft for bringing to my attention the references cited in the remark.

Let $\Delta_n(\theta)$ be given by (1.1) and define the sequence $\{\phi_n\}$ as follows

$$\phi_n = \phi_n[\Delta_n(\theta_0)] = \begin{cases} 1 & \text{if} \quad \Delta_n(\theta_0) > c_n \\ \gamma_n & \text{if} \quad \Delta_n(\theta_0) = c_n \\ 0 & \text{otherwise,} \end{cases} \qquad (3.2)$$

where the sequences $\{c_n\}$ and $\{\gamma_n\}$ are determined by the requirement

$$\mathscr{E}_{\theta_0}\phi_n = \alpha \quad \text{for all } n. \qquad (3.3)$$

Then we have the following result.

THEOREM 3.1 Under assumptions (A1–4), (A5), the sequence of tests $\{\phi_n\}$ defined by (3.2) and (3.3) is AUMP for testing $H_0 \colon \theta = \theta_0$ against $A \colon \theta \in \omega$, where ω is given by (1.2).

Proof The proof is by contradiction. Suppose that $\{\phi_n\}$ is not AUMP. Then (3.1) is violated for some sequence of level α tests $\{\omega_n\}$, and let the left-hand side of (3.1) have the value $\delta > 0$. That is,

$$\limsup [\sup (\mathscr{E}_\theta \omega_n - \mathscr{E}_\theta \phi_n; \theta \in \omega)] = \delta. \qquad (3.4)$$

Then there exists a subsequence $\{m\} \subseteq \{n\}$ and a sequence $\{\theta_m\}$ with $\theta_m \in \omega$ for all m such that

$$\mathscr{E}_{\theta_m}\omega_m - \mathscr{E}_{\theta_m}\phi_m \to \delta. \qquad (3.5)$$

From hereafter, the main steps in the proof may be summarized as follows. We first show that (3.5) cannot happen if the sequence $\{\theta_m\}$ is such that $\{m^{\frac{1}{2}}(\theta_m - \theta_0)\}$ is unbounded. For this, we employ assumption (A5), Theorem 4.2 in Chapter 2 and Lemma 2.1 herein. Next we show that (3.5) cannot occur either if the sequence $\{\theta_m\}$ is such that $\{m^{\frac{1}{2}}(\theta_m - \theta_0)\}$ remains bounded. For this purpose, we employ Theorems 4.1, 4.5, 4.6 of Chapter 2, Lemmas 2.1, 2.2 herein and the Neyman–Pearson fundamental lemma. Consider the sequence $\{m^{\frac{1}{2}}(\theta_m - \theta_0)\}$ and suppose first that it is unbounded. Then there exists a subsequence $\{r\} \subseteq \{m\}$ such that $r^{\frac{1}{2}}(\theta_r - \theta_0) \to \infty$. Then assumption (A5) implies

$$\Delta_r(\theta_0) \to \infty \quad \text{in } P_{\theta_r}\text{-probability} \tag{3.6}$$

(this is the only instance where (A5) is employed). Now by setting for simplicity $\sigma^2(\theta_0) = 4\mathscr{E}_{\theta_0}[\dot{\phi}_1(\theta_0)]^2$, Theorem 4.2 in Chapter 2 implies that

$$\mathscr{L}[\Delta_n(\theta_0)|P_{\theta_0}] \Rightarrow N(0, \sigma^2(\theta_0)).$$

Then Lemma 2.1 applies with

$$Y_n = \Delta_n(\theta_0), \quad Q_n = P_{\theta_0}, \quad \mu = 0 \quad \text{and} \quad \sigma^2 = \sigma^2(\theta_0)$$

and gives that $\{c_n\}$ stays bounded. This fact together with (3.6) implies that

$$P_{\theta_r}[\Delta_r(\theta_0) > c_r] \to 1 \quad \text{and} \quad P_{\theta_r}[\Delta_r(\theta_0) = c_r] \to 0,$$

so that

$$\mathscr{E}_{\theta_r}\phi_r \to 1.$$

This result together with (3.5) (with m replaced by r) implies that (3.4) cannot occur.

Now consider the case that $\{m^{\frac{1}{2}}(\theta_m - \theta_0)\}$ is bounded. Then there exists a subsequence $\{s\} \subseteq \{m\}$ such that $s^{\frac{1}{2}}(\theta_s - \theta_0) \to h \geqslant 0$. First we deal with the case that $h > 0$. To this end, set

$$h_s = s^{\frac{1}{2}}(\theta_s - \theta_0),$$

so that $\theta_s = \theta_0 + h_s s^{-\frac{1}{2}}$ with $h_s \to h$. Then by Theorem 4.6 in Chapter 2, one gets

$$\mathscr{L}[h\Delta_s(\theta_0)|P_{\theta_s}] \Rightarrow N(h^2\sigma^2(\theta_0), h^2\sigma^2(\theta_0)).$$

Then Lemma 2.1 applies again with $\{n\}$ replaced by $\{s\}$ and with

$$Y_s = h\Delta_s(\theta_0), \quad Q_s = P_{\theta_s} \quad \text{and} \quad \mu = \sigma^2 = h^2\sigma^2(\theta_0).$$

It follows then that

$$P_{\theta_s}[h\Delta_s(\theta_0) = hc_s] \to 0. \qquad (3.7)$$

Also by the same lemma applied with Y_s as above, $Q_s = P_{\theta_0}$ and $\mu = 0$, we get $hc_s \to h\sigma(\theta_0)\xi_\alpha$, so that

$$P_{\theta_s}[h\Delta_s(\theta_0) > hc_s] = 1 - P_{\theta_s}\left[\frac{h\Delta_s(\theta_0) - h^2\sigma^2(\theta_0)}{h\sigma(\theta_0)} \leqslant \frac{hc_s - h^2\sigma^2(\theta_0)}{h\sigma(\theta_0)}\right]$$

$$\to 1 - \Phi\left[\frac{h\xi_\alpha\sigma(\theta_0) - h^2\sigma^2(\theta_0)}{h\sigma(\theta_0)}\right]$$

$$= 1 - \Phi[\xi_\alpha - h\sigma(\theta_0)];$$

i.e.
$$P_{\theta_s}[h\Delta_s(\theta_0) > hc_s] \to 1 - \Phi[\xi_\alpha - h\sigma(\theta_0)]. \qquad (3.8)$$

Relations (3.7) and (3.8) imply that

$$\mathscr{E}_{\theta_s}\phi_s \to 1 - \Phi[\xi_\alpha - h\sigma(\theta_0)]. \qquad (3.9)$$

Next define the sequence $\{\psi_s\}$ as follows

$$\psi_s = \psi_s(X_0, X_1, ..., X_s)$$

$$= \begin{cases} 1 & \text{if } \Lambda(\theta_0, \theta_s) + \tfrac{1}{2}h^2\sigma^2(\theta_0) > d_s \\ \delta_s & \text{if } \Lambda(\theta_0, \theta_s) + \tfrac{1}{2}h^2\sigma^2(\theta_0) = d_s \\ 0 & \text{otherwise,} \end{cases} \qquad (3.10)$$

where the sequences $\{d_s\}$ and $\{\delta_s\}$ are determined by the requirement
$$\mathscr{E}_{\theta_0}\psi_s = \alpha \quad \text{for all } s, \qquad (3.11)$$

and recall that $\Lambda(\theta_0, \theta_s) = \log[dP_{s,\theta_s}/dP_{s,\theta_0}]$.

By Theorem 4.5 in Chapter 2,

$$\mathscr{L}[\Lambda(\theta_0, \theta_s) + \tfrac{1}{2}h^2\sigma^2(\theta_0)|P_{\theta_s}] \Rightarrow N(h^2\sigma^2(\theta_0), h^2\sigma^2(\theta_0)),$$

so that in a way entirely analogous to the one by which (3.9) was obtained, one has

$$\mathscr{E}_{\theta_s}\psi_s \to 1 - \Phi[\xi_\alpha - h\sigma(\theta_0)]. \qquad (3.12)$$

Thus by virtue of (3.9) and (3.12),

$$\mathscr{E}_{\theta_s}\phi_s - \mathscr{E}_{\theta_s}\psi_s \to 0. \qquad (3.13)$$

Replacing m by s in (3.5) (which can be done since $\{s\} \subseteq \{m\}$), we have
$$\mathscr{E}_{\theta_s}\omega_s - \mathscr{E}_{\theta_s}\phi_s \to \delta. \qquad (3.14)$$

Thus, on account of (3.13) and (3.14),

$$\mathscr{E}_{\theta_s}\omega_s - \mathscr{E}_{\theta_s}\psi_s \to \delta.$$

Hence for all sufficiently large s, $s \geqslant s_1$, say, one has

$$\mathscr{E}_{\theta_s}\psi_s \leqslant \mathscr{E}_{\theta_s}\omega_s - \tfrac{1}{2}\delta. \tag{3.15}$$

For each $s \geqslant s_1$, consider the problem of testing the simple hypothesis $H_0: \theta = \theta_0$ against the simple alternative $A_s: \theta = \theta_s$ at level α. Then by the Neyman–Pearson fundamental lemma the test defined by (3.10) and (3.11) is the most powerful one. Accordingly, relation (3.15) cannot occur. We arrived at (3.15) by assuming (3.5) and the existence of $\{s\} \subseteq \{m\}$ and $\theta_s \epsilon \omega$ such that $s^{\frac{1}{2}}(\theta_s - \theta_0) \to h > 0$. Therefore (3.5) and hence (3.4) cannot be true.

Finally, it remains for us to show that (3.4) is also not true for the case that $\{m^{\frac{1}{2}}(\theta_m - \theta_0)\}$ is bounded and the only convergent subsequence converges to zero. To this effect, let $\{s\} \subseteq \{m\}$ and $\theta_s \epsilon \omega$ for all s be such that $s^{\frac{1}{2}}(\theta_s - \theta_0) = h_s \to 0$. Then Theorem 4.1 in Chapter 2 implies that

$$\Lambda(\theta_0, \theta_s) \to 0 \quad \text{in } P_{\theta_s}\text{-probability},$$

so that, for every $\epsilon > 0$,

$$P_{\theta_0}[|\Lambda(\theta_0, \theta_s)| > \epsilon] \to 0. \tag{3.16}$$

Now Lemma 2.2 applies with

$$P = P_{\theta_0}, \quad Q = P_{\theta_s} \quad \text{and} \quad Z = \Lambda(\theta_0, \theta_s)$$

and gives

$$\|P_{\theta_0} - P_{\theta_s}\| \leqslant 2[1 - \exp(-\epsilon)] + 2P_{\theta_0}[|\Lambda(\theta_0, \theta_s)| > \epsilon].$$

This result together with (3.16) implies that

$$\|P_{\theta_0} - P_{\theta_s}\| \to 0. \tag{3.17}$$

Next

$$
\begin{aligned}
|\alpha - \mathscr{E}_{\theta_s}\phi_s| &= |P_{\theta_0}[\Delta_s(\theta_0) > c_s] + \gamma_s P_{\theta_0}[\Delta_s(\theta_0) = c_s] \\
&\quad - P_{\theta_s}[\Delta_s(\theta_0) > c_s] - \gamma_s P_{\theta_s}[\Delta_s(\theta_0) = c_s]| \\
&\leqslant 2\|P_{\theta_0} - P_{\theta_s}\| \to 0 \quad \text{by (3.17)},
\end{aligned}
$$

so that

$$\mathscr{E}_{\theta_s}\phi_s \to \alpha. \tag{3.18}$$

In a manner entirely similar to the one used for obtaining (3.18), one also concludes that

$$\mathscr{E}_{\theta_s}\psi_s \to \alpha,$$

so that

$$\mathscr{E}_{\theta_s}\phi_s - \mathscr{E}_{\theta_s}\psi_s \to 0. \tag{3.19}$$

But relation (3.19) is the same as (3.13) which led us to a contradiction to (3.4). Therefore, (3.19) implies that (3.4) cannot hold and the proof of the theorem is concluded. ∎

As was pointed out in the course of the proof of Theorem 3.1, assumption (A5) was only used for the purpose of arriving at a contradiction to (3.4) in the case that the alternatives θ_n either do not approach θ_0 sufficiently fast or stay away from it, in the sense that $\{n^{\frac{1}{2}}(\theta_n - \theta_0)\}$ is unbounded. Therefore, if we restrict ourselves to alternatives θ_n which tend to θ_0 at the appropriate rate, in the sense that $\{n^{\frac{1}{2}}(\theta_n - \theta_0)\}$ stays bounded, the proof of Theorem 3.1 has the following corollary.

COROLLARY 3.1 Under the assumptions (A1–4), the sequence of tests $\{\phi_n\}$ defined by (3.2) and (3.3) is asymptotically locally most powerful for testing $H_0: \theta = \theta_0$ against $A: \theta \in \omega$. That is, the sequence $\{\phi_n\}$ satisfies (3.1) when the parameter values are restricted to those θs in ω for which $\{n^{\frac{1}{2}}(\theta - \theta_0)\}$ remains bounded from above.

When the alternatives lie on the other side of the hypothesis being tested, a theorem and a corollary analogous to Theorem 3.1 and Corollary 3.1 also hold true; their proof is entirely analogous to the one given. More precisely, define $\{\phi_n'\}$ by

$$\phi_n' = \phi_n'[\Delta_n(\theta_0)] = \begin{cases} 1 & \text{if} \quad \Delta_n(\theta_0) < c_n' \\ \gamma_n' & \text{if} \quad \Delta_n(\theta_0) = c_n' \\ 0 & \text{otherwise,} \end{cases} \tag{3.20}$$

where the sequences $\{c_n'\}$ and $\{\gamma_n'\}$ are determined by the requirement

$$\mathscr{E}_{\theta_0}\phi_n' = \alpha \quad \text{for all } n. \tag{3.21}$$

Then we have

THEOREM 3.2 Under assumptions (A1–4), (A5′), the sequence of tests $\{\phi_n'\}$ defined by (3.20) and (3.21) is AUMP for testing $H_0: \theta = \theta_0$ against $A': \theta \in \omega'$, where ω' is given by (1.2).

COROLLARY 3.2 Under the assumptions (A1–4), the sequence of tests $\{\phi'_n\}$ defined by (3.20) and (3.21) is asymptotically locally most powerful for testing $H_0: \theta = \theta_0$ against $A': \theta \in \omega'$. That is, the sequence $\{\phi'_n\}$ satisfies (3.1) when the parameter values are restricted to those θs in ω' for which $\{n^{\frac{1}{2}}(\theta - \theta_0)\}$ remains bounded from below.

A discussion of the testing problem just concluded is found in Johnson and Roussas [1].

4 AUMP tests for the examples of Section 1

In [2], Wald considered the problem of constructing an AUMP test when the underlying process consists of i.i.d. r.v.s. The tests are based on the maximum likelihood estimate (MLE) and the regularity conditions used are quite stringent. Wald's result has not been carried over to the Markov case to the best of our knowledge.

In the present framework, the Markov case is also covered and no reference whatsoever is made to the MLE; its role is played here by the r.v. $\Delta_n(\theta_0)$. Furthermore the assumptions employed are substantially weaker than those used by Wald.

For the purpose of comparing tests, it should be mentioned that for testing $H_0: \theta = \theta_0$ against $A: \theta \in \omega$ at level α, Wald's tests $\{w_n\}$ are defined as follows

$$w_n = w_n(X_1, ..., X_n) = \begin{cases} 1 & \text{if } \hat{\theta}_n > \bar{c}_n \\ \bar{\gamma}_n & \text{if } \hat{\theta}_n = \bar{c}_n \\ 0 & \text{otherwise,} \end{cases} \qquad (4.1)$$

where $\hat{\theta}_n$ is the MLE of θ and the sequences $\{\bar{c}_n\}$ and $\{\bar{\gamma}_n\}$ are defined by the requirement

$$\mathscr{E}_{\theta_0} w_n = \alpha \quad \text{for all } n. \qquad (4.2)$$

In the following examples only tests for the alternatives A are considered since the tests for the alternatives A' are constructed symmetrically.

EXAMPLES

4.1   Refer to Example 1.1. In this case, $\Delta_n(\theta_0)$ is given by
(1.3) and is as follows

$$\Delta_n(\theta_0) = \frac{1}{\sigma^2} n^{-\frac{1}{2}} \sum_{j=1}^{n} (X_j - \theta_0) = \frac{1}{\sigma^2} n^{\frac{1}{2}} (\bar{X}_n - \theta_0).$$

Therefore the test in (3.2) rejects H_0 whenever $\Delta_n(\theta_0) > c_n$, or
equivalently, whenever $n^{\frac{1}{2}}(\bar{X}_n - \theta_0) > c_n^*$, where $c_n^* = \sigma^2 c_n$. Since
$\bar{X}_n = \hat{\theta}_n$ the present test is the same as that of Wald and both
coincide with the existing UMP test for all n.

4.2   Refer to Example 1.2. Here by virtue of (1.4), one has

$$\Delta_n(\theta_0) = n^{-\frac{1}{2}} \sum_{j=1}^{n} \left[\frac{1}{2\theta_0^2}(X_j - \mu)^2 - \frac{1}{2\theta_0} \right] = \frac{n^{\frac{1}{2}}}{2\theta_0^2}(S_n^2 - \theta_0),$$

where $S_n^2 = n^{-1} \sum_{j=1}^{n} (X_j - \mu)^2$. Therefore the test in (3.2) rejects H_0
whenever $\Delta_n(\theta_0) > c_n$, or equivalently, whenever $n^{\frac{1}{2}}(S_n^2 - \theta_0) > c_n^*$,
where $c_n^* = 2\theta_0^2 c_n$. Again $S_n^2 = \hat{\theta}_n^2$ and therefore the present test
is the same as that of Wald and both coincide with the existing
UMP test for all n.

4.3   Refer to Example 1.3. From (1.9), one has with P_{θ_0}-
probability equal to one

$$\Delta_n(\theta_0) = 2n^{-\frac{1}{2}} \sum_{j=1}^{n} [I_{(\theta_0, \infty)}(X_j) - \tfrac{1}{2}]$$

and therefore the test in (3.2) rejects H_0 whenever $\Delta_n(\theta_0) > c_n$, or
equivalently, whenever

$$\sum_{j=1}^{n} I_{(\theta_0, \infty)}(X_j) > c_n^*, \qquad (4.3)$$

where $$c_n^* = \frac{n}{2} + \frac{n}{2} \frac{c_n}{\sqrt{n}}.$$

Now $$P_{\theta_0}[I_{(\theta_0, \infty)}(X_j) = 1] = P_{\theta_0}(X_j > \theta_0) = \tfrac{1}{2}.$$

Therefore in (4.3) the constant c_n^* can be determined exactly by
means of the binomial tables for a moderate sample size, since,
under P_{θ_0}, $\sum_{j=1}^{n} I_{(\theta_0, \infty)}(X_j)$ is distributed as $B(n, \tfrac{1}{2})$. Of course, both

here and in all other examples the cut-off points can always be determined approximately, since $\Delta_n(\theta_0)$ is asymptotically normal, under P_{θ_0}. Actually, here $\Delta_n(\theta_0)$ is asymptotically $N(0, 1)$, under P_{θ_0}, because $\Gamma(\theta) = 1$ by (3.13) in Chapter 2.

By taking into consideration the fact that θ_0 is the population median (mean), the test in (4.3) has an intuitive interpretation; namely, it rejects H_0 if too many Xs exceed θ_0.

It is well known that the MLE of θ in the present case is the sample median. Therefore the test based on the MLE rejects H_0 whenever $n^{\frac{1}{2}}(\theta_n - \theta_0) > \bar{c}_n$, or equivalently, whenever

$$\theta_n > \theta_0 + \bar{c}_n n^{-\frac{1}{2}}.$$

Furthermore, it is shown (see, e.g. Rao [3], p. 355) that

$$\mathscr{L}[n^{\frac{1}{2}}(\theta_n - \theta_0)|P_{\theta_0}] \Rightarrow N(0, 1),$$

so that the two tests are asymptotically equivalent. For any n, it can be shown (see Exercise 2) that

$$\sum_{j=1}^{n} I_{(\theta_0, \infty)}(X_j) > \tfrac{1}{2}n \quad \text{if and only if} \quad \theta_n > \theta_0; \qquad (4.4)$$

that is, the two tests are equivalent, provided the adjusting constants $\dfrac{n}{2}\dfrac{c_n}{\sqrt{n}}$ and $\dfrac{c_n}{\sqrt{n}}$ are omitted.

4.4 Refer to Example 1.4. Here $\Delta_n[p(\theta_0)]$ is given by (1.21) and the test in (3.2) can be carried out explicitly. In the present case, Wald's test based on the MLE is not directly applicable, since the relevant theory has not been carried over to the Markov case (to the best of our knowledge). Also the MLE is not easy to obtain (see Exercise 3).

5 Testing a simple hypothesis against two-sided alternatives

Consider the sets ω and ω' defined by (1.2) and set

$$\omega'' = \omega \cup \omega'.$$

In the present section, the problem of testing the hypothesis $H_0: \theta = \theta_0(\in \Theta)$ against the alternative $A'': \theta \in \omega''$ is taken up.

One would not possibly expect the existence of an AUMP test. However, an AUMP unbiased (or AMP unbiased in Wald's [2] terminology), according to the following definition, does exist, as is shown herein.

DEFINITION 5.1   For testing $H_0: \theta = \theta_0$ against
$$A'': \theta \in \omega'' = \{\theta \in \Theta; \theta \neq \theta_0\},$$
the sequence of tests $\{\lambda_n\}$ is said to be *asymptotically unbiased* if $\liminf [\inf (\mathscr{E}_\theta \lambda_n; \theta \in \omega'')] \geqslant \alpha$ and is said to be AUMP *unbiased* (AUMPU) of asymptotic level of significance α if it is asymptotically unbiased and of asymptotic level α and

$$\limsup [\sup (\mathscr{E}_\theta \omega_n - \mathscr{E}_\theta \lambda_n; \theta \in \omega'')] \leqslant 0 \qquad (5.1)$$

for any sequence of asymptotically unbiased and of asymptotic level of significance α tests $\{\omega_n\}$.

When the sample size n is fixed, the unbiasedness of a test requires that its power remains $\geqslant \alpha$. When n is allowed to increase indefinitely, it seems natural to replace the concept of unbiasedness of a test by asymptotic unbiasedness, as is done in Definition 5.1. The approach used in this section (by employing results of Chapter 3) also makes it possible to consider sequences of tests of asymptotic level of significance α rather than of exact level α.

The main result in this section is Theorem 5.1. The auxiliary results presented below will facilitate the proof of the theorem.

The following lemma is a version of the Helly–Bray theorem (see Loève [1], p. 182) and is of some interest in its own right.

LEMMA 5.1   Let $\{F_n\}$ be a sequence of d.f.s on R and let $\{f_n\}$ be a sequence of real-valued, measurable functions defined on R. Also let F be a d.f. on R and let f and g be real-valued, continuous functions defined on R. It is assumed that

(i) $F_n \Rightarrow F$,

(ii) $|f_n(x)| < g(x)$ for all n and all $x \in R$,

(iii) $f_n(x) \to f(x)$ uniformly on finite intervals,

(iv) $\int g \, dF_n \to \int g \, dF$ and $\int g \, dF$ is finite.

Then $\int f_n \, dF_n \to \int f \, dF$ and $\int f \, dF$ is finite.

Proof For any $c, d \in R$ with $c < d$, one clearly has

$$\left| \int f_n \, \mathrm{d}F_n - \int f \, \mathrm{d}F \right| \leqslant \int_{(-\infty, c]} |f_n| \, \mathrm{d}F_n + \int_{(d, \infty)} |f_n| \, \mathrm{d}F_n$$
$$+ \int_{(-\infty, c]} |f| \, \mathrm{d}F + \int_{(d, \infty)} |f| \, \mathrm{d}F$$
$$+ \left| \int_{(c, d]} f_n \, \mathrm{d}F_n - \int_{(c, d]} f \, \mathrm{d}F \right|. \qquad (5.2)$$

On account of (ii) and the fact that

$$\left| \int_{(c, d]} f_n \, \mathrm{d}F_n - \int_{(c, d]} f \, \mathrm{d}F \right| \leqslant \int_{(c, d]} |f_n - f| \, \mathrm{d}F_n$$
$$+ \left| \int_{(c, d]} f \, \mathrm{d}F_n - \int_{(c, d]} f \, \mathrm{d}F \right|,$$

inequality (5.2) becomes as follows

$$\left| \int f_n \, \mathrm{d}F_n - \int f \, \mathrm{d}F \right| \leqslant \int_{(-\infty, c]} g \, \mathrm{d}F_n + \int_{(d, \infty)} g \, \mathrm{d}F_n$$
$$+ \int_{(-\infty, c]} |f| \, \mathrm{d}F + \int_{(d, \infty)} |f| \, \mathrm{d}F$$
$$+ \int_{(c, d]} |f_n - f| \, \mathrm{d}F_n + \left| \int_{(c, d]} f \, \mathrm{d}F_n - \int_{(c, d]} f \, \mathrm{d}F \right|,$$

or

$$\left| \int f_n \, \mathrm{d}F_n - \int f \, \mathrm{d}F \right| \leqslant \int g \, \mathrm{d}F_n - \int_{(c, d]} g \, \mathrm{d}F_n + \int_{(-\infty, c]} |f| \, \mathrm{d}F$$
$$+ \int_{(d, \infty)} |f| \, \mathrm{d}F + \int_{(c, d]} |f_n - f| \, \mathrm{d}F_n$$
$$+ \left| \int_{(c, d]} f \, \mathrm{d}F_n - \int_{(c, d]} f \, \mathrm{d}F \right|. \qquad (5.3)$$

Now by means of (ii)–(iv), it follows that $\int |f| \, \mathrm{d}F < \infty$. Thus, for every $\epsilon > 0$, we may choose c and d to be continuity points of F and such that

$$\int_{(-\infty, c]} |f| \, \mathrm{d}F < \frac{\epsilon}{6}, \quad \int_{(d, \infty)} |f| \, \mathrm{d}F < \frac{\epsilon}{6}$$

and
$$\int g \, dF - \int_{(c,\,d]} g \, dF < \frac{\epsilon}{6}. \qquad (5.4)$$

By (iv) and the Helly–Bray lemma (see Loève [1], p. 180), one has

$$\int g \, dF_n - \int_{(c,\,d]} g \, dF_n \to \int g \, dF - \int_{(c,\,d]} g \, dF.$$

Therefore by means of the last inequality in (5.4), one has for all sufficiently large n

$$\int g \, dF_n - \int_{(c,\,d]} g \, dF_n < \frac{2\epsilon}{6}. \qquad (5.5)$$

Furthermore, for all sufficiently large n, one also has

$$\int_{(c,\,d]} |f_n - f| \, dF_n < \frac{\epsilon}{6} \quad \text{and} \quad \left| \int_{(c,\,d]} f \, dF_n - \int_{(c,\,d]} f \, dF \right| < \frac{c}{6} \quad (5.6)$$

because of (iii) and the Helly–Bray lemma, respectively. By virtue of (5.4)–(5.6) and for all sufficiently large n, inequality (5.3) becomes $\left| \int f_n \, dF_n - \int f \, dF \right| < \epsilon$, as was to be shown.]

COROLLARY 5.1   Let $f_n^* = f_n I_{(a_n,\, b_n)}$ and $f^* = f I_{(a,\, b)}$, where $a_n \to a, b_n \to b$ and a, b are continuity points of F in the case they are finite. Then under the assumptions of the foregoing lemma, we have

$$\int f_n^* \, dF_n \to \int f^* \, dF.$$

Proof   Since $|f_n^*(x)| \leqslant |f_n(x)|$, so that (ii) is satisfied, it is clear that (5.3) holds true with f_n and f being replaced by f_n^* and f^*, respectively. Also $|f^*(x)| \leqslant |f(x)|$ and hence (5.4) is valid and so is (5.5). Thus it only remains for us to establish relations (5.6). If $a = -\infty$ and $b = \infty$, then $f^* = f$ and hence the second inequality in (5.6) holds true. Also, for all sufficiently large n, $a_n < c$ and $b_n > d$, so that $f_n^* = f_n$ on $(c, d]$. Hence the first inequality in (5.6) is also valid and the conclusion follows. Suppose next that a is finite and $b = \infty$. In this case, we assume that $a > c$, since otherwise the previous argument applies again. Without loss of generality, we also assume that $a < d$. There is $\delta_1 > 0$ such that for all sufficiently large n, $c < a - \delta_1 < a_n < a + \delta_1 < d$ and $a - \delta_1$, $a + \delta_1$ are continuity points of F. Therefore

$$\int_{(c,\,d]} |f_n^* - f^*| \, \mathrm{d}F_n = \int_{(a-\delta_1,\,d]} |f_n^* - f^*| \, \mathrm{d}F_n$$

$$= \int_{I_1} |f_n^* - f^*| \, \mathrm{d}F_n + \int_{(a+\delta_1,\,d]} |f_n - f| \, \mathrm{d}F_n,$$

where $I_1 = (a - \delta_1, a + \delta_1]$; f_n^* was replaced by f_n above because $b_n > d$ for all sufficiently large n.

Since $\int_{(a+\delta_1,\,d]} |f_n - f| \, \mathrm{d}F_n < \epsilon$ for all sufficiently large n by means of (iii), we need only be concerned with $\int_{I_1} |f_n^* - f^*| \, \mathrm{d}F_n$. In connection with this, we have

$$\int_{I_1} |f_n^* - f^*| \, \mathrm{d}F_n \leqslant \int_{I_1} |f_n^*| \, \mathrm{d}F_n + \int_{I_1} |f^*| \, \mathrm{d}F_n$$

$$\leqslant \int_{I_1} |f_n| \, \mathrm{d}F_n + \int_{I_1} |f| \, \mathrm{d}F_n$$

$$\leqslant 2 \int_{I_1} g \, \mathrm{d}F_n \quad \text{by means of (ii).}$$

Furthermore, $\int_{I_1} g \, dF_n \to \int_{I_1} g \, dF$ and by the fact that a is a continuity point of F, one can choose δ_1 as before and also such that $\int_{I_1} g \, \mathrm{d}F < \epsilon$. It follows that for all sufficiently large n,

$$\int_{(c,\,d]} |f_n^* - f^*| \, \mathrm{d}F_n < 5\epsilon.$$

Next

$$\left| \int_{(c,\,d]} f^* \, \mathrm{d}F_n - \int_{(c,\,d]} f^* \, \mathrm{d}F \right| = \left| \int_{(a,\,d]} f \, \mathrm{d}F_n - \int_{(a,\,d]} f \, \mathrm{d}F \right| < \epsilon$$

for all sufficiently large n. It follows that the two relations in (5.6) hold true (with $\frac{\epsilon}{6}$ replaced by 5ϵ and ϵ, respectively) and hence the corollary itself.

The case that $a = -\infty$ and b is finite, as well as the case that both a and b are finite, are treated in an entirely similar way. ▉

LEMMA 5.2   Let Δ_n^* be defined by (1.5) in Chapter 3 when $\theta = \theta_0$. Then one has

$$\mathscr{E}_{\theta_0}\Delta_n^* \to 0 \quad \text{and} \quad \mathscr{E}_{\theta_0}|\Delta_n^*| \to \int |x| \, dN(0, \Gamma).$$

Proof   In the inequality $1 + x \leqslant \exp x$, $x \geqslant 0$, replace x by $\frac{1}{2}\lambda_0|\Delta_n^*|$, where λ_0 is as in (1.4) of Chapter 3. Then

$$|1 + \tfrac{1}{2}\lambda_0\Delta_n^*| \leqslant 1 + \tfrac{1}{2}\lambda_0|\Delta_n^*| \leqslant \exp\left(\tfrac{1}{2}\lambda_0|\Delta_n^*|\right),$$

and hence

$$(1 + \tfrac{1}{2}\lambda_0\Delta_n^*)^2 \leqslant (1 + \tfrac{1}{2}\lambda_0|\Delta_n^*|)^2 \leqslant \exp\left(\lambda_0|\Delta_n^*|\right).$$

It follows that

$$\mathscr{E}_{\theta_0}(1 + \tfrac{1}{2}\lambda_0\Delta_n^*)^2 \leqslant \mathscr{E}_{\theta_0}(1 + \tfrac{1}{2}\lambda_0|\Delta_n^*|)^2$$
$$\leqslant \mathscr{E}_{\theta_0}\exp\left(\lambda_0|\Delta_n^*|\right) \leqslant M(< \infty) \quad \text{for all } n$$

by Propositions 1.2 and 2.2 in Chapter 3.

By setting $F_n = \mathscr{L}(\Delta_n^*|P_{\theta_0})$, one has then

$$\mathscr{E}_{\theta_0}(1 + \tfrac{1}{2}\lambda_0\Delta_n^*)^2 = \int (1 + \tfrac{1}{2}\lambda_0x)^2 \, dF_n \leqslant M \quad \text{for all } n \quad (5.7)$$

and

$$\mathscr{E}_{\theta_0}(1 + \tfrac{1}{2}\lambda_0|\Delta_n^*|)^2 = \int (1 + \tfrac{1}{2}\lambda_0|x|)^2 \, dF_n \leqslant M \quad \text{for all } n. \quad (5.7')$$

Also
$$\mathscr{L}(\Delta_n^*|P_{\theta_0}) = F_n \Rightarrow N(0, \Gamma) \quad (5.8)$$

by Proposition 2.4 (i) in Chapter 3.

By means of (5.7), (5.7') and (5.8), the corollary on p. 184 in Loève[1] applies and gives

$$\int (1 + \tfrac{1}{2}\lambda_0x) \, dF_n \to \int (1 + \tfrac{1}{2}\lambda_0x) \, dN(0, \Gamma) \quad (5.9)$$

and
$$\int (1 + \tfrac{1}{2}\lambda_0|x|) \, dF_n \to \int (1 + \tfrac{1}{2}\lambda_0|x|) \, dN(0, \Gamma). \quad (5.9')$$

Since $\int x \, dN(0, \Gamma) = 0$, relations (5.9) and (5.9') are equivalent to

$$\int x \, dF_n = \mathscr{E}_{\theta_0}\Delta_n^* \to 0$$

and
$$\int |x|\, dF_n = \mathscr{E}_{\theta_0} |\Delta_n^*| \to \int |x|\, dN(0, \Gamma),$$

respectively, as was to be shown. ∎

LEMMA 5.3 Let the sequence of tests, $\{\phi_n^*(\Delta_n^*)\}$, be defined as follows

$$\phi_n^*(\Delta_n^*) = \begin{cases} 1 & \text{if} \quad \Delta_n^* < a_n^* \text{ or } \Delta_n^* > b_n^* \\ & \quad (a_n^* < b_n^*), \\ \gamma_{1,n}^* \text{ or } \gamma_{2,n}^* & \text{if} \quad \Delta_n^* = a_n^* \text{ or } \Delta_n^* = b_n^*, \\ & \quad \text{respectively,} \\ 0 & \text{if} \quad a_n^* < \Delta_n^* < b_n^*, \end{cases} \quad (5.10)$$

where the constants a_n^*, b_n^*, $\gamma_{1,n}^*$ and $\gamma_{2,n}^*$ are determined by the requirements
$$\mathscr{E}_{\theta_0} \phi_n^*(\Delta_n^*) = \alpha$$
and
$$\mathscr{E}_{\theta_0}[\Delta_n^* \phi_n^*(\Delta_n^*)] = \alpha \mathscr{E}_{\theta_0} \Delta_n^* \quad \text{for all } n \quad (0 < \alpha < 1). \quad (5.11)$$

Then $a_n^* \to -\xi_{\frac12\alpha}$ and $b_n^* \to \xi_{\frac12\alpha}$, where ξ_p is the upper pth quantile of the $N(0, \Gamma)$.

Proof We have

$$\mathscr{E}_{\theta_0} \phi_n^*(\Delta_n^*) = P_{\theta_0}(\Delta_n^* < a_n^*) + P_{\theta_0}(\Delta_n^* > b_n^*) + \gamma_{1,n}^* P_{\theta_0}(\Delta_n^* = a_n^*)$$
$$+ \gamma_{2,n}^* P_{\theta_0}(\Delta_n^* = b_n^*). \quad (5.12)$$

Now (5.8) and Lemma 2.1 (i) yield

$$P_{\theta_0}(\Delta_n^* = a_n^*) \to 0 \quad \text{and} \quad P_{\theta_0}(\Delta_n^* = b_n^*) \to 0. \quad (5.13)$$

Next we assert that $a_n^* \leqslant M_1(< \infty)$ and $(-\infty <) M_2 \leqslant b_n^*$ for all n. In fact, suppose that $\{a_n^*\}$ is not bounded above. Then there exists $\{a_m^*\} \subseteq \{a_n^*\}$ such that $a_m^* \to \infty$ and since $a_m^* < b_m^*$, one also has $b_m^* \to \infty$. But then relations (5.11)–(5.13) imply that

$$\alpha = \mathscr{E}_{\theta_0} \phi_m^*(\Delta_m^*) \to 1,$$

a contradiction. The case in which $\{b_n^*\}$ is assumed to be unbounded from below is treated symmetrically.

Furthermore we claim that both $\{a_n^*\}$ and $\{b_n^*\}$ are bounded from both below and above. In fact, suppose that $\{a_n^*\}$ is not bounded from below. Then there exists $\{a_m^*\} \subseteq \{a_n^*\}$ such that

$a_m^* \to -\infty$. Look at $\{b_m^*\}$ and suppose first that it is unbounded from above. Then there exists $\{b_r^*\} \subseteq \{b_m^*\}$ such that $b_r^* \to \infty$. Thus $a_r^* \to -\infty$ and $b_r^* \to \infty$. Then relations (5.11)–(5.13) imply that

$$\alpha = \mathscr{E}_{\theta_0} \phi_r^*(\Delta_r^*) \to 0,$$

a contradiction. Next suppose that $\{b_m^*\}$ is bounded. Then there exists $\{b_r^*\} \subseteq \{b_m^*\}$ such that $b_r^* \to b(<\infty)$. Thus $a_r^* \to -\infty$ and $b_r^* \to b(<\infty)$. But by (5.13), one has

$$\lim \mathscr{E}_{\theta_0}[\Delta_r^* \phi_r^*(\Delta_r^*)]$$

$$= \lim \left[\int_{(-\infty,\, a_r^*)} x \, d\mathscr{L}(\Delta_r^* | P_{\theta_0}) + \int_{(b_r^*,\, \infty)} x \, d\mathscr{L}(\Delta_r^* | P_{\theta_0}) \right.$$
$$\left. + a_r^* P_{\theta_0}(\Delta_r^* = a_r^*) \right].$$

$$(5.14)$$

Furthermore the assumptions of Lemma 5.1 are fulfilled with

$$F_r = \mathscr{L}(\Delta_r^* | P_{\theta_0}), \quad F = N(0,\Gamma), \quad f_n(x) = f(x) = x$$

and
$$g(x) = |x|;$$

the convergence in (iv) is satisfied because of Lemma 5.2. Then Corollary 5.1 applies in an obvious manner and gives

$$\int_{(-\infty,\, a_r^*)} x \, d\mathscr{L}(\Delta_r^* | P_{\theta_0}) \to 0$$

and
$$\int_{(b_r^*,\, \infty)} x \, d\mathscr{L}(\Delta_r^* | P_{\theta_0}) \to \int_{(b,\, \infty)} x \, dN(0,\Gamma). \qquad (5.15)$$

Also

$$\mathscr{E}_{\theta_0} \Delta_r^* = \int_{(-\infty,\, a_r^*)} x \, dF_r + \int_{(a_r^*,\, \infty)} x \, dF_r + a_r^* P_{\theta_0}(\Delta_r^* = a_r^*)$$

and hence (5.15) and Lemma 5.2 imply that $a_r^* P_{\theta_0}(\Delta_r^* = a_r^*) \to 0$. Therefore taking the limits on both sides of the second relation in (5.11) and utilizing the last result obtained above along with Lemma 5.2 and relations (5.14) and (5.15), one has

$$0 = \alpha \int_{(b,\, \infty)} x \, dN(0,\Gamma),$$

a contradiction. The case that $\{b_n^*\}$ is assumed to be unbounded from above is treated symmetrically.

Therefore both $\{a_n^*\}$ and $\{b_n^*\}$ are bounded. Then there exists $\{m\} \subseteq \{n\}$ such that $a_m^* \to a$ and $b_m^* \to b$. From (5.14), with r replaced by m, it follows then as above that

$$\lim \mathscr{E}_{\theta_0}[\Delta_m^* \phi_m^*(\Delta_m^*)] = \int_{[a,\,b]^c} x \, dN(0, \Gamma). \qquad (5.16)$$

Then Lemma 5.2 together with (5.11) and (5.16) implies

$$\int_{[a,\,b]^c} x \, dN(0, \Gamma) = 0.$$

Hence $a = -b$. Taking this into consideration and letting $n \to \infty$ through the subsequence $\{m\}$ in the first equation in (5.11), one obtains $\int_{[-b,\,b]} dN(0, \Gamma) = 1 - \alpha$; thus, $a_m^* \to -\xi_{\frac{1}{2}\alpha}$ and $b_m^* \to \xi_{\frac{1}{2}\alpha}$. Since all convergent subsequences converge to the same limits, it follows that $a_n^* \to -\xi_{\frac{1}{2}\alpha}$ and $b_n^* \to \xi_{\frac{1}{2}\alpha}$, as was to be shown. ∎

COROLLARY 5.2 Consider the sequence $\{h_n\}$ such that $h_n \to h \in R$ and set $\theta_n = \theta_0 + h_n n^{-\frac{1}{2}}$. Then

$$\mathscr{E}_{\theta_n} \phi_n^*(\Delta_n^*) \to \int_{(|x| \geqslant \xi_{\frac{1}{2}\alpha})} dN(\Gamma h, \Gamma).$$

Proof From Proposition 2.4 (ii) in Chapter 3, one has

$$\mathscr{L}(\Delta_n^* | P_{\theta_n}) \Rightarrow N(\Gamma h, \Gamma).$$

Then the corollary is an immediate consequence of Lemma 2.1 (i), the fact that $a_n^* \to -\xi_{\frac{1}{2}\alpha}$, $b_n^* \to \xi_{\frac{1}{2}\alpha}$ and Polya's lemma. ∎

Let $\Delta_n = \Delta_n(\theta_0)$ be given by (1.1) for $\theta = \theta_0$ and define the sequence $\{\phi_n\}$ as follows.

$$\phi_n = \phi_n(\Delta_n) = \left.\begin{cases} 1 & \text{if} \quad \Delta_n < a_n \text{ or } \Delta_n > b_n \quad (a_n < b_n) \\ 0 & \text{if} \quad a_n \leqslant \Delta_n \leqslant b_n, \end{cases}\right\} \quad (5.17)$$

where the sequences $\{a_n\}$ and $\{b_n\}$ are chosen, so that

$$a_n \to -\xi_{\frac{1}{2}\alpha} \quad \text{and} \quad b_n \to \xi_{\frac{1}{2}\alpha}, \qquad (5.18)$$

and ξ_p is the upper pth quantile of the $N(0, \Gamma)$.

Then we have the following result.

THEOREM 5.1 Under assumptions (A1-5), (A5'), the

sequence of tests $\{\phi_n(\Delta_n)\}$ defined by (5.17) and (5.18) is AUMPU of asymptotic level of significance α for testing $H_0: \theta = \theta_0$ against $A'': \theta \in \omega''$, where ω'' is given in Definition 5.1.

Proof The proof of the theorem is presented in the following three stages. We first show that $\{\phi_n(\Delta_n)\}$ is of asymptotic level of significance α. Next it is proved that $\{\phi_n(\Delta_n)\}$ satisfies the condition of asymptotic unbiasedness; this is done by contradiction. Finally, it is shown that $\{\phi_n(\Delta_n)\}$ is AUMP within the class of asymptotically unbiased and of asymptotic level α tests. The proof of this fact is also by contradiction and consists of two steps. First we take care of those parameter values θ_n for which $\{n^{\frac{1}{2}}(\theta_n - \theta_0)\}$ is unbounded. For those θ_n for which $\{n^{\frac{1}{2}}(\theta_n - \theta_0)\}$ remains bounded, we employ the exponential approximation discussed in Chapter 3 in order to replace the given family by an exponential family. For this latter family, we set up the appropriate UMPU test which exists and then, by utilizing available results, we return to the original family and show the required optimal character of the sequence of tests $\{\phi_n(\Delta_n)\}$.

The details of the proof are as follows. In the first place, the level of significance is asymptotically α because of (5.18) and the fact that $\mathscr{L}(\Delta_n | P_{\theta_0}) \Rightarrow N(0, \Gamma)$ by Theorem 4.2 in Chapter 2.

Next we show that the condition of asymptotic unbiasedness holds for $\{\phi(\Delta_n)\}$, namely

$$\liminf\{\inf[\mathscr{E}_\theta \phi_n(\Delta_n); \theta \in \omega'']\} \geqslant \alpha. \qquad (5.19)$$

For contradiction, suppose (5.19) is not true and let the expression on the left-hand side of (5.19) be equal to $\delta < \alpha$. Then there exists a subsequence $\{m\} \subseteq \{n\}$ and a sequence $\{\theta_m\}$ with $\theta_m \in \omega''$ for all m such that

$$\mathscr{E}_{\theta_m}\phi_m(\Delta_m) \to \delta. \qquad (5.20)$$

First suppose that $\{m^{\frac{1}{2}}(\theta_m - \theta_0)\}$ is unbounded. Then there exists $\{r\} \subseteq \{m\}$ with $\theta_r < \theta_0$ for all r or $\theta_r > \theta_0$ for all r such that, respectively,

$$r^{\frac{1}{2}}(\theta_r - \theta_0) \to -\infty \quad \text{or} \quad r^{\frac{1}{2}}(\theta_r - \theta_0) \to \infty. \qquad (5.21)$$

But

$$\mathscr{E}_{\theta_r}\phi_r(\Delta_r) = P_{\theta_r}(\Delta_r < a_r) + P_{\theta_r}(\Delta_r > b_r) \qquad (5.22)$$

and therefore (5.18) together with (5.21) and (A5) or (A5'), respectively, implies that $\mathscr{E}_{\theta_r}\phi_r(\Delta_r) \to 1$ which contradicts (5.20).

Next suppose that $\{m^{\frac{1}{2}}(\theta_m - \theta_0)\}$ remains bounded. Then there exists $\{s\} \subseteq \{m\}$ such that $s^{\frac{1}{2}}(\theta_s - \theta_0) \to h$. Setting $h_s = s^{\frac{1}{2}}(\theta_s - \theta_0)$, we have then $\theta_s = \theta_0 + h_s s^{-\frac{1}{2}}$ with $h_s \to h$. But then $\{P_{s,\theta_0}\}$ and $\{P_{s,\theta_s}\}$ are contiguous by Proposition 6.1 in Chapter 2, so that Theorem 4.6 in the same chapter applies and gives

$$\mathscr{L}(\Delta_s | P_{\theta_s}) \Rightarrow N(\Gamma h, \Gamma). \tag{5.23}$$

If $h = 0$, then (5.22) with (r replaced by s) implies by means of (5.18) and (5.23) that $\mathscr{E}_{\theta_s}\phi_s(\Delta_s) \to \alpha$ which contradicts (5.20); if $h \neq 0$, then

$$\mathscr{E}_{\theta_s}\phi_s(\Delta_s) \to P[N(\Gamma h, \Gamma) < -\xi_{\frac{1}{2}\alpha}] + P[N(\Gamma h, \Gamma) > \xi_{\frac{1}{2}\alpha}] > \alpha.$$

Once more we have a contradiction to (5.20). Therefore (5.19) holds true.

It remains for us to show that $\{\phi_n(\Delta_n)\}$ is UMP within the class of asymptotically unbiased tests with asymptotic level equal to α. So let $\{\omega_n\}$ be any sequence of such tests. To show that

$$\limsup \{\sup [\mathscr{E}_\theta \omega_n - \mathscr{E}_\theta \phi_n(\Delta_n); \theta \in \omega'']\} \leqslant 0. \tag{5.24}$$

The proof is by contradiction. Suppose that (5.24) is not true and let the left-hand side in (5.24) be equal to 8δ for some $\delta > 0$. Then there exists a subsequence $\{m\} \subseteq \{n\}$ and a sequence $\{\theta_m\}$ with $\theta_m \in \omega''$ for all m such that

$$\mathscr{E}_{\theta_m}\omega_m - \mathscr{E}_{\theta_m}\phi_m(\Delta_m) \to 8\delta. \tag{5.25}$$

Consider the sequence $\{m^{\frac{1}{2}}(\theta_m - \theta_0)\}$ and suppose first that it is unbounded. Then there exists $\{r\} \subseteq \{m\}$ and $\{\theta_r\}$ with $\theta_r < \theta_0$ for all r or $\theta_r > \theta_0$ for all r such that, respectively,

$$r^{\frac{1}{2}}(\theta_r - \theta_0) \to -\infty \quad \text{or} \quad r^{\frac{1}{2}}(\theta_r - \theta_0) \to \infty. \tag{5.26}$$

In this case, repeating the arguments employed in connection with (5.21) above, we conclude that $\mathscr{E}_{\theta_r}\phi_r(\Delta_r) \to 1$, so that (5.25) cannot be true. Next suppose that $\{m^{\frac{1}{2}}(\theta_m - \theta_0)\}$ is bounded. Then there exists $\{s\} \subseteq \{m\}$ such that $s^{\frac{1}{2}}(\theta_s - \theta_0) \to h$. Setting

$$h_s = s^{\frac{1}{2}}(\theta_s - \theta_0),$$

we have then $\theta_s = \theta_0 + h_s s^{-\frac{1}{2}}$ with $h_s \to h$. Once more the sequences $\{P_{s,\theta_0}\}$ and $\{P_{s,\theta_s}\}$ are contiguous as was mentioned and justified above. Now the test ω_n is a function of the Xs. However, from an asymptotic power point of view, we can replace it by $\overline{\omega}_n(\Delta_n)$, where $\overline{\omega}_n(\Delta_n) = \mathscr{E}_{\theta_0}(\omega_n|\Delta_n)$. This is so by Theorem 5.1 in Chapter 3 according to which

$$\mathscr{E}_{\theta_s}\overline{\omega}_s(\Delta_s) - \mathscr{E}_{\theta_s}\omega_s \to 0 \quad \text{and} \quad \mathscr{E}_{\theta_0}\overline{\omega}_s(\Delta_s) - \mathscr{E}_{\theta_0}\omega_s \to 0. \quad (5.27)$$

Thus (5.25) (with m replaced by s) and (5.27) imply that

$$\mathscr{E}_{\theta_s}\overline{\omega}_s(\Delta_s) - \mathscr{E}_{\theta_s}\phi_s(\Delta_s) > 7\delta \quad \text{for all } s \geqslant n_1, \quad \text{say.} \quad (5.28)$$

Next for any sequence of tests $\{\psi_n\}$ defined on R, one has

$$|\mathscr{E}_{\theta_s}\psi_s(\Delta_s) - \mathscr{E}_{\theta_s}\psi_s(\Delta_s^*)| = \left| \int_{(\Delta_s = \Delta_s^*)} [\psi_s(\Delta_s) - \psi_s(\Delta_s^*)] \, \mathrm{d}P_{\theta_s} \right.$$

$$\left. + \int_{(\Delta_s \neq \Delta_s^*)} [\psi_s(\Delta_s) - \psi_s(\Delta_s^*)] \, \mathrm{d}P_{\theta_s} \right|$$

$$\leqslant 2P_{\theta_s}(\Delta_s \neq \Delta_s^*) \to 0, \quad (5.29)$$

the convergence to zero being true because of Proposition 2.3 (ii) of Chapter 3. So

$$\mathscr{E}_{\theta_s}\psi_s(\Delta_s) - \mathscr{E}_{\theta_s}\psi_s(\Delta_s^*) \to 0,$$

and, in particular,

$$\mathscr{E}_{\theta_s}\overline{\omega}_s(\Delta_s^*) - \mathscr{E}_{\theta_s}\overline{\omega}_s(\Delta_s) \to 0 \quad \text{and} \quad \mathscr{E}_{\theta_s}\phi_s(\Delta_s) - \mathscr{E}_{\theta_s}\phi_s(\Delta_s^*) \to 0,$$

so that for all sufficiently large s, $s \geqslant n_2$, say, one has

$$\mathscr{E}_{\theta_s}\overline{\omega}_s(\Delta_s^*) - \mathscr{E}_{\theta_s}\overline{\omega}_s(\Delta_s) > -\delta \quad \text{and} \quad \mathscr{E}_{\theta_s}\phi_s(\Delta_s) - \mathscr{E}_{\theta_s}\phi_s(\Delta_s^*) > -\delta. \quad (5.30)$$

Adding up (5.28) and (5.30), and letting $n_3 = \max(n_1, n_2)$, one has

$$\mathscr{E}_{\theta_s}\overline{\omega}_s(\Delta_s^*) - \mathscr{E}_{\theta_s}\phi_s(\Delta_s^*) > 5\delta \quad \text{for all } s \geqslant n_3. \quad (5.31)$$

Next, as it was mentioned in the proof of Corollary 5.2, one has that

$$\mathscr{L}(\Delta_s^*|P_{\theta_s}) \Rightarrow N(\Gamma h, \Gamma)$$

and consequently

$$\mathscr{E}_{\theta_s}\phi_s(\Delta_s^*) \to \int_{(|x| > \xi_{\frac{1}{2}\alpha})} \mathrm{d}N(\Gamma h, \Gamma).$$

Thus

$$\int_{(|x| > \xi_{\frac{1}{2}\alpha})} \mathrm{d}N(\Gamma h, \Gamma) - \mathscr{E}_{\theta_s}\phi_s(\Delta_s^*) < \delta \quad \text{for all } s \geqslant n_4, \quad \text{say.} \quad (5.32)$$

Now the convergence in (5.19) is still true if $\{\theta_s\} = \{\theta_0\}$, on account of Proposition 2.3 (i). Therefore

$$\mathcal{E}_{\theta_s}\overline{\omega}_s(\Delta_s^*) - \mathcal{E}_{\theta_s}\overline{\omega}_s(\Delta_s) \to 0$$

and hence, by means of this result and also (5.27), one has

$$\mathcal{E}_{\theta_s}\overline{\omega}_s(\Delta_s^*) = \alpha_s \to \alpha. \tag{5.33}$$

At this point, look at the probability measure R_{s,h_s} defined by

$$\frac{dR_{s,h_s}}{dP_{s,\theta_0}} = \exp\left[h_s\Delta_s^* - B_s(h_s)\right], \quad \exp\left[B_s(h_s)\right] = \mathcal{E}_{\theta_0}\exp\left(h_s\Delta_s^*\right) \tag{5.34}$$

(see also (1.6), (1.7) and Theorem 3.1 in Chapter 3). In the family of probability measures defined by (5.34), consider the problem of testing $H_{0,h}\colon h_s = 0$ against $A_h''\colon h_s \neq 0$ at level α_s for all sufficiently large s, $s \geqslant n_5$, say, so that $0 < \alpha_s < 1$. Then the test defined below is a UMPU test. (See, e.g. Lehmann [3], p. 126.) Namely,

$$\overline{\phi}_s(\Delta_s^*) = \begin{cases} 1 & \text{if} \quad \Delta_s^* < \overline{a}_s \text{ or } \Delta_s^* > \overline{b}_s \quad (\overline{a}_s < \overline{b}_s) \\ \overline{\gamma}_{1,s} \text{ or } \overline{\gamma}_{2,s} & \text{if} \quad \Delta_s^* = \overline{a}_s \text{ or } \Delta_s^* = \overline{b}_s, \quad \text{respectively,} \\ 0 & \text{if} \quad \overline{a}_s < \Delta_s^* < b_s^*, \end{cases} \tag{5.35}$$

where the constants $\overline{a}_s$, $\overline{b}_s$, $\overline{\gamma}_{1,s}$ and $\overline{\gamma}_{2,s}$ are defined by the requirements

$$\left.\begin{array}{ll} & \mathcal{E}[\overline{\phi}_s(\Delta_s^*)|R_{s,0}] = \alpha_s \\ \text{and} & \mathcal{E}[\Delta_s^*\overline{\phi}_s(\Delta_s^*)|R_{s,0}] = \alpha_s\mathcal{E}(\Delta_s^*|R_{s,0}) \quad \text{for all } s \geqslant n_5. \end{array}\right\} \tag{5.36}$$

From (5.34) it follows that $R_{s,0} = P_{s,\theta_0}$. Thus (5.36) becomes

$$\left.\begin{array}{ll} & \mathcal{E}_{\theta_0}\overline{\phi}_s(\Delta_s^*) = \alpha_s \\ \text{and} & \mathcal{E}_{\theta_0}[\Delta_s^*\overline{\phi}_s(\Delta_s^*)] = \alpha_s\mathcal{E}_{\theta_0}(\Delta_s^*) \quad \text{for all } s \geqslant n_5. \end{array}\right\} \tag{5.36'}$$

Since $\alpha_s \to \alpha$ by (5.33), it follows that

$$\int_{(|x|>\xi_{\frac{1}{2}\alpha_s})} dN(\Gamma h, \Gamma) \to \int_{(|x|>\xi_{\frac{1}{2}\alpha})} dN(\Gamma h, \Gamma),$$

so that

$$\int_{(|x|>\xi_{\frac{1}{2}\alpha_s})} dN(\Gamma h, \Gamma) - \int_{(|x|>\xi_{\frac{1}{2}\alpha})} dN(\Gamma h, \Gamma) < \delta \quad \text{for all } s \geqslant n_6, \quad \text{say.} \tag{5.37}$$

Also
$$\mathscr{E}_{\theta_s}\overline{\phi}_s(\Delta_s^*) - \int_{(|x|>\xi_{\frac{1}{2}\alpha_s})} dN(\Gamma h, \Gamma) \to 0 \quad \text{by Corollary 5.3,}$$
so that
$$\mathscr{E}_{\theta_s}\overline{\phi}_s(\Delta_s^*) - \int_{(|x|>\xi_{\frac{1}{2}\alpha_s})} dN(\Gamma h, \Gamma) < \delta \quad \text{for all } s \geqslant n_7, \quad \text{say.}$$
$$(5.38)$$

Letting now $s \geqslant n_8 = \max(n_4, n_6, n_7)$ and adding up (5.32), (5.37) and (5.38), one has
$$\mathscr{E}_{\theta_s}\overline{\phi}_s(\Delta_s^*) - \mathscr{E}_{\theta_s}\phi_s(\Delta_s^*) < 3\delta. \qquad (5.39)$$
Since $\mathscr{E}_{\theta_s}\overline{\phi}_s(\Delta_s^*) = \mathscr{E}_{\theta_0}\overline{\omega}_s(\Delta_s^*) = \alpha_s$ and $\overline{\phi}_s(\Delta_s^*)$ is UMPU, we have
$$\mathscr{E}[\overline{\omega}_s(\Delta_s^*)|R_{s, h_s}] - \mathscr{E}[\overline{\phi}_s(\Delta_s^*)|R_{s, h_s}] < \delta \quad \text{for all } s \geqslant n_9, \quad \text{say.}$$
$$(5.40)$$

The detailed and rigorous justification of (5.40) is given in Lemma 5.4 below in order not to interrupt the proof of the theorem. But
$$|\mathscr{E}_{\theta_s}\overline{\omega}_s(\Delta_s^*) - \mathscr{E}[\overline{\omega}_s(\Delta_s^*)|R_{s, h_s}]| \leqslant \|P_{s, \theta_s} - R_{s, h_s}\| \to 0$$
by Theorem 3.1 in Chapter 3, and therefore
$$\mathscr{E}_{\theta_s}\overline{\omega}_s(\Delta_s^*) - \mathscr{E}[\overline{\omega}_s(\Delta_s^*)|R_{s, h_s}] \to 0. \qquad (5.41)$$
Similarly $\quad \mathscr{E}[\overline{\phi}_s(\Delta_s^*)|R_{s, h_s}] - \mathscr{E}_{\theta_s}\overline{\phi}_s(\Delta_s^*) \to 0, \qquad (5.42)$
so that by means of (5.40)–(5.42) and for all $s \geqslant n_{10}$, say, one has
$$\mathscr{E}_{\theta_s}\overline{\omega}_s(\Delta_s^*) - \mathscr{E}_{\theta_s}\overline{\phi}_s(\Delta_s^*)$$
$$< \mathscr{E}[\overline{\omega}_s(\Delta_s^*)|R_{s, h_s}] - \mathscr{E}[\overline{\phi}_s(\Delta_s^*)|R_{s, h_s}] + \delta \leqslant 2\delta. \quad (5.43)$$
From (5.39) and (5.43) and for all $s \geqslant n_{11} = \max(n_8, n_{10})$, we obtain
$$\mathscr{E}_{\theta_s}\overline{\omega}_s(\Delta_s^*) - \mathscr{E}_{\theta_s}\phi_s(\Delta_s^*) < 5\delta.$$

However, this contradicts (5.31). The proof of the theorem is completed. ∎

As in the case of Theorem 3.1 (and also Theorem 3.2), from the proof of Theorem 5.1 it is immediate that, as long as one restricts himself to alternatives θ for which $\{n^{\frac{1}{2}}(\theta-\theta_0)\}$ stays bounded, one has

COROLLARY 5.3 Under assumptions (A 1–4), the sequence of tests $\{\phi_n\}$ defined by (5.17) and (5.18) is asymptotically locally most powerful unbiased for testing $H_0: \theta = \theta_0$ against $A'': \theta \in \omega''$. That is, the sequence $\{\phi_n\}$ satisfies (5.1) when the parameter values are restricted to those θs in ω'' for which $\{n^{\frac{1}{2}}(\theta - \theta_0)\}$ remains bounded from below for $\theta < \theta_0$ and bounded from above for $\theta > \theta_0$.

REMARK 5.1 In order to see the explicit form of the test in the case of the concrete examples that we have been dealing with, the reader is referred to Section 4.

LEMMA 5.4 In connection with the family of probability measures $\{R_{s,h_s}\}$ defined by (5.34), consider the sequence of tests $\{\overline{\omega}_s(\Delta_s^*)\}$ such that $\mathscr{E}_{\theta_0}\overline{\omega}_s(\Delta_s^*) = \alpha_s \to \alpha$ (see (5.33)). Then

$$\mathscr{E}[\overline{\omega}_s(\Delta_s^*)|R_{s,h_s}] - \mathscr{E}[\overline{\phi}_s(\Delta_s^*)|R_{s,h_s}] < \delta$$

$$\text{for all sufficiently large } s, \quad (5.44)$$

where δ is the one employed in the proof of Theorem 5.1.

Proof For each $s \geqslant n_5$, where n_5 appears after relationship (5.34), consider the test $\overline{\omega}_s(\Delta_s^*)$ with $\mathscr{E}_{\theta_0}\overline{\omega}_s(\Delta_s^*) = \alpha_s$. Then there exists a test $\tilde{\omega}_s(\Delta_s^*)$ of the form

$$\tilde{\omega}_s(\Delta_s^*) = \begin{cases} 1 & \text{if} \quad \Delta_s^* < \tilde{a}_s \text{ or } \Delta_s^* > \tilde{b}_s \quad (\tilde{a}_s < \tilde{b}_s) \\ \tilde{\gamma}_{1,s} \text{ or } \tilde{\gamma}_{2,s} & \text{if} \quad \Delta_s^* = \tilde{a}_s \text{ or } \Delta_s^* = \tilde{b}_s, \quad \text{respectively,} \\ 0 & \text{if} \quad \tilde{a}_s < \Delta_s^* < \tilde{b}_s \end{cases}$$

such that $\qquad\qquad \mathscr{E}_{\theta_0}\tilde{\omega}_s(\Delta_s^*) = \alpha_s$

and $\qquad\quad \mathscr{E}[\tilde{\omega}_s(\Delta_s^*)|R_{s,h_s^*}] \geqslant \mathscr{E}[\tilde{\omega}_s(\Delta_s^*)|R_{s,h_s^*}],$ $\qquad\qquad$ (5.45)

where R_{s,h_s^*} is defined by (5.34) with h_s replaced by $h_s^* \to h^* \in R$. This is so by Theorem 2, p. 220 in Ferguson [1]. Now the fact that $\lim\inf[\inf(\mathscr{E}_\theta\omega_n; \theta \neq \theta_0)] \geqslant \alpha$, implies that

$$\lim\inf \mathscr{E}_{\theta_s^*}\omega_s \geqslant \alpha \quad \text{for any sequence } \{\theta_s^*\} \text{ with } \theta_s^* \neq \theta_0 \text{ for all } s. \quad (5.46)$$

In fact, $\mathscr{E}_{\theta_s^*}\omega_s \geqslant \inf(\mathscr{E}_\theta\omega_s; \theta \neq \theta_0)$ and hence

$$\lim\inf \mathscr{E}_{\theta_s^*}\omega_s \geqslant \lim\inf[\inf(\mathscr{E}_\theta\omega_s; \theta \neq \theta_0)] \geqslant \alpha.$$

Let now $\theta_s^* = \theta_0 + h_s^* s^{-\frac{1}{2}}$. Then by Theorem 5.1, Chapter 3, one has $\mathscr{E}_{\theta_s^*}\bar{\omega}_s(\Delta_s) - \mathscr{E}_{\theta_s^*}\omega_s \to 0$. This result together with (5.46) implies in an obvious manner that

$$\liminf \mathscr{E}_{\theta_s^*}\bar{\omega}_s(\Delta_s) \geqslant \alpha. \tag{5.47}$$

Next, as in (5.29),

$$\left| \mathscr{E}_{\theta_s^*}\bar{\omega}_s(\Delta_s) - \mathscr{E}_{\theta_s^*}\bar{\omega}_s(\Delta_s^*) \right| \leqslant 2P_{\theta_s^*}(\Delta_s \neq \Delta_s^*) \to 0,$$

so that $\mathscr{E}_{\theta_s^*}\bar{\omega}_s(\Delta_s^*) - \mathscr{E}_{\theta_s^*}\bar{\omega}_s(\Delta_s) \to 0$. This result together with (5.47) implies as above that

$$\liminf \mathscr{E}_{\theta_s^*}\bar{\omega}_s(\Delta_s^*) \geqslant \alpha. \tag{5.48}$$

But $\left| \mathscr{E}_{\theta_s^*}\bar{\omega}_s(\Delta_s^*) - \mathscr{E}[\bar{\omega}_s(\Delta_s^*)|R_{s,h_s^*}] \right| \leqslant \left\| P_{\theta_s^*} - R_{s,h_s^*} \right\| \to 0$

by Theorem 3.1, Chapter 3, so that

$$\mathscr{E}[\bar{\omega}_s(\Delta_s^*)|R_{s,h_s^*}] - \mathscr{E}_{\theta_s^*}\bar{\omega}_s(\Delta_s^*) \to 0.$$

This result together with (5.48) gives

$$\liminf \mathscr{E}[\bar{\omega}_s(\Delta_s^*)|R_{s,h_s^*}] \geqslant \alpha. \tag{5.49}$$

Now inequality (5.49) and the inequality in (5.45) provide us with the following relationship

$$\liminf \mathscr{E}[\tilde{\omega}_s(\Delta_s^*)|R_{s,h_s^*}] \geqslant \alpha. \tag{5.50}$$

Next $\left| \mathscr{E}_{\theta_s^*}\tilde{\omega}_s(\Delta_s^*) - \mathscr{E}[\tilde{\omega}_s(\Delta_s^*)|R_{s,h_s^*}] \right| \leqslant \left\| P_{\theta_s^*} - R_{s,h_s^*} \right\| \to 0,$

so that $\mathscr{E}_{\theta_s^*}\tilde{\omega}_s(\Delta_s^*) - \mathscr{E}[\tilde{\omega}_s(\Delta_s^*)|R_{s,h_s^*}] \to 0.$

This result, together with (5.50), gives

$$\liminf \mathscr{E}_{\theta_s^*}\tilde{\omega}_s(\Delta_s^*) \geqslant \alpha. \tag{5.51}$$

We shall now show that $\tilde{a}_s \to -\xi_{\frac{1}{2}\alpha}$ and $\tilde{b}_s \to \xi_{\frac{1}{2}\alpha}$. To this end, there exists a subsequence $\{r\} \subseteq \{s\}$ such that $\tilde{a}_r \to a$ and $\tilde{b}_r \to b$, where $-\infty \leqslant a \leqslant b \leqslant \infty$. By taking into consideration the fact that $\mathscr{L}(\Delta_r^*|P_{\theta_0}) \Rightarrow N(0,\Gamma)$, $\mathscr{E}_{\theta_0}\tilde{\omega}_r(\Delta_r^*) \to \alpha$,

Lemma 2.1 (i) and Polya's lemma, one has that

$$\mathscr{E}_{\theta_0}\tilde{\omega}_r(\Delta_r^*) \to \Phi(a) + 1 - \Phi(b) = \alpha,$$

where Φ is the d.f. of $N(0,\Gamma)$. It follows that the equalities $a = b = -\infty$ or $a = b = \infty$ are excluded. Let $a = -\infty$, so that

$b = \xi_\alpha$. Then by the fact that $\mathscr{L}(\Delta_r^* | P_{\theta_r^*}) \Rightarrow N(\Gamma h^*, \Gamma)$ and considerations similar to the ones employed above, one has that $\mathscr{E}_{\theta_r^*} \tilde{\omega}_r(\Delta_r^*) \to 1 - \Phi(\xi_\alpha - \Gamma h^*)$. By choosing $h^* < 0$, we get $1 - \Phi(\xi_\alpha - \Gamma h^*) < 1 - \Phi(\xi_\alpha) = \alpha$, so that $\liminf \mathscr{E}_{\theta_r^*} \tilde{\omega}_r(\Delta_r^*) < \alpha$ which contradicts (5.51). Thus $a > -\infty$, and in a similar fashion $b < \infty$.

We now claim that $a = -\xi_{\frac{1}{2}\alpha}$ and $b = \xi_{\frac{1}{2}\alpha}$. For contradiction, assume that $b < \xi_{\frac{1}{2}\alpha}$, so that $a < -\xi_{\frac{1}{2}\alpha}$. Then as before,

$$\mathscr{E}_{\theta_0} \tilde{\omega}_r(\Delta_r^*) \to \Phi(a) + 1 - \Phi(b) = \alpha,$$

whereas

$$\mathscr{E}_{\theta_r^*} \tilde{\omega}_r(\Delta_r^*) \to \Phi(a - \Gamma h^*) + 1 - \Phi(b - \Gamma h^*) < \Phi(a) + 1 - \Phi(b)$$

by choosing $h^* < 0$. Thus $\liminf \mathscr{E}_{\theta_r^*} \tilde{\omega}_r(\Delta_r^*) < \alpha$ which contradicts (5.51). Therefore $b \geqslant \xi_{\frac{1}{2}\alpha}$, and b may not be $> \xi_{\frac{1}{2}\alpha}$ by a similar argument, so that $b = \xi_{\frac{1}{2}\alpha}$. Similarly $a = -\xi_{\frac{1}{2}\alpha}$: It follows that all convergent subsequences of $\{\tilde{a}_s\}$ and $\{\tilde{b}_s\}$ converge to $-\xi_{\frac{1}{2}\alpha}$ and $\xi_{\frac{1}{2}\alpha}$, respectively, so that $\tilde{a}_s \to -\xi_{\frac{1}{2}\alpha}$ and $\tilde{b}_s \to \xi_{\frac{1}{2}\alpha}$. Consequently, replacing θ_s^* by $\theta_s = \theta_0 + h_s s^{-\frac{1}{2}}$, $h_s \to h$, one has

$$\mathscr{E}_{\theta_s} \tilde{\omega}_s(\Delta_s^*) - \mathscr{E}_{\theta_s} \overline{\phi}_s(\Delta_s^*) \to 0$$

and hence $\mathscr{E}[\tilde{\omega}_s(\Delta_s^*) | R_{s, h_s}] - \mathscr{E}[\overline{\phi}_s(\Delta_s^*) | R_{s, h_s}] \to 0.$

Thus $\mathscr{E}[\tilde{\omega}_s(\Delta_s^*) | R_{s, h_s}] - \mathscr{E}[\overline{\phi}_s(\Delta_s^*) | R_{s, h_s}] < \delta$

for all sufficiently large s, and hence by means of (5.45),

$$\mathscr{E}[\overline{\omega}_s(\Delta_s^*) | R_{s, h_s}] - \mathscr{E}[\overline{\phi}_s(\Delta_s^*) | R_{s, h_s}] < \delta \quad \text{for all sufficiently large } s,$$

as was to be seen. ▮

6   Testing a one-sided hypothesis against one-sided alternatives

In Section 3, it was shown that the sequence of tests $\{\phi_n\}$ defined by (3.2) and (3.3) was AUMP for testing $H_0: \theta = \theta_0$ against $A: \theta \in \omega = \{\theta \in \Theta; \theta > \theta_0 \in \Theta\}$. In establishing this result, only the material discussed in Chapter 2 was employed.

Now by utilizing the theory of exponential approximation developed in Chapter 3, one can show that the before-mentioned sequence of tests is also AUMP for testing the one-sided hypothesis $H_1\colon \theta \in \overline{\omega} = \{\theta \in \Theta;\ \theta \leqslant \theta_0\} = \omega' \cup \{\theta_0\}$ against A. The proof of this fact is along the same lines with the proof of Theorem 5.1. However, we consider it appropriate that an outline of the proof be presented.

More precisely, we shall establish the following result. In this, $\Delta_n = \Delta_n(\theta_0)$.

THEOREM 6.1 Under assumptions (A1–5), (A5′), the sequence of tests $\{\phi_n(\Delta_n)\}$ defined by (3.2) and (3.3) is AUMP for testing $H_1\colon \theta \in \overline{\omega}$ against $A\colon \theta \in \omega$ within the class of all sequences of tests $\{\lambda_n\}$ for which

$$\limsup\,[\sup\,(\mathscr{E}_\theta \lambda_n;\ \theta \in \overline{\omega})] \leqslant \alpha. \tag{6.1}$$

Proof In the first place, we show that $\{\phi_n(\Delta_n)\}$ is in the class under consideration; that is, we show that

$$\limsup\,\{\sup[\mathscr{E}_\theta \phi_n(\Delta_n);\ \theta \in \overline{\omega}]\} \leqslant \alpha. \tag{6.2}$$

The proof is by contradiction. Suppose that (6.2) is not true. Then there exists a subsequence $\{m\} \subseteq \{n\}$ and $\{\theta_m\}$ with $\theta_m \in \overline{\omega}$ for all m such that
$$\mathscr{E}_{\theta_m}\phi_m(\Delta_m) \to \delta > \alpha. \tag{6.3}$$
Look at the sequence $\{m^{\frac{1}{2}}(\theta_m - \theta_0)\}$ and suppose first that it is unbounded. Then there exists $\{r\} \subseteq \{m\}$ such that $r^{\frac{1}{2}}(\theta_r - \theta_0) \to -\infty$. At this point we recall that $\mathscr{L}(\Delta_n|P_{\theta_0}) \to N(0, \Gamma)$ and therefore by Lemma 2.1 (ii), $c_n \to \xi_\alpha$. This was also seen in the proof of Theorem 3.1 in a slightly different context. Then by (A5)′ it follows that

$$\mathscr{E}_{\theta_r}\phi_r(\Delta_r) = P_{\theta_r}(\Delta_r > c_r) + \gamma_r P_{\theta_r}(\Delta_r = c_r) \leqslant P_{\theta_r}(\Delta_r \geqslant c_r) \to 0$$

and this contradicts (6.3).

Next suppose $\{m^{\frac{1}{2}}(\theta_m - \theta_0)\} = \{h_m\}$ remains bounded. Then there exists $\{s\} \subseteq \{m\}$ such that $h_s \to h \leqslant 0$. Thus we have $\theta_s = \theta_0 + h_s s^{-\frac{1}{2}}$ with $h_s \to h$. At this point we recall that
$$\mathscr{L}(\Delta_s|P_{\theta_s}) \Rightarrow N(\Gamma h, \Gamma),$$
so that

$$\mathscr{E}_{\theta_s}\phi_s(\Delta_s) \to \int_{(x > \xi_\alpha)} \mathrm{d}N(\Gamma h, \Gamma) \leqslant \alpha \quad \text{since } \Gamma h \leqslant 0.$$

Thus we have obtained a contradiction to (6.3) again. It follows that (6.2) is valid.

We now show the AUMP property of $\{\phi_n(\Delta_n)\}$. This is also done by contradiction. To this end, suppose that for a sequence of tests $\{\omega_n\}$ satisfying (6.1), we have

$$\limsup \{\sup [\mathscr{E}_\theta \omega_n - \mathscr{E}_\theta \phi_n(\Delta_n); \theta \in \omega]\} = 7\delta > 0. \quad (6.4)$$

Then there exists $\{m\} \subseteq \{n\}$ and $\{\theta_m\}$ with $\theta_m \in \omega$ for all m such that

$$\mathscr{E}_{\theta_m} \omega_m - \mathscr{E}_{\theta_m} \phi_m(\Delta_m) \to 7\delta. \quad (6.5)$$

For the case that $\{m^{\frac{1}{2}}(\theta_m - \theta_0)\}$ is unbounded, one has by (A5) that $\mathscr{E}_{\theta_r} \phi_r(\Delta_r) \to 1$ for a subsequence $\{\theta_r\} \subseteq \{\theta_m\}$ and this shows that (6.5) and hence (6.4) cannot occur. It remains then for us to arrive at a contradiction to (6.5) for the case that

$$\{m^{\frac{1}{2}}(\theta_m - \theta_0)\} = \{h_m\}$$

stays bounded. In this case there exists $\{s\} \subseteq \{m\}$ such that $h_s \to h$, so that $\theta_s = \theta_0 + h_s s^{-\frac{1}{2}}$ with $h_s \to h$. Introducing

$$\bar{\omega}_n(\Delta_n) = \mathscr{E}_{\theta_0}(\omega_n | \Delta_n),$$

we obtain as in the proof of Theorem 5.1 that

$$\mathscr{E}_{\theta_s} \bar{\omega}_s(\Delta_s^*) - \mathscr{E}_{\theta_s} \phi_s(\Delta_s^*) > 4\delta \quad \text{for all } s \geqslant n_1^*, \quad \text{say.} \quad (6.6)$$

Again, as in the proof of Theorem 5.1, one has

$$\int_{(x > \xi_\alpha)} \mathrm{d}N(\Gamma h, \Gamma) - \mathscr{E}_{\theta_s} \phi_s(\Delta_s^*) < \delta \quad \text{for all } s \geqslant n_2^*, \quad \text{say,} \quad (6.7)$$

and, by passing to a subsequence of $\{s\}$ if necessary,

$$\mathscr{E}_{\theta_0} \bar{\omega}_s(\Delta_{s1}^*) = \alpha_s \to \alpha. \quad (6.8)$$

In the family of probability measures R_{s, h_s} defined by (5.34), consider now the problem of testing $H_{1, h}: h_s \leqslant 0$ against $A_h: h_s > 0$ at level α_s for each s. Then the test defined by (6.9) and (6.10) below is a UMP test. (See, e.g. Lehmann [3], p. 70.) Namely,

$$\bar{\phi}_s(\Delta_s^*) = \begin{cases} 1 & \text{if } \Delta_s^* > \bar{c}_s \\ \bar{\gamma}_s & \text{if } \Delta_s^* = \bar{c}_s \\ 0 & \text{otherwise,} \end{cases} \quad (6.9)$$

where the constants $\bar{c}_s$ and $\bar{\gamma}_s$ are determined by the requirement

$$\mathscr{E}[\bar{\phi}_s(\Delta_s^*) | R_{s, 0}] = \mathscr{E}_{\theta_0} \bar{\phi}_s(\Delta_s^*) = \alpha_s \quad \text{for all } s. \quad (6.10)$$

Because of (6.8) one also sees as in the proof of Theorem 5.1 that

$$\int_{(x>\xi_{\alpha_s})} dN(\Gamma h, \Gamma) - \int_{(x>\xi_\alpha)} dN(\Gamma h, \Gamma) < \delta \quad \text{for all } s \geq n_3^*, \quad \text{say,}$$

(6.11)

and

$$\mathscr{E}_{\theta_s} \overline{\phi}_s(\Delta_s^*) - \int_{(x>\xi_{\alpha_s})} dN(\Gamma h, \Gamma) < \delta \quad \text{for all } s \geq n_4^*, \quad \text{say.} \quad (6.12)$$

From (6.7), (6.11) and (6.12), it follows that

$$\mathscr{E}_{\theta_s} \overline{\phi}_s(\Delta_s^*) - \mathscr{E}_{\theta_s} \phi_s(\Delta_s^*) < 3\delta \quad \text{for all } s \geq n_5^* = \max(n_2, n_3, n_4).$$

(6.13)

By the fact that $\overline{\phi}_s(\Delta_s^*)$ is UMP at level α_s and $\mathscr{E}_{\theta_0} \overline{\omega}_s(\Delta_s^*) = \alpha_s$, one has

$$\mathscr{E}[\overline{\omega}_s(\Delta_s^*) | R_{s, h_s}] - \mathscr{E}[\overline{\phi}_s(\Delta_s^*) | R_{s, h_s}] \leq 0. \quad (6.14)$$

From now on, one simply repeats the arguments employed after relationship (5.40) in order to conclude that

$$\mathscr{E}_{\theta_s} \overline{\omega}_s(\Delta_s^*) - \mathscr{E}_{\theta_s} \phi_s(\Delta_s^*) < 4\delta \quad \text{for all } s \geq n_6^*, \quad \text{say.}$$

However, this contradicts (6.6) and completes the proof of the theorem. ∎

To Theorem 6.1, there is the following corollary.

COROLLARY 6.1 Under assumptions (A1–4), the sequence of tests $\{\phi_n(\Delta_n)\}$ defined by (3.2) and (3.3) is asymptotically locally most powerful for testing $H_1: \theta \in \overline{\omega}$ against $A: \theta \in \omega$ within the class of all sequences of tests $\{\lambda_n\}$ satisfying (6.1) when the parameter values are restricted to those θs in ω for which $\{n^{\frac{1}{2}}(\theta - \theta_0)\}$ remains bounded from above. That is, within the class of sequences of tests just described, $\{\phi_n(\Delta_n)\}$ satisfies (3.1) when the parameter values are restricted to those θs in ω for which $\{n^{\frac{1}{2}}(\theta - \theta_0)\}$ remains bounded from above.

For the symmetric case, we have the following results which are established in a way entirely analogous to the one employed for proving Theorem 6.1. Namely,

THEOREM 6.2 Under assumptions (A1–5), (A5′), the sequence of tests $\{\phi_n'(\Delta_n)\}$ defined by (3.20) and (3.21) is AUMP for testing $H_1': \theta \in \overline{\omega}' = \{\theta \in \Theta; \theta \geq \theta_0\} = \omega \cup \{\theta_0\}$ against $A': \theta \in \omega'$ within the class of all sequences of tests $\{\lambda_n\}$ such that

$$\limsup [\sup(\mathscr{E}_\theta \lambda_n; \theta \in \overline{\omega}')] \leq \alpha. \quad (6.15)$$

COROLLARY 6.2 Under assumptions (A 1–4), the sequence of tests $\{\phi_n'(\Delta_n)\}$ defined by (3.20) and (3.21) is asymptotically locally most powerful for testing $H_1': \theta \in \bar{\omega}'$ against $A': \theta \in \omega'$ within the class of all sequences of tests $\{\lambda_n\}$ satisfying (6.15) when the parameter values are restricted to those θs in ω' for which $\{n^{\frac{1}{2}}(\theta - \theta_0)\}$ remains bounded from below. That is, within the class of sequences of tests just described, $\{\phi_n'(\Delta_n)\}$ satisfies (3.1) when the parameter values are restricted to those θs in ω' for which $\{n^{\frac{1}{2}}(\theta - \theta)\}$ remains bounded from below.

REMARK 6.1 For the explicit form of the tests in the case of the concrete examples that we have been dealing with, the reader is referred to Section 4.

Exercises

1. Verify condition (A 5)' in connection with Example 1.3.
2. Verify the equivalence asserted in relation (4.4).
3. In connection with Example 4.4, show that an MLE of $p(\theta) = \exp(-\theta), \theta > 0$, is a root of the equation

$$p^3(\theta) + a_2 p^2(\theta) + a_1 p(\theta) + a_0 = 0,$$

where

$$a_0 = -\frac{1}{n} \sum_{j=1}^{n} X_{j-1} X_j, \quad a_1 = \frac{1}{n} \sum_{j=1}^{n} (X_{j-1}^2 + X_j^2) - 1 \quad \text{and} \quad a_2 = a_0.$$

5. *Some statistical applications: asymptotic efficiency of estimates*

Summary

In this chapter, we apply some of the basic results obtained in Chapter 2 to the problem of the asymptotic efficiency of a sequence of estimates of the unknown parameter $\theta \in \Theta$. The asymptotic efficiency adopted in this chapter is the one proposed by Wolfowitz [1] (see Definition 1.1). The main results obtained herein are in the nature of establishing a certain upper bound for the limiting probability of concentration of various estimates under consideration. All these results (Theorems 4.1, 5.1 and 5.2), except for one (Theorem 4.2), are established for the case that Θ is an open subset of R. In Sections 6 and 7, we consider the asymptotic efficiency from the classical point of view and show that standard results for both the one-dimensional and multi-dimensional case (see Theorems 6.1 and 7.1) are obtained either as special cases of, or are closely related to, the main results mentioned above.

1 W-efficiency – preliminaries

The classical approach of proving asymptotic efficiency (a.eff.) of estimates has been geared towards showing that an MLE is a.eff. More specifically, under suitable regularity conditions, an MLE, properly normalized, is asymptotically normal with mean zero and variance the inverse of Fisher's information number. One then considers the class of all estimates which, properly normalized, are asymptotically normal with mean zero, and calls the one with the smallest variance (of the limiting normal distribution) an a.eff. estimate, if such an estimate exists. Again, under appropriate conditions, an MLE is shown to be

[128]

a.eff. except possibly on a subset of the parameter space whose Lebesgue measure is zero (see, e.g. LeCam [1], [3] and Bahadur [1]). However, an example has been constructed by Hodges (see LeCam [1]), where on this exceptional set the variance of the limiting normal distribution of an estimate is actually smaller than the corresponding variance of an MLE. Such estimates are known as *superefficient* estimates.

Several legitimate objections have been raised against the classical approach just outlined (see, e.g. Wolfowitz [1]). The existence of superefficient estimates is disturbing enough; but what is entirely unnecessary is to confine ourselves only to those estimates which, properly normalized, are asymptotically normal. Of course, this is necessary if a.eff. is to be judged on the basis of the variance. However, there is no reason that this should be the criterion of a.eff.

Arguing along these lines, Wolfowitz [1] was led to introducing an alternative more general and quite reasonable criterion of a.eff. For the formulation of this criterion, some additional notation is required.

As usual, $X_0, X_1, ..., X_n$ are $n + 1$ observations from the Markov process under consideration and Θ is an open subset of R. Let $\mathscr{C}^*$ be a class (to be specified later on) of sequences of estimates (of θ) $\{T_n\} = \{T_n(X_0, X_1, ..., X_n)\}$, and let

$$\{V_n\} = \{V_n(X_0, X_1, ..., X_n)\}$$

be a sequence of estimates not necessarily belonging in $\mathscr{C}^*$. Also for each n and θ, consider two classes of intervals in R (to be specified later on) of the following form.

$$I_n(\theta) = \{\text{certain intervals } i_n(\theta; t_1, t_2); t_1, t_2 > 0\}; \theta \in \Theta. \quad (1.1)$$

$$J_n(\theta; T) = \{\text{certain intervals } j_n(\theta; t_1, t_2, T); t_1, t_2 > 0\}; \theta \in \Theta. \quad (1.2)$$

The presence of T in the definition of $J_n(\theta; T)$ indicates that the j-intervals also depend on a quantity associated with the sequence $\{T_n\}$.

DEFINITION 1.1 Let $\{T_n\}$, $\{V_n\}$ and $\mathscr{C}^*$ be as above and for each n and each θ, let $I_n(\theta)$ and $J_n(\theta; T)$ be defined by (1.1) and (1.2), respectively. Suppose that $\lim P_\theta[V_n \in i_n(\theta; t_1, t_2)]$ exists for all $t_1, t_2 > 0$ and every $\theta \in \Theta$. Then the sequence of estimates $\{V_n\}$ is said to be *asymptotically efficient in Wolfowitz sense* (*W-efficient*, for short) with respect to the class $\mathscr{C}^*$ if the inequality

$$\limsup P_\theta[T_n \in j_n(\theta; t_1, t_2, T)] \leqslant \lim P_\theta[V_n \in i_n(\theta; t_1, t_2)] \quad (1.3)$$

holds for all $\{T_n\}$ in $\mathscr{C}^*$, every $t_1, t_2 > 0$ and all (or almost all with respect to Lebesgue measure l in R, a.a. $[l]$) θ in Θ.

The optimality of $\{V_n\}$ consists in that for every $t_1, t_2 > 0$ and all (a.e. $[l]$) $\theta \in \Theta$, asymptotically, V_n is concentrated in $i_n(\theta; t_1, t_2)$ at least as much as any T_n, such that $\{T_n\} \in \mathscr{C}^*$, is concentrated in $j_n(\theta; t_1, t_2, T)$.

Actually, Wolfowitz' programme was to show W-efficiency of the MLE in the i.i.d. case. More specifically, what he does is this. He takes Θ to be a finite interval with or without its endpoints and defines the class $\mathscr{C}^*$, to be denoted here by $\mathscr{C}_W$, as follows

$$\mathscr{C}_W = \{\{T_n\}; P_\theta[(n^{\frac{1}{2}}(T_n - \theta) \leqslant x] \to F_T(x; \theta),$$
$$\text{a d.f., } uniformly \text{ in } R \times \Theta\}.$$

Of course, the presence of T in the limiting d.f. simply indicates that this limiting d.f. depends on $\{T_n\}$. For each $\theta \in \Theta$, let $l_T(\theta)$, $u_T(\theta)$ be the 'smallest' and 'largest' median of $F_T(\cdot; \theta)$ and define $I_n(\theta), J_n(\theta; T)$ as follows:

$$I_n(\theta) = \{i_n(\theta; t_1, t_2); i_n(\theta; t_1, t_2)$$
$$= [\theta - t_1 n^{-\frac{1}{2}}, \theta + t_2 n^{-\frac{1}{2}}], t_1, t_2 > 0\},$$
$$J_n(\theta; T) = \{j_n(\theta; t_1, t_2, T); j_n(\theta; t_1, t_2, T)$$
$$= [\theta - t_1 n^{-\frac{1}{2}} + w_T(\theta) n^{-\frac{1}{2}}, \theta + t_2 n^{-\frac{1}{2}} + W_T(\theta) n^{-\frac{1}{2}}], t_1, t_2 > 0\},$$

where w_T and W_T are certain specified functions defined on Θ and such that $l_T(\theta) \leqslant w_T(\theta) \leqslant W_T(\theta) \leqslant u_T(\theta)$, $\theta \in \Theta$.

With the above notation and under suitable regularity conditions, Wolfowitz shows that (1.3) holds true (with $\limsup$ replaced by $\lim$) for all but countably many θs when V_n is equal to the MLE.

Prominent among his assumptions is that

$$P_\theta[n^{\frac{1}{2}}(V_n - \theta) \leqslant x] \to \Phi(x; \theta) \quad uniformly \text{ in } R \times \Theta,$$

where $\Phi(\cdot; \theta)$ is the d.f. of $N(0, 1/I(\theta))$, $I(\theta)$ being Fisher's information number.

Kaufman [1] has generalized Wolfowitz' result for the case that Θ is k-dimensional.

Now the problem of establishing W-efficiency can be split into two parts. First one concentrates on determining an upper bound, $B(\theta; t_1, t_2)$, say, such that

$$
\begin{aligned}
\limsup P_\theta[T_n \in j_n(\theta; t_1, t_2, T)] \\
\leqslant B(\theta; t_1, t_2) \quad \text{for all } \{T_n\} \in \mathscr{C}^*, \\
\text{every } t_1, t_2 > 0 \quad \text{and all (a.a. } [l]),\ \theta \in \Theta,
\end{aligned} \right\} \quad (1.4)
$$

and then one attempts to find a sequence of estimates $\{V_n\}$ for which the upper bound is attained, in the sense that

$$
\lim P_\theta[V_n \in i_n(\theta; t_1, t_2)]
$$

$$
\to B(\theta; t_1, t_2) \quad \text{for all } t_1, t_2 > 0 \text{ and every } \theta \in \Theta, \quad (1.5)
$$

where $i_n(\theta; t_1, t_2) \in I_n(\theta)$ for a certain specification of $I_n(\theta)$. In other words, one would use inequality (1.4) in much the same way that the Cramér–Rao inequality is employed (only in reversed order). However, the optimality of $\{V_n\}$ is not defined by (1.5) but rather by (1.3), the reason being that (1.3) may be fulfilled for some $\{V_n\}$ while (1.5) may fail to occur.

Schmetterer [1], Roussas [5] and Pfanzagl [1] have modified Wolfowitz' assumptions and have established inequality (1.3) for several specifications of the class $\mathscr{C}^*$. The first two of these authors also established their results in a Markov process framework.

Our objective here is to show that assumptions (A 1–4) are sufficient for establishing inequality (1.4) for certain specifications of the classes $\mathscr{C}^*$ and $J_n(\theta; T)$. Accordingly, we shall not concern ourselves, with the problem of proving (1.5) for some specific $\{V_n\}$.

2 Some lemmas

The lemmas below are needed in the proof of the results in the remainder of this chapter and occasionally elsewhere, too.

DEFINITION 2.1 For each n, let f_n and f be functions defined on $E \subseteq R^k$ into R. We say that $\{f_n\}$ converges *con-*

tinuously to f in E if for each $x \in E$, $f_n(x_n) \to f(x)$ whenever $x_n \to x$.

The following result then holds true.

LEMMA 2.1 If $\{f_n\}$ converges continuously to f in E, then f is continuous on E.

Proof In order to establish continuity at an arbitrary $x \in E$, it suffices to show that $f(x_n) \to f(x)$ whenever $x_n \to x$. For contradiction, suppose that f is not continuous at x. Then there exists a sequence $\{x_n\}$ such that

$$|f(x_n) - f(x)| > \epsilon \quad \text{for some } \epsilon > 0. \qquad (2.1)$$

Now from the assumption of continuous convergence it follows that $f_n(x) \to f(x)$. (Just take $\{x_n\} = \{x\}$.) Then for each fixed n, there exists an integer m_n such that

$$|f_{m_n}(x_n) - f(x_n)| < \tfrac{1}{2}\epsilon. \qquad (2.2)$$

Clearly, the m_ns can be chosen so that $m_n < m_{n'}$ if $n < n'$. Thus $\{m_n\} \subseteq \{n\}$. Setting $y_{m_n} = x_n$, we have that $y_{m_n} \to x$ and relations (2.1) and (2.2) imply

$$|f(y_{m_n}) - f(x)| > \epsilon, \quad |f_{m_n}(y_{m_n}) - f(y_{m_n})| < \tfrac{1}{2}\epsilon. \qquad (2.3)$$

From (2.3), it follows that $|f_{m_n}(y_{m_n}) - f(x)| \geqslant \tfrac{1}{2}\epsilon$, which contradicts the assumption of the lemma. The proof is completed. ∎

The concepts of continuous convergence and uniform convergence are related as in the following lemma.

LEMMA 2.2 (i) If $\{f_n\}$ converges to f uniformly on E, then the convergence is continuous, provided f is continuous on E.

(ii) If E is compact, then continuous convergence of $\{f_n\}$ to f implies uniform convergence.

Proof (i) Let $x, x_n \in E$ for all n be such that $x_n \to x$. Then we have

$$|f_n(x_n) - f(x)| \leqslant |f_n(x_n) - f(x_n)| + |f(x_n) - f(x)| \to 0$$

by continuity of f and uniform convergence of $\{f_n\}$ to f.

(ii) In the first place, f is continuous on E by Lemma 2.1. For contradiction, assume that $\{f_n\}$ does not converge uniformly to f.

Then there exists a subsequence $\{m\} \subseteq \{n\}$ and a sequence $\{x_m\}$ with $x_m \in E$ for all m, such that

$$|f_m(x_m) - f(x_m)| \geqslant \epsilon \quad \text{for some } \epsilon > 0. \qquad (2.4)$$

By the compactness of E, there exists a subsequence $\{x_r\} \subseteq \{x_m\}$ such that $x_r \to x \in E$. Replacing m by r in (2.4), we have

$$\epsilon \leqslant |f_r(x_r) - f(x_r)| \leqslant |f_r(x_r) - f(x)| + |f(x) - f(x_r)|. \qquad (2.5)$$

But $|f(x) - f(x_r)| < \tfrac{1}{2}\epsilon$ for all sufficiently large r, by the continuity of f. Hence from (2.5), one has $|f_r(x_r) - f(x)| \geqslant \tfrac{1}{2}\epsilon$ for all sufficiently large r. However, this contradicts the assumption of continuous convergence. The proof is completed.]

The lemma stated below generalizes the dominated convergence theorem and its proof can be found in Loève [1], pp. 162–3 and also Pratt [1].

LEMMA 2.3 For each n, let h_n, g_n, G_n and h, g, G be $\mathscr{B}^k$-measurable, real-valued functions defined on R^k such that

(i) $h_n(x) \to h(x)$, $g_n(x) \to g(x)$, $G_n(x) \to G(x)$, $x \in R^k$,

(ii) $g_n(x) \leqslant h_n(x) \leqslant G_n(x)$ for all n and $x \in R^k$, and

(iii) $\int g_n \, dl^k \to \int g \, dl^k$, $\int G_n \, dl^k \to \int G \, dl^k$ with $\int g \, dl^k$ and $\int G \, dl^k$ finite. Then $\int h_n \, dl^k \to \int h \, dl^k$ and $\int h \, dl^k$ is finite, where l^k is the Lebesgue measure on $\mathscr{B}^k$.

A slight variation of the following result is proved in Bahadur [1].

LEMMA 2.4 Let $\{f_n\}$ be a sequence of measurable functions defined on an open subset E of R^k into R and let $f_n(x) \to 0$, $x \in E$ and $|f_n(x)| \leqslant M(< \infty)$ for all n and $x \in E$. Then for any sequence $\{x_n\}$ with $x_n \to 0$ there exists a subsequence $\{m\} \subseteq \{n\}$ and an l^k-null subset of E (both of which may depend on $\{x_n\}$) such that for all $x \in E$ outside this null set, one has

$$f_m(x + x_m) \to 0 \quad \text{pointwise.}$$

Proof For any $x_n \to 0$, consider the integral

$$\int |f_n(x + x_n)| \, d\Phi^{(k)}(x),$$

where $\Phi^{(k)}$ is the d.f. of $N(0, I)$ (I is the $k \times k$ unit matrix), and perform the transformation $x + x_n = y$. We have then

$$\int |f_n(x + x_n)| \, \mathrm{d}\Phi^{(k)}(x)$$

$$= \int |f_n(x + x_n)| \, (2\pi)^{-k/2} \exp\left(-\tfrac{1}{2} x' x\right) \mathrm{d}l^k(x)$$

$$= \int |f_n(y)| \, (2\pi)^{-k/2} \exp\left[-\tfrac{1}{2}(y - x_n)' (y - x_n)\right] \mathrm{d}l^k(y).$$

On the basis of our assumptions, the conditions of Lemma 2.3 are fulfilled with

$$h_n(y) = |f_n(y)| \, (2\pi)^{-k/2} \exp\left[-\tfrac{1}{2}(y - x_n)' (y - x_n)\right], \quad g_n(y) = 0,$$

$$G_n(y) = (2\pi)^{-k/2} M \exp\left[-\tfrac{1}{2}(y - x_n)' (y - x_n)\right]$$

and $\quad h(y) = g(y) = 0, \quad G(y) = (2\pi)^{-k/2} M \exp\left(-\tfrac{1}{2} y' y\right).$

Therefore

$$\int h_n(y) \, \mathrm{d}l^k$$

$$= \int |f_n(y)| \, (2\pi)^{-k/2} \exp\left[-\tfrac{1}{2}(y - x_n)' (y - x_n)\right] \mathrm{d}l^k(y) \to 0,$$

so that $\qquad \int |f_n(x + x_n)| \, \mathrm{d}\Phi^{(k)}(x) \to 0.$

Therefore the Markov inequality implies that

$$f_n(x + x_n) \to 0 \quad \text{in } \Phi^{(k)}\text{-measure}.$$

Hence there exists a subsequence $\{m\} \subseteq \{n\}$ and a $\Phi^{(k)}$-null set (both of which may depend on $\{x_n\}$) such that for all $x \in E$ outside this null set, one has

$$f_m(x + x_m) \to 0 \quad \text{pointwise}.$$

Since $\Phi^{(k)} \approx l^k$, one has finally that for all $x \in E$ outside the before-mentioned l^k-null set,

$$f_m(x + x_m) \to 0 \quad \text{pointwise},$$

as was to be seen. ∎

The lemma to be presented below follows from Lemma 2.4 above (for $k = 1$ and $E = R$) and is discussed in Pfanzagl [1].

LEMMA 2.5 Let $\{f_n\}$ be a sequence of $\mathscr{B}$-measurable, real-valued functions defined on R and being such that

$$\liminf f_n(x) \geqslant 0,\ x \in R, \quad \text{and} \quad |f_n(x)| \leqslant M(< \infty)$$

for all n and $x \in R$. Then for any sequence $\{x_n\}$ with $x_n \to 0$ there exists an l-null set (which may depend on $\{x_n\}$) outside of which one has

$$\limsup f_n(x + x_n) \geqslant 0.$$

Proof Let $f_n^* = \inf(f_n, 0)$. Then the sequence $\{f_n^*\}$ satisfies the requirement of Lemma 2.4. Therefore for $x_n \to 0$, there exists a subsequence $\{m\} \subseteq \{n\}$ and an l^k-null set (both of which may depend on $\{x_n\}$) outside of which one has

$$f_m^*(x + x_m) \to 0 \quad \text{pointwise.}$$

From the definition of f_n^*, one has

$$f_n(x + x_n) \geqslant f_n^*(x + x_n) \quad \text{for all } n \text{ and } x \in R.$$

Therefore

$$\limsup f_n(x + x_n) \geqslant \limsup f_n^*(x + x_n) \geqslant \lim f_m^*(x + x_m) = 0,$$

as was to be shown.]

In closing this section, we would like to point out that Lemmas 2.1, 2.2 and 2.4 are valid in a more general set-up than the one considered here (see Schmetterer [1], Section 1), but their present formulation is sufficient for our purposes.

3 A representation theorem

In the present section, we consider a class of sequences of estimates (of θ) defined by (3.1) below such that when each estimate of a given sequence in the class under consideration is properly normalized, the resulting sequence of their distributions converges (weakly) to a probability measure $\mathscr{L}(\theta)$, say. Then it is shown (see Theorem 3.1) that $\mathscr{L}(\theta)$ assumes a certain representation as the convolution of two specified probability measures. This result is employed in the next section in connection with the asymptotic efficiency of estimates and, of course, is of considerable interest in its own right.

For an arbitrary $\theta \in \Theta$ and any $h \in R^k$, set $\theta_n = \theta + hn^{-\frac{1}{2}}$, so that $\theta_n \in \Theta$ for sufficiently large n. Define the class $\mathscr{C}_H$ of sequences of estimates of θ, $\{T_n\}$, as follows

$$\mathscr{C}_H = \{\{T_n\}; \mathscr{L}[n^{\frac{1}{2}}(T_n - \theta_n)|P_{\theta_n}]$$
$$\Rightarrow \mathscr{L}(\theta), \text{ a probability measure}\}, \qquad (3.1)$$

where $\mathscr{L}(\theta)$, in general, depends on $\{T_n\}$. The probability measure $\mathscr{L}(\theta)$ can be represented as the convolution of two probability measures. This result is due to Hájek [2]. However, the proof to be presented below is based on an idea of Bickel [1]; see also Roussas and Soms [9]. Finally, an alternative proof can be obtained from some general results of LeCam [7]. The precise formulation of the theorem is as follows.

THEOREM 3.1 Consider the class $\mathscr{C}_H$ defined by (3.1). Then one has that

$$\mathscr{L}(\theta) = \mathscr{L}_1(\theta) * \mathscr{L}_2(\theta),$$

where $\mathscr{L}_1(\theta) = N(0, \Gamma^{-1}(\theta))$ and $\mathscr{L}_2(\theta)$ is defined by (3.15) below.

Proof Since θ is kept fixed throughout the proof, we shall omit it from our notation for simplicity. Thus we will write $\mathscr{L}, \mathscr{L}_1, \mathscr{L}_2, \Gamma$, etc. instead of $\mathscr{L}(\theta), \mathscr{L}_1(\theta), \mathscr{L}_2(\theta), \Gamma(\theta)$, etc. Now let $\Delta_n^* = \Delta_n^*(\theta)$ be the truncated version of Δ_n defined by (1.5) in Chapter 3 and consider the sequence

$$\mathscr{L}_n^* = \mathscr{L}_n^*(\theta) = \mathscr{L}[(n^{\frac{1}{2}}(T_n - \theta), \Delta_n^*)|P_\theta].$$

Then by the weak compactness theorem there exists $\{m\} \subseteq \{n\}$ such that $\mathscr{L}_m^* \Rightarrow \mathscr{L}^*$, a measure. Then the marginal measures $[\mathscr{L}m^{\frac{1}{2}}(T_m - \theta)|P_\theta]$ and $\mathscr{L}(\Delta_m^*|P_\theta)$ of $\mathscr{L}[(m^{\frac{1}{2}}(T_m - \theta), \Delta_m^*)|P_\theta]$ converge (weakly) to the corresponding marginal measures of $\mathscr{L}^*$. These latter marginal measures are then probability measures since both $\mathscr{L}[m^{\frac{1}{2}}(T_m - \theta)|P_\theta]$ and $\mathscr{L}(\Delta_m^*|P_\theta)$ converge to probability measures. It follows that $\mathscr{L}^*$ itself is a probability measure. Setting (T, Δ) for the identity mapping on $R^k \times R^k$, we have then that $\mathscr{L}[(T, \Delta)|\mathscr{L}^*] = \mathscr{L}^*$, so that

$$\mathscr{L}[(m^{\frac{1}{2}}(T_m - \theta), \Delta_m^*)|P_\theta] \Rightarrow \mathscr{L}[(T, \Delta)|\mathscr{L}^*]. \qquad (3.2)$$

It is shown in Lemma 3.1 below that

$$\mathscr{E}_\theta \exp\left[iu'm^{\frac{1}{2}}(T_m - \theta) + \Lambda_m\right]$$
$$- \mathscr{E}_\theta \exp\left[iu'm^{\frac{1}{2}}(T_m - \theta) + h'\Delta_m^* - \tfrac{1}{2}h'\Gamma h\right] \to 0 \quad (3.3)$$

and

$$\mathscr{E}_\theta \exp\left[iu'm^{\frac{1}{2}}(T_m - \theta) + h'\Delta_m^* - \tfrac{1}{2}h'\Gamma h\right]$$
$$\to \mathscr{E}_{\mathscr{L}*} \exp\left(iu'T + h'\Delta - \tfrac{1}{2}h'\Gamma h\right). \quad (3.4)$$

Therefore by setting

$$\phi(u, h) = \mathscr{E}_{\mathscr{L}*} \exp\left(iu'T + ih'\Delta\right) \quad (3.5)$$

and replacing h by zero, we obtain by means of (3.4) that

$$\mathscr{E}_\theta \exp\left[iu'm^{\frac{1}{2}}(T_m - \theta)\right] \to \mathscr{E}_{\mathscr{L}*} \exp iu'T = \mathscr{E}_{\mathscr{L}} \exp iu'T = \phi(u, 0).$$
$$(3.6)$$

Also from (3.1) and (3.5), it follows that

$$\mathscr{E}_{\theta_m} \exp\left[iu'm^{\frac{1}{2}}(T_m - \theta_m)\right] \to \phi(u, 0). \quad (3.7)$$

Next
$$\mathscr{E}_{\theta_m} \exp\left[iu'm^{\frac{1}{2}}(T_m - \theta_m)\right]$$
$$= \mathscr{E}_\theta \exp\left[iu'm^{\frac{1}{2}}(T_m - \theta_m) + \Lambda_m\right]$$
$$= \mathscr{E}_\theta \exp\left[iu'm^{\frac{1}{2}}(T_m - \theta) - iu'h + \Lambda_m\right]$$
$$= \exp\left(-iu'h\right) \mathscr{E}_\theta \exp\left[iu'm^{\frac{1}{2}}(T_m - \theta) + \Lambda_m\right]$$

and this last expression converges to

$$\exp\left(-iu'h\right) \mathscr{E}_{\mathscr{L}*} \exp\left(iu'T + h'\Delta - \tfrac{1}{2}h'\Gamma h\right)$$

by means of (3.3) and (3.4). That is

$$\mathscr{E}_{\theta_m} \exp\left[iu'm^{\frac{1}{2}}(T_m - \theta_m)\right]$$
$$\to \exp\left(-iu'h\right) \mathscr{E}_{\mathscr{L}*} \exp\left(iu'T + h'\Delta - \tfrac{1}{2}h'\Gamma h\right). \quad (3.8)$$

From (3.7) and (3.8), we obtain

$$\phi(u, 0) = \exp\left(-iu'h\right) \mathscr{E}_{\mathscr{L}*} \exp\left(iu'T + h'\Delta - \tfrac{1}{2}h'\Gamma h\right). \quad (3.9)$$

It is shown in Lemma 3.2 below that in (3.9) we may replace h by ih. By doing so, we obtain

$$\phi(u, 0) = \exp u'h \, \mathscr{E}_{\mathscr{L}*} \exp\left(iu'T + ih'\Delta + \tfrac{1}{2}h'\Gamma h\right). \quad (3.10)$$

By means of (3.5), relation (3.10) may also be written as follows

$$\phi(u, 0) = (\exp u'h)\, \phi(u, h) \exp\left(\tfrac{1}{2}h'\Gamma h\right),$$

so that $\quad \phi(u, h) = \phi(u, 0) \exp(-u'h) \exp(-\tfrac{1}{2}h'\Gamma h).$ (3.11)

Next, it is easily seen that

$$-u'h - \tfrac{1}{2}h'\Gamma h = -\tfrac{1}{2}(h' + u'\Gamma^{-1})\,\Gamma(h + \Gamma^{-1}u) + \tfrac{1}{2}u'\Gamma^{-1}u,$$

so that (3.11) becomes

$$\phi(u, h) = \exp[-\tfrac{1}{2}(h' + u'\Gamma^{-1})\,\Gamma(h + \Gamma^{-1}u)]\,\phi(u, 0)\exp(\tfrac{1}{2}u'\Gamma^{-1}u).$$
   (3.12)

Setting successively $h = 0$ and $h = -\Gamma^{-1}u$ in (3.12), we obtain

$$\phi(u, 0) = \exp(-\tfrac{1}{2}u'\Gamma^{-1}u)\,\phi(u, 0)\exp(\tfrac{1}{2}u'\Gamma^{-1}u)$$ (3.13)

and $\qquad \phi(u, -\Gamma^{-1}u) = \phi(u, 0)\exp(\tfrac{1}{2}u'\Gamma^{-1}u),$ (3.14)

respectively.

From (3.5) and (3.14), it follows that $\phi(u, 0)\exp(\tfrac{1}{2}u'\Gamma^{-1}u)$ is a characteristic function (under $\mathscr{L}^*$), namely, the characteristic function of the random vector $T - \Gamma^{-1}\Delta$. Define $\mathscr{L}_2$ by

$$\mathscr{L}_2 = \mathscr{L}(T - \Gamma^{-1}\Delta \,|\, \mathscr{L}^*).$$ (3.15)

Also $\exp(-\tfrac{1}{2}u'\Gamma^{-1}u)$ is a characteristic function (under $\mathscr{L}^*$), namely, the characteristic function of the random vector $\Gamma^{-1}\Delta$ which is distributed as $N(0, \Gamma^{-1})$. Set $\mathscr{L}_1 = N(0, \Gamma^{-1})$. Then from relation (3.13) and the composition theorem (see, e.g. Loève [1], p. 193), it follows that $\mathscr{L} = \mathscr{L}_1 * \mathscr{L}_2$, as was to be seen. ▮

We now proceed to establish the lemmas used in the proof of the theorem.

LEMMA 3.1   With the notation employed in the theorem, one has that

(i) $\mathscr{E}_\theta \exp[iu'm^{\frac{1}{2}}(T_m - \theta) + \Lambda_m] - \mathscr{E}_\theta \exp[iu'm^{\frac{1}{2}}(T_m - \theta) + h'\Delta_m^*$
$\quad - \tfrac{1}{2}h'\Gamma h] \to 0,$

and

(ii) $\mathscr{E}_\theta \exp[iu'm^{\frac{1}{2}}(T_m - \theta) + h'\Delta_m^* - \tfrac{1}{2}h'\Gamma h]$
$\qquad \to \mathscr{E}_{\mathscr{L}^*} \exp(iu'T + h'\Delta - \tfrac{1}{2}h'\Gamma h).$

Proof (i) For simplicity, set

$$U_n = \exp\Lambda_n,$$

$$V_n = \exp(h'\Delta_n^* - \tfrac{1}{2}h'\Gamma h)$$

and $$Q^* = N(-\tfrac{1}{2}h'\Gamma h, h'\Gamma h).$$

Then the convergence $\mathscr{L}(\Lambda_n|P_\theta) \Rightarrow \mathscr{L}(\Delta|Q^*)$, implies that

$$\mathscr{L}(U_n|P_\theta) \Rightarrow \mathscr{L}(\exp\Delta|Q^*),$$

whereas $\mathscr{E}_\theta U_n = \mathscr{E}_{Q^*}\exp\Delta = 1$. Also the convergence

$$\mathscr{L}(\Delta_n^*|P_\theta) \Rightarrow N(0,\Gamma)$$

implies that $\quad \mathscr{L}(h'\Delta_n^* - \tfrac{1}{2}h'\Gamma h|P_\theta) \Rightarrow \mathscr{L}(\Delta|Q^*),$

so that $$\mathscr{L}(V_n|P_\theta) \Rightarrow \mathscr{L}(\exp\Delta|Q^*).$$

Furthermore, $\mathscr{E}_\theta \exp h'\Delta_n^* \to \mathscr{E}_Q \exp h'\Delta = \exp\tfrac{1}{2}h'\Gamma h$ by (3.8) in Chapter 3, where $Q = N(0,\Gamma)$. Thus $\mathscr{E}_\theta V_n \to 1$, so that Lemma 2.1 in Chapter 3 applies and gives that $U_n, V_n, n \geqslant 1$, are uniformly integrable. Next $U_n - V_n \to 0$ in P_θ-probability by arguing as in (3.5) of Chapter 3 and using the fact that $\Lambda_n - h'\Delta_n^* \to -\tfrac{1}{2}h'\Gamma h$ in P_θ-probability. Then, by Lemma 2.2 in Chapter 3, it follows that $|U_n - V_n|, n \geqslant 1$, are uniformly integrable and Corollary 2.1 of the same chapter gives that $\mathscr{E}_\theta|U_n - V_n| \to 0$. The proof of (i) is then completed by observing that the left-hand side of (i) is bounded in absolute value by $\mathscr{E}_\theta|U_n - V_n|$.

(ii) Clearly, it suffices to show that

$$\mathscr{E}_\theta[\exp iu'm^{\frac{1}{2}}(T_m - \theta) + h'\Delta_m^*] \to \mathscr{E}_{\mathscr{L}^*}\exp(iu'T + h'\Delta). \quad (3.16)$$

Set $A_m = (|h'\Delta_m^*| > c)$. Then one has that

$$|\mathscr{E}_\theta \exp[iu'm^{\frac{1}{2}}(T_m - \theta) + h'\Delta_m^*] - \mathscr{E}_{\mathscr{L}^*}\exp(iu'T + h'\Delta)|$$

$$\leqslant \int_{A_m} \exp h'\Delta_m^* \, dP_\theta + \int_{A_m} \exp h'\Delta \, d\mathscr{L}^*$$

$$+ \left| \int_{A_m^c} \exp[iu'm^{\frac{1}{2}}(T_m - \theta) + h'\Delta_m^*] \, dP_\theta \right.$$

$$\left. - \int_{A_m^c} \exp(iu'T + h'\Delta) \, d\mathscr{L}^* \right|. \quad (3.17)$$

As was seen in the proof of (i), $\mathscr{E}_\theta \exp(h'\Delta_m^*) \to \exp(\tfrac{1}{2}h'\Gamma h)$. Also $\mathscr{L}(h'\Delta_m^*|P_\theta) \Rightarrow \mathscr{L}(h'\Delta|Q)$ implies that

$$\mathscr{L}(\exp h'\Delta_m^*|P_\theta) \Rightarrow \mathscr{L}(\exp h'\Delta|Q).$$

Therefore Lemma 2.1 in Chapter 3 gives that $\exp h'\Delta_m^*$, for all m, are uniformly integrable. Thus one may choose c sufficiently large, so that

$$\int_{A_m} \exp h'\Delta_m^* \, dP_\theta < \epsilon \quad \text{and} \quad \int_{A_m} \exp h'\Delta \, d\mathscr{L}^*$$

$$= \int_{A_m} \exp h'\Delta \, dQ < \epsilon. \quad (3.18)$$

Furthermore

$$\int_{A_m^c} \exp\left[iu'm^{\frac{1}{2}}(T_m - \theta) + h'\Delta_m^*\right] dP_\theta$$

$$= \int_{(|h'\Delta| \leqslant c)} \exp(iu'T + h'\Delta) \, d\mathscr{L}_m^*$$

$$\to \int_{(|h'\Delta| \leqslant c)} \exp(iu'T + h'\Delta) \, d\mathscr{L}^* \quad (3.19)$$

by the fact that $\mathscr{L}_m^* \Rightarrow \mathscr{L}^*$ and $\mathscr{L}^*(R^k \times B) = Q(B) = 0$, where $B = \{\Delta \in R^k; |h'\Delta| = c\}$. Then relations (3.17)–(3.19) complete the proof of (3.16) and hence that of (ii). ∎

LEMMA 3.2 With the notation employed in the theorem, consider the expectation $\mathscr{E}_{\mathscr{L}^*} \exp(iu'T + h'\Delta)$ as a function of h, call it $g(h)$, where $h = (h_1, \ldots, h_k)'$, $h_j \in R$, $j = 1, \ldots, k$. Then, for $j = 1, \ldots, k$ $g(h)$ is analytic in the jth coordinate h_j when the first $j - 1$ coordinates h_r, $r = 1, \ldots, j - 1$ are any complex numbers and the last $k - j$ coordinates h_r, $r = j + 1, \ldots, k$ are any real numbers.

Proof By setting $\bar{h} = (h_2, \ldots, h_k)'$, $\Delta = (\Delta_1, \ldots, \Delta_k)'$ and $\bar{\Delta} = (\Delta_2, \ldots, \Delta_k)'$, one has that

$$\mathscr{E}_{\mathscr{L}^*} \exp(iu'T + h'\Delta) = \mathscr{E}_{\mathscr{L}^*} \exp\left[(iu'T + \bar{h}'\bar{\Delta}) + h_1\Delta_1\right]$$

$$= \mathscr{E}_{\mathscr{L}^*}\left[\exp(iu'T + \bar{h}'\bar{\Delta}) \sum_{j=0}^{\infty} \frac{(h_1\Delta_1)^j}{j!}\right]$$

$$= \mathscr{E}_{\mathscr{L}^*} \sum_{j=0}^{\infty} \left[\exp(iu'T + \bar{h}'\bar{\Delta}) \frac{\Delta_1^j}{j!}\right] h_1^j.$$

Now at this point we observe that for $n \geqslant 0$, one has

$$\left| \sum_{j=0}^{n} \left[\exp\left(iu'T + \bar{h}'\bar{\Delta}\right) \frac{\Delta_1^j}{j!} \right] h_1^j \right| = \exp\bar{h}'\bar{\Delta} \left| \sum_{j=0}^{n} \frac{\Delta_1^j}{j!} h_1^j \right|$$

$$\leqslant \exp\bar{h}'\bar{\Delta} \sum_{j=0}^{n} \frac{|\Delta_1|^j}{j!} |h_1|^j$$

$$\leqslant \exp\left(\bar{h}'\bar{\Delta} + |h_1\Delta_1|\right) \quad (3.20)$$

and this last expression is, clearly, Q-integrable and hence $\mathscr{L}^*$-integrable. Then by the dominated convergence theorem, we get

$$\mathscr{E}_{\mathscr{L}^*} \exp\left(iu'T + h'\Delta\right) = \sum_{j=0}^{\infty} \left\{ \mathscr{E}_{\mathscr{L}^*} \left[\exp\left(iu'T + \bar{h}'\bar{\Delta}\right) \frac{\Delta_1^j}{j!} \right] \right\} h_1^j$$

which shows that $g(h) = g(h_1, ..., h_k)$ is analytic in h_1 when the other coordinates are kept fixed. Thus, h_1 may be replaced by a complex variable. Making this replacement and working as above, we have that the left-hand side in (3.20) is

$$\sum_{j=0}^{n} \left[\exp\left(iu'T + ih_{12}\Delta_1 + h_{11}\Delta_1 + \bar{\bar{h}}'\bar{\bar{\Delta}}\right) \frac{\Delta_2^j}{j!} \right] h_2^j,$$

where $h_1 = h_{11} + ih_{12}$, $\bar{\bar{h}} = (h_3, ..., h_k)'$ and $\bar{\bar{\Delta}} = (\Delta_3, ..., \Delta_k)'$, whereas the last bound on the right-hand side of the same relationship is equal to $\exp\left(h_{11}\Delta_1 + \bar{\bar{h}}'\bar{\bar{\Delta}} + |h_2\Delta_2|\right)$ which is Q-integrable and hence $\mathscr{L}^*$-integrable. Therefore h_2 may be replaced by a complex variable. In a similar fashion each one of the remaining coordinates may be replaced by complex variables and this completes the proof of the lemma.]

4   W-efficiency of estimates: upper bounds via Theorem 3.1

The main purpose of the present section is that of establishing the inequalities in Theorems 4.1 and 4.2 for two specifications of a class of sequences of estimates (of θ). The first specification, which is due to Schmetterer [1], is given in (4.8) and is restricted to the case that θ is real-valued. The result associated with it is presented as Theorem 4.1. A certain multiparameter version of this result is presented as Theorem 4.2. The corresponding class of sequences of estimates is given by (4.10) and is the multi-parameter analogue of the class $\mathscr{C}_S$ given by (4.8). For the proof of the above-mentioned theorems, one needs an auxiliary result which is formulated as Proposition 4.1.

For $a \in R^k$ with $a \neq 0$ and $v \in R$, consider the half space

$$H_a(v) = \{x \in R^k; a'x \leqslant v\}.$$

Denote by L the d.f. corresponding to a probability measure $\mathscr{L}$ in R^k, set $Q_1 = N(0, \Gamma^{-1})$ and let $\Phi(x; \Gamma^{-1})$ be the corresponding d.f. Then it is clear that there exists $v^* \in R$ such that

$$\int_{H_a(v)} \mathrm{d}\mathscr{L} = \int_{H_a(v)} \mathrm{d}L(x) = \int_{H_a(v^*)} \mathrm{d}\Phi(x; \Gamma^{-1}) = \int_{H_a(v^*)} \mathrm{d}Q_1. \quad (4.1)$$

Next, denote by μ and G the probability measure and the d.f., respectively, corresponding to the characteristic function

$$\phi(u, 0) \exp\left(\tfrac{1}{2} u' \Gamma^{-1} u\right),$$

and let (X, Y) be two random vectors distributed as $Q_1 \times \mu$. Also let $Q_2 \times \mu$ be an alternative distribution of (X, Y), where Q_2 is the measure induced by $\Phi(x + \lambda \Gamma^{-1} a; \Gamma^{-1})$, and consider the problem of testing the hypothesis that (X, Y) are distributed as $Q_1 \times \mu$ against the alternative that they are distributed as $Q_2 \times \mu$. If $f(x; \Gamma^{-1})$ is the density of Q_1 with respect to the k-dimensional Lebesgue measure l^k, it follows that $f(x; \Gamma^{-1})$ is also the density of $Q_1 \times \mu$ with respect to $l^k \times \mu$. This is so because for any $B_1, B_2 \in \mathscr{B}$, one has that

$$\int_{B_1 \times B_2} f(x; \Gamma^{-1}) \, \mathrm{d}(l^k \times \mu) = \mu(B_2) \, Q_1(B_1) = (Q_1 \times \mu)(B_1 \times B_2).$$

Similarly, $f(x + \lambda \Gamma^{-1} a; \Gamma^{-1})$ is the density of $Q_2 \times \mu$ with respect to $l^k \times \mu$. Then for the hypothesis testing problem above the most powerful test rejects the hypothesis when

$$\frac{f(x + \lambda \Gamma^{-1} a; \Gamma^{-1})}{f(x; \Gamma^{-1})} \geqslant c,$$

which is equivalent to $a'x \leqslant c^*$ after some simplifications, provided $\lambda > 0$. Thus the test which rejects the hypothesis in question whenever $a'X \leqslant v^*$ is at least as powerful as the test which rejects it whenever $a'(X+Y) \leqslant v$. These two tests have the same level of significance because

$$\int_{H_a(v^*)} \mathrm{d}(Q_1 \times \mu) = \int_{H_a(v^*)} \mathrm{d}[\Phi(x;\Gamma^{-1}) \times G(y)]$$
$$= \int_{H_a(v^*)} \mathrm{d}\Phi(x;\Gamma^{-1}) = \int_{H_a(v)} \mathrm{d}L(x),$$

the last equality being true because of (4.1), and

$$\int_{[a'(x+y) \leqslant v]} \mathrm{d}(Q_1 \times \mu) = \int_{[a'(x+y) \leqslant v]} \mathrm{d}[\Phi(x;\Gamma^{-1}) \times G(y)]$$
$$= \int_{(a'z \leqslant v)} \mathrm{d}L(z) = \int_{H_a(v)} \mathrm{d}L(x).$$

The second equality from the right-hand side above is true because the characteristic function of the sum $X+Y$ of the independent random vectors X and Y is equal to

$$\exp\left(-\tfrac{1}{2}u'\Gamma^{-1}u\right)\phi(u,0)\exp\left(\tfrac{1}{2}u'\Gamma^{-1}u\right) = \phi(u,0)$$

whose corresponding distribution is $\mathscr{L}$.

It follows that the powers of the above tests satisfy the following inequality

$$\int_{H_a(v^*)} \mathrm{d}(Q_1 \times \mu) \geqslant \int_{[a'(x+y) \leqslant v]} \mathrm{d}(Q_1 \times \mu). \qquad (4.2)$$

But

$$\int_{[a'(x+y) \leqslant v]} \mathrm{d}(Q_2 \times \mu) = \int_{[a'(x+y) \leqslant v]} \mathrm{d}[\Phi(x+\lambda\Gamma^{-1}a;\Gamma^{-1}) \times G(y)]$$
$$= \int_{[a'(z+y-\lambda\Gamma^{-1}a) \leqslant v]} \mathrm{d}[\Phi(z;\Gamma^{-1}) \times G(y)]$$
$$= \int_{\{a'[(x+y)-\lambda\Gamma^{-1}a] \leqslant v\}} \mathrm{d}[\Phi(x;\Gamma^{-1}) \times G(y)]$$
$$= \int_{[a'(z-\lambda\Gamma^{-1}a) \leqslant v]} \mathrm{d}L(z).$$

That is, $\displaystyle\int_{[a'(x+y) \leqslant v]} \mathrm{d}(Q_2 \times \mu) = \int_{[a'(z-\lambda\Gamma^{-1}a) \leqslant v]} \mathrm{d}L(z). \qquad (4.3)$

Also

$$\int_{H_a(v^*)} \mathrm{d}(Q_2 \times \mu) = \int_{(a'x \leqslant v^*)} \mathrm{d}[\Phi(x + \lambda\Gamma^{-1}a \,;\, \Gamma^{-1}) \times G(y)]$$

$$= \int_{[a'(z - \lambda\Gamma^{-1}a) \leqslant v^*]} \mathrm{d}[\Phi(z \,;\, \Gamma^{-1}) \times G(y)]$$

$$= \int_{[a'(z - \lambda\Gamma^{-1}a) \leqslant v^*]} \mathrm{d}\Phi(z \,;\, \Gamma^{-1}).$$

In the last expression above, we set $\lambda = \dfrac{s}{a'\Gamma^{-1}a}$, $s > 0$. Then

$$a'(z - \Gamma^{-1}a) = a'\left(z - \frac{s\Gamma^{-1}a}{a'\Gamma^{-1}a}\right) = a'z - \frac{sa'\Gamma^{-1}a}{a'\Gamma^{-1}a} = a'z - s,$$

so that

$$\int_{[a'(z - \Gamma^{-1}a) \leqslant v^*]} \mathrm{d}\Phi(z \,;\, \Gamma^{-1}) = \int_{(a'z \leqslant v^* + s)} \mathrm{d}\Phi(z \,;\, \Gamma^{-1}).$$

Thus $$\int_{H_a(v^*)} \mathrm{d}(Q_2 \times \mu) = \int_{(a'z \leqslant v^* + s)} \mathrm{d}\Phi(z \,;\, \Gamma^{-1}). \qquad (4.4)$$

On the other hand, for the above choice of λ, relation (4.3) becomes as follows

$$\int_{[a'(x+y) \leqslant v]} \mathrm{d}(Q_2 \times \mu) = \int_{(a'z \leqslant v^* + s)} \mathrm{d}L(z). \qquad (4.5)$$

Then relations (4.2), (4.4) and (4.5) give

$$\int_{(a'x \leqslant v^* + s)} \mathrm{d}\Phi(x \,;\, \Gamma^{-1}) \geqslant \int_{(a'x \leqslant v^* + s)} \mathrm{d}L(x) \quad (s > 0). \qquad (4.6)$$

For $s < 0$, we arrive at

$$\int_{(a'x \leqslant v^* + s)} \mathrm{d}\Phi(x \,;\, \Gamma^{-1}) \leqslant \int_{(a'x \leqslant v + s)} \mathrm{d}L(x) \qquad (4.6')$$

by considering as alternative distribution of (X, Y) the distribution $Q_2' \times \mu$, where Q_2' is the measure induced by $\Phi(x - \lambda\Gamma^{-1}a \,;\, \Gamma^{-1})$, $\lambda = -\dfrac{s}{a'\Gamma a}$, $s < 0$. Since also $\displaystyle\int_{H_a(v^*)} \mathrm{d}\Phi(x \,;\, \Gamma^{-1}) = \int_{H_a(v)} \mathrm{d}L(x)$ by

(4.1), considering the two cases that $s > 0$ and $s < 0$, relations (4.6) and (4.6′) provide easily the following inequality

$$\int_{H_a(v+s)\Delta H_a(v)} dL(x) \leqslant \int_{H_a(v^*+s)\Delta H_a(v^*)} d\Phi(x;\Gamma^{-1}) \quad (s \in R).$$

Thus we have established the following result.

PROPOSITION 4.1 For $a \in R^k$ with $a \neq 0$ and $a \in R$, consider the hyperplane $H_a(\alpha) = \{x \in R^k; a'x \leqslant \alpha\}$ and let $L(x)$ and $\Phi(x;\Gamma^{-1})$ be the d.f.s of the (probability) measures $\mathscr{L}$ (in R^k) and $N(0,\Gamma^{-1})$, respectively. Furthermore, for any $v \in R$, let v^* be defined by $\int_{H_a(v)} dL(x) = \int_{H_a(v^*)} d\Phi(x;\Gamma^{-1})$. Then for any $v, s \in R$, one has that

$$\int_{H_a(v+s)\Delta H_a(v)} dL(x) \leqslant \int_{H_a(v^*+s)\Delta H_a(v^*)} d\Phi(x;\Gamma^{-1}). \qquad (4.7)$$

We are now going to derive the main results in Schmetterer [1] and Roussas [5] as special cases of (4.7). To this end, suppose that Θ is an open subset of R (i.e. $k = 1$) and define the class $\mathscr{C}_S$ of sequences of estimates (of θ) by

$$\mathscr{C}_S = \{\{T_n\}; P_\theta[n^{\frac{1}{2}}(T_n - \theta) \leqslant x] \to F_T(x;\theta), \text{ a d.f.},$$

continuously in Θ for each fixed $x \in R$ and

also *continuously* in R for each fixed $\theta \in \Theta\}$.   (4.8)

From (4.8), we have that for each $\theta \in \Theta$,

$$P_\theta[n^{\frac{1}{2}}(T_n - \theta) \leqslant x] \to F_T(x;\theta)$$

continuously in R. Therefore, by Lemma 2.1, $F_T(\cdot\,;\theta)$ is continuous and hence the 'smallest' and 'largest' median $l_T(\theta)$ and $u_T(\theta)$, respectively, are such that $F_T(l_T;\theta) = F_T(u_T;\theta) = \frac{1}{2}$. Then consider the following definition of the class $J_n(\theta)$ mentioned in (1.2).

$$J_n(\theta;T) = \{j_n(\theta;t_1,t_2,T); j_n(\theta;t_1,t_2,T)$$
$$= (\theta - t_1 n^{-\frac{1}{2}} + l_T(\theta)n^{-\frac{1}{2}}, \theta + t_2 n^{-\frac{1}{2}} + u_T(\theta)n^{-\frac{1}{2}}), t_1, t_2 > 0\}.$$
$$\qquad (4.9)$$

Finally, for each $\theta \in \Theta$, let $B(\theta; t_1, t_2)$ be defined by

$$B(\theta; t_1, t_2) = \Phi[t_2 \sigma(\theta)] - \Phi[t_1 \sigma(\theta)], \; t_1, t_2 > 0. \qquad (4.10)$$

Then one has the following result.

THEOREM 4.1   Let $\mathcal{C}_S$, $J_n(\theta; T)$ and $B(\theta; t_1, t_2)$ be defined by (4.8), (4.9) and (4.10), respectively. Then we have

$$\lim P_\theta[T_n \in j_n(\theta; t_1, t_2, T)]$$

$$\leqslant B(\theta; t_1, t_2) \quad \text{for all } t_1, t_2 > 0 \text{ and every } \theta \in \Theta.$$

Proof   In the first place, it is clear that

$$\mathcal{C}_S \subseteq \mathcal{C}_H \quad \text{and} \quad F_T(\cdot; \theta) = L(\cdot; \theta).$$

Next $P_\theta[n^{\frac{1}{2}}(T_n - \theta) \leqslant l_T(\theta)] \to \frac{1}{2} = \displaystyle\int_{H_1[l_T(\theta)]} dL(x; \theta)$

$$= \int_{H_1(0)} d\Phi[x; \sigma^{-2}(\theta)] \, (= \tfrac{1}{2}),$$

and similarly

$$P_\theta[n^{\frac{1}{2}}(T_n - \theta) \leqslant u_T(\theta)] \to \tfrac{1}{2} = \int_{H_1[u_T(\theta)]} dL(x; \theta)$$

$$= \int_{H_1(0)} d\Phi[x; \sigma^{-2}(\theta)] \, (= \tfrac{1}{2}).$$

Therefore for the present case, if we take v equal to $l_T(\theta)$ and $u_T(\theta)$ in Proposition 4.1, then the corresponding v^*s are equal to 0. By also taking s to be equal to $-t_1$ and t_2, relation (4.7) gives, respectively,

$$\int_{H_1[l_T(\theta) - t_1] \Delta H_1[l_T(\theta)]} dL(x; \theta) \leqslant \int_{H_1(0 - t_1) \Delta H_1(0)} d\Phi[x; \sigma^{-2}(\theta)]$$

and

$$\int_{H_1[u_T(\theta) + t_2] \Delta H_1[u_T(\theta)]} dL(x; \theta) \leqslant \int_{H_1(0 + t_2) \Delta H_1(0)} d\Phi[x; \sigma^{-2}(\theta)].$$

Equivalently,

$$\lim P_\theta[l_T(\theta) - t_1 < n^{\frac{1}{2}}(T_n - \theta) < l_T(\theta)] \leqslant \tfrac{1}{2} - \Phi[-t_1 \sigma(\theta)]$$

and

$$\lim P_\theta[u_T(\theta) < n^{\frac{1}{2}}(T_n - \theta) < u_T(\theta) + t_2] \leqslant \Phi[t_2 \sigma(\theta)] - \tfrac{1}{2}.$$

Hence $\lim P_\theta[T_n \in j_n(\theta; t_1, t_2, T)]$

$$= \lim P_\theta[l_T(\theta) - t_1 < n^{\frac{1}{2}}(T_n - \theta) < u_T(\theta) + t_2]$$

$$\leqslant \Phi[t_2 \sigma(\theta)] - \Phi[-t_1 \sigma(\theta)] = B(\theta; t_1, t_2),$$

as was to be seen.]

Now we do not assume that $k = 1$. Then by interpreting the various inequalities below in the coordinatewise sense, we consider the class $\mathscr{C}_R$ of sequences of estimates defined as in (4.8), namely

$$\mathscr{C}_R = \{\{T_n\}; P_\theta[n^{\frac{1}{2}}(T_n - \theta) \leqslant x] \to F_T(x; \theta), \text{ a d.f.},$$

$$\textit{continuously in } \Theta \text{ for each fixed } x \in R^k \text{ and also}$$

$$\textit{continuously in } R^k \text{ for each fixed } \theta \in \Theta\}. \qquad (4.11)$$

Then for each $h \in R^k$, it is clear that the d.f. of $n^{\frac{1}{2}}h'(T_n - \theta)$, under P_θ, converges to a d.f. $F_T(x; \theta, h)$, say, continuously in Θ for each fixed $x \in R^k$ and also continuously in R^k for each fixed $\theta \in \Theta$. Furthermore, the d.f. $F_T(\cdot; \theta, h)$ is continuous, so that the 'smallest' and 'largest' median $l_T(\theta, h)$ and $u_T(\theta, h)$, respectively, are such that $F_T(l_T; \theta) = F_T(u_T; \theta) = \frac{1}{2}$. It is also clear that $\mathscr{C}_R \subseteq \mathscr{C}_H$. Now working as in the proof of the previous theorem, one obtains the following result.

THEOREM 4.2 Let $\mathscr{C}_R$ be obtained by (4.11) and for each $h \in R^k$, let $l_T(\theta, h)$ and $u_T(\theta, h)$ be as above. Then for each $h \in R^k$, one has

$$\lim P_\theta[-t_1 + l_T(\theta, h) < n^{\frac{1}{2}}h'(T_n - \theta) < t_2 + u_T(\theta, h)]$$

$$\leqslant \Phi[t_2 \sigma^{-1}(\theta, h)] - \Phi[-t_1 \sigma^{-1}(\theta, h)] \quad \text{for all } t_1, t_2 > 0$$

$$\text{and every } \theta \in \Theta,$$

where $\sigma^2(\theta, h) = h'\Gamma^{-1}(\theta) h$.

5 W-efficiency of estimates: upper bounds

In this section, we establish upper bounds for two additional specifications of the class $\mathscr{C}^*$ for the case that Θ is an open subset of R. For the proof of the relevant theorems, we shall require the relations (5.1) and (5.2) below. In order to formulate them, let

$\theta \in \Theta$, let $0 \neq h \in R$ and set $\theta_n = \theta + hn^{-\frac{1}{2}}$. Then from Theorems 4.3 and 4.5 in Chapter 2, it follows immediately that for $x \in R$,

$$P_\theta \left[\frac{\Lambda(\theta, \theta_n) + \frac{1}{2}h^2\sigma^2(\theta)}{|h|\,\sigma(\theta)} \leqslant x \right] \to \Phi(x) \qquad (5.1)$$

and
$$P_{\theta_n} \left[\frac{\Lambda(\theta, \theta_n) + \frac{1}{2}h^2\sigma^2(\theta)}{|h|\,\sigma(\theta)} \leqslant x \right] \to \Phi[x - |h|\,\sigma(\theta)], \qquad (5.2)$$

where
$$\sigma^2(\theta) = \Gamma(\theta) = 4\mathscr{E}_\theta[\dot{\phi}_1(\theta)]^2 \qquad (5.3)$$

and, recall that,
$$\Lambda(\theta, \theta_n) = \log\left(dP_{n,\theta_n}/dP_{n,\theta}\right).$$

In the sequel, we shall write Λ_n rather than $\Lambda(\theta, \theta_n)$ for the sake of simplicity.

In the definition of the class $\mathscr{C}_S$, Pfanzagl [1] removes the requirement of continuous convergence and assumes instead that the limiting distribution has zero as its median. More precisely, he defines the class $\mathscr{C}^*$, to be denoted here by $\mathscr{C}_P$, as follows.

$$\mathscr{C}_P = \{\{T_n\}\,;\mathscr{L}[n^{\frac{1}{2}}(T_n - \theta)|P_\theta] \Rightarrow \mathscr{L}_{T,\theta}, \text{ a probability}$$

measure, such that $\mathscr{L}_{T,\theta}((-\infty, 0]) \geqslant \frac{1}{2}$ and

$$\mathscr{L}_{T,\theta}([0, \infty)) \geqslant \tfrac{1}{2} \text{ for all } \theta \in \Theta\}. \qquad (5.4)$$

Thus if $0 \in [l_T(\theta), u_T(\theta)]$ for all $\theta \in \Theta$, then $\mathscr{C}_P \supseteq \mathscr{C}_S$. The requirement that 0 be a median for $\mathscr{L}_{T,\theta}$ for all $\theta \in \Theta$, is a very natural one, as the following argument shows. Suppose that 0 is not a median for $\mathscr{L}_{T,\theta}$ for some θ and let $l_T(\theta) > 0$. Choose a continuity point $x = x(\theta, T)$ of $\mathscr{L}_{T,\theta}$ such that $0 < x < l_T(\theta)$. Then

$$\tfrac{1}{2} > \mathscr{L}_{T,\theta}((-\infty, x])$$
$$= \lim P_\theta[n^{\frac{1}{2}}(T_n - \theta) < x] \geqslant \limsup P_\theta[n^{\frac{1}{2}}(T_n - \theta) \leqslant 0]$$
$$= \limsup P_\theta(T_n \leqslant \theta).$$

In other words, $\limsup P_\theta(T_n \leqslant \theta) < \frac{1}{2}$ and in a similar fashion, $\limsup P_\theta(T_n \geqslant \theta) < \frac{1}{2}$ if $u_T(\theta) < 0$. Such an asymptotic behaviour of an estimate of θ perhaps would be judged by many as not being satisfactory.

For the proof of the first theorem in this chapter, the following lemma presented in Pfanzagl [1] is required.

LEMMA 5.1 Let $\{T_n\}$ be a sequence of estimates such that $\mathscr{L}[n^{\frac{1}{2}}(T_n - \theta)|P_\theta] \Rightarrow \mathscr{L}_{T,\theta}$, a probability measure,

$$\liminf P_\theta[n^{\frac{1}{2}}(T_n - \theta) \geqslant x] \geqslant \tfrac{1}{2} \quad \text{for all } x < 0$$

and $\quad \liminf P_\theta[n^{\frac{1}{2}}(T_n - \theta) \leqslant x] \geqslant \tfrac{1}{2} \quad \text{for all } x > 0.$

Then, under (5.1) and (5.2), one has

$$\liminf P_\theta[T_n \in j_n(\theta; t_1, t_2)]$$
$$\leqslant B(\theta; t_1, t_2) \quad \text{for all } t_1, t_2 > 0 \text{ and a.a. } [l]\ \theta \in \Theta; \quad (5.5)$$

the quantity $B(\theta; t_1, t_2)$ is given by (4.10) and $j_n(\theta; t_1, t_2) \in J_n(\theta)$, where $J_n(\theta)$ is defined by (5.6) below.

$$J_n(\theta) = \{j_n(\theta; t_1, t_2); j_n(\theta; t_1, t_2) = (\theta - t_1 n^{-\frac{1}{2}}, \theta + t_2 n^{-\frac{1}{2}}), t_1, t_2 > 0\}.$$
$$(5.6)$$

Proof For each fixed $x < 0$, define $\{f_n(x; \cdot)\}$ as follows

$$f_n(x; \theta) = P_\theta[n^{\frac{1}{2}}(T_n - \theta) \geqslant x] - \tfrac{1}{2}, \theta \in \Theta.$$

Then, clearly, the conditions of Lemma 2.5 are satisfied. Thus, if for $r \in R$ we consider the sequence $x_n(r) = rn^{-\frac{1}{2}}$, it follows that there exists an l-null set which depends on $\{x(r)\}$ and perhaps also on x, call it $N(x, r)$, such that

$$\limsup f_n(x; \theta + rn^{-\frac{1}{2}}) \geqslant 0 \quad \text{for all } \theta \in N^c(x, r).$$

Let R_0 stand for the set of rational numbers in R and set $N_1 = \cup [N(x, r); x \in (-\infty, 0) \cap R_0, r \in (0, \infty) \cap R_0]$. Then $l(N_1) = 0$ and $\limsup f_n(x; \theta + rn^{-\frac{1}{2}}) \geqslant 0$ for all $\theta \in N_1^c, x \in (-\infty, 0) \cap R_0$ and $r \in (0, \infty) \cap R_0$, or equivalently,

$$\limsup P[n^{\frac{1}{2}}(T_n - \theta) \geqslant x + r | P_{\theta + rn^{-\frac{1}{2}}}] \geqslant \tfrac{1}{2}$$

$$\text{for all } \theta \in N_1^c, x \in (-\infty, 0) \cap R_0 \text{ and } r \in (0, \infty) \cap R_0. \quad (5.7)$$

Set s for the left-hand side of (5.7). Then there exists a subsequence $\{m\} \subseteq \{n\}$ such that

$$\lim P[m^{\frac{1}{2}}(T_m - \theta) \geqslant x + r | P_{\theta + rn^{-\frac{1}{2}}}] = s.$$

Applying (3.2) with $\theta_m = \theta + rm^{-\frac{1}{2}}$ and $\delta > 0$, we get

$$P_{\theta_m}\left(\frac{\Lambda_m + \frac{1}{2}r^2\sigma^2}{r\sigma} \leqslant -\delta r\sigma\right) \to \Phi[-(1+\delta)r\sigma] < \tfrac{1}{2},$$

so that for all sufficiently large m, $m \geqslant m_1$, say, one has

$$P_{\theta_m}\left(\frac{\Lambda_m + \frac{1}{2}r^2\sigma^2}{r\sigma} \leqslant -\delta r\sigma\right)$$

$$< P_{\theta_m}[m^{\frac{1}{2}}(T_m - \theta) \geqslant x + r] \quad \text{for all } \theta \in N_1^c,$$

$$x \in (-\infty, 0) \cap R_0 \text{ and } r \in (0, \infty) \cap R_0. \quad (5.8)$$

At this point, an ingenious use of the Neyman–Pearson fundamental lemma, due to Wald [1], provides the following inequality

$$P_\theta\left(\frac{\Lambda_m + \frac{1}{2}r^2\sigma^2}{r\sigma} \leqslant -\delta r\sigma\right) < P_\theta[m^{\frac{1}{2}}(T_m - \theta) \geqslant x + r]$$

$$\text{for all } \theta \in N_1^c, \quad x \in (-\infty, 0) \cap R_0 \text{ and } r \in (0, \infty) \cap R_0. \quad (5.9)$$

Namely, define C_m and D_m by

$$C_m = \left(\frac{\Lambda_m + \frac{1}{2}r^2\sigma^2}{r\sigma} \leqslant -\delta r\sigma\right), \quad D_m = [m^{\frac{1}{2}}(T_m - \theta) \geqslant x + r].$$

Then for each $m \geqslant m_1$, C_m and D_m are alternative critical regions for testing the simple hypothesis H: the underlying probability measure is $P_{m,\theta}$ against the simple alternative A: the underlying probability measure is P_{m,θ_m}. Since C_m is optimal, by the Neyman–Pearson fundamental lemma, and its power $P_{\theta_m}(C_m)$ is less than $P_{\theta_m}(D_m)$, the power of D_m, by (5.8), the level of C_m must also be less than the level of D_m. That is, $P_\theta(C_m) < P_\theta(D_m)$, or, equivalently,

$$P_\theta\left(\frac{\Lambda_m + \frac{1}{2}r^2\sigma^2}{r\sigma} \leqslant -\delta r\sigma\right) < P_\theta[m^{\frac{1}{2}}(T_m - \theta) \geqslant x + r]$$

which is inequality (5.9).

In inequality (5.9), we first take the limits as $m \to \infty$ and utilize (5.1) and then let $\delta \to 1$. One has then

$$\limsup P_\theta[n^{\frac{1}{2}}(T_n - \theta) \geqslant x + r] \geqslant \Phi(-r\sigma)$$

$$\text{for all } \theta \in N_1^c, \quad x \in (-\infty, 0) \cap R_0 \text{ and } r \in (0, \infty) \cap R_0. \quad (5.10)$$

Next, for each fixed $x > 0$, define $\{f_n(x\,;\cdot)\}$ by

$$f_n(x\,;\theta) = P_\theta[n^{\frac{1}{2}}(T_n - \theta) \leqslant x] - \tfrac{1}{2} \quad (\theta \in \Theta).$$

Then once again the conditions of Lemma 2.5 are fulfilled and working as above with θ_m now defined by $\theta_m = \theta - rm^{-\frac{1}{2}}$, one obtains

$$\limsup P_\theta[n^{\frac{1}{2}}(T_n - \theta) \leqslant x - r] \geqslant \Phi(-r\sigma) \quad \text{for all } \theta \in N_2^c$$

with $\quad l(N_2) = 0, x \in (0, \infty) \cap R_0 \quad \text{and} \quad r \in (0, \infty) \cap R_0.$ (5.11)

Now, for $\epsilon > 0$, let $y = y(\epsilon, \theta) > 0$ be defined by

$$\Phi(y\sigma) - \Phi(-y\sigma) = \epsilon$$

and set $\eta = \eta(\epsilon, \theta) = 2y\sigma$. Then for any x_1, x_2 such that

$$|x_1 - x_2| \leqslant 2y$$

(equivalently, $|x_1\sigma - x_2\sigma| \leqslant \eta$), we clearly have

$$|\Phi(x_1\sigma) - \Phi(x_2\sigma)| \leqslant \epsilon. \tag{5.12}$$

Let $t > 0$ be arbitrary and choose

$$r \in (t, t + 2y) \cap R_0, \quad x \in (t - r, 0) \cap R_0.$$

Then $|r - t| \leqslant 2y$, or equivalently, $|(-t)\sigma - (-r)\sigma| \leqslant \eta$ and hence relation (5.12) implies

$$-\epsilon \leqslant \Phi(-r\sigma) - \Phi(-t\sigma) \leqslant \epsilon. \tag{5.13}$$

Thus for $\theta \in N_1^c$, relation (5.10) gives

$$\limsup P_\theta[n^{\frac{1}{2}}(T_n - \theta) \geqslant x + r] \geqslant \Phi(-r\sigma). \tag{5.14}$$

Since also $x + r > t$ by the choice of x, one has

$$\limsup P_\theta[n^{\frac{1}{2}}(T_n - \theta) \geqslant t] \geqslant \limsup P_\theta[n^{\frac{1}{2}}(T_n - \theta) \geqslant x + r]$$
$$\geqslant \Phi(-r\sigma) \geqslant \Phi(-t\sigma) - \epsilon$$

on account of (5.14) and (5.13). Since this is true for every $\epsilon > 0$, one concludes that

$$\limsup P_\theta[n^{\frac{1}{2}}(T_n - \theta) \geqslant t] \geqslant \Phi(-t\sigma)$$

$$\text{for all } t > 0 \text{ and all } \theta \in N_1^c,$$

or equivalently,

$$\liminf P_\theta[n^{\frac{1}{2}}(T_n - \theta) < t] \leqslant \Phi(t\sigma)$$

$$\text{for all } t > 0 \text{ and all } \theta \in N_1^c.$$

Finally, replacing t by t_2, one has

$$\liminf P_\theta[n^{\frac{1}{2}}(T_n - \theta) < t_2] \leqslant \Phi(t_2\sigma)$$

$$\text{for all } t_2 > 0 \text{ and all } \theta \in N_1^c. \quad (5.15)$$

For an arbitrary $t > 0$, choose

$$r \in (t, t + 2y) \cap R_0 \quad \text{and} \quad x \in (0, r - t) \cap R_0.$$

Then (5.12) holds true again. For $\theta \in N_2^c$, relation (5.11) implies

$$\limsup P_\theta[n^{\frac{1}{2}}(T_n - \theta) \leqslant x - r] \geqslant \Phi(-r\sigma). \quad (5.16)$$

Since also $x - r < -t$ by the choice of x, one has

$$\limsup P_\theta[n^{\frac{1}{2}}(T_n - \theta) \leqslant -t] \geqslant \limsup P_\theta[n^{\frac{1}{2}}(T_n - \theta) \leqslant x - r]$$

$$\geqslant \Phi(-r\sigma) \geqslant \Phi(-t\sigma) - \epsilon$$

on account of (5.16) and (5.13). From this, it follows as above that

$$\limsup P_\theta[n^{\frac{1}{2}}(T_n - \theta) \leqslant -t] \geqslant \Phi(-t\sigma)$$

$$\text{for all } t > 0 \text{ and all } \theta \in N_2^c.$$

Upon replacing t by t_1, one has

$$\limsup P_\theta[n^{\frac{1}{2}}(T_n - \theta) \leqslant -t_1] \geqslant \Phi(-t_1\sigma)$$

$$\text{for all } t_1 > 0 \text{ and all } \theta \in N_2^c. \quad (5.17)$$

Now for any $t_1, t_2 > 0$, let $-t_1', t_2'$ be continuity points of $\mathscr{L}_{T,\theta}$ such that $-t_1' < -t_1, t_2' > t_2$. Then

$$\liminf P_\theta[-t_1 < n^{\frac{1}{2}}(T_n - \theta) < t_2]$$

$$\leqslant \lim P_\theta[-t_1' < n^{\frac{1}{2}}(T_n - \theta) < t_2']$$

$$= \lim P_\theta[n^{\frac{1}{2}}(T_n - \theta) < t_2'] - \lim P_\theta[n^{\frac{1}{2}}(T_n - \theta) \leqslant -t_1']$$

$$= \liminf P_\theta[n^{\frac{1}{2}}(T_n - \theta) < t_2'] - \limsup P_\theta[n^{\frac{1}{2}}(T_n - \theta) \leqslant -t_1']$$

and this is bounded above by $\Phi(t_2'\sigma) - \Phi(-t_1'\sigma)$ by means of (5.15) and (5.17), provided $\theta \in N_0^c = N_1^c \cap N_2^c$, so that $l(N_0) = 0$. Thus

$$\liminf P_\theta[-t_1 < n^{\frac{1}{2}}(T_n - \theta) < t_2] \leqslant \Phi(t_2'\sigma) - \Phi(-t_1'\sigma).$$

Letting now $-t_1' \to -t_1$ and $t_2' \to t_2$ through continuity points of $\mathscr{L}_{T,\theta}$, we get that for all $\theta \in N_0^c$ and all $t_1, t_2 > 0$,

$$\liminf P_\theta[-t_1 < n^{\frac{1}{2}}(T_n - \theta) < t_2] \leqslant \Phi(t_2\sigma) - \Phi(-t_1\sigma) = B(\theta; t_1, t_2),$$

as was to be shown. ▮

Now we can formulate the first main result in this section.

THEOREM 5.1   Let $\mathscr{C}_P$, $J_n(\theta\,;t_1,t_2)$ and $B(\theta\,;t_1,t_2)$ be defined by (5.4), (5.6) and (4.10), respectively. Then one has

$$\limsup P_\theta[T_n \in j_n(\theta\,;t_1,t_2)] \leqslant B(\theta\,;t_1,t_2)$$

$$\text{for all } t_1, t_2 > 0 \text{ and a.a. } [l] \ \theta \in \Theta.$$

Proof   First we show that the assumptions of Lemma 5.1 are satisfied. Suppose $x < 0$. Then for each $\theta \in \Theta$, there exists $y = y(T,\theta)$ in $[x,0)$ such that $\mathscr{L}_{T,\theta}(\{y\}) = 0$. Thus

$$P_\theta[n^{\frac{1}{2}}(T_n - \theta) \geqslant x] \geqslant P_\theta[n^{\frac{1}{2}}(T_n - \theta) \geqslant y]$$

and hence

$$\liminf P_\theta[n^{\frac{1}{2}}(T_n - \theta) \geqslant x] \geqslant \liminf P_\theta[n^{\frac{1}{2}}(T_n - \theta) \geqslant y]$$
$$= \lim P_\theta[n^{\frac{1}{2}}(T_n - \theta) \geqslant y]$$
$$= \mathscr{L}_{T,\theta}([y,\infty)) \geqslant \mathscr{L}_{T,\theta}([0,\infty)) \geqslant \tfrac{1}{2}.$$

That is,

$$\liminf P_\theta[n^{\frac{1}{2}}(T_n - \theta) \geqslant x] \geqslant \tfrac{1}{2} \quad \text{for all } x < 0,$$

and in a similar fashion

$$\liminf P_\theta[n^{\frac{1}{2}}(T_n - \theta) \leqslant x] \geqslant \tfrac{1}{2} \quad \text{for all } x > 0.$$

Next we proceed as follows. For $\in\, > 0$, define y and η as in the proof of Lemma 5.1. Let $t_1, t_2 > 0$ be arbitrary and choose $s_1 \in (t_1, t_1 + 2y]$, $s_2 \in (t_2, t_2 + 2y]$, so that

$$\mathscr{L}_{T,\theta}(\{s_1\}) = \mathscr{L}_{T,\theta}(\{s_2\}) = 0.$$

Then $s_1 - t_1 \leqslant 2y$, $s_2 - t_2 \leqslant 2y$, or equivalently, $s_1\sigma - t_1\sigma \leqslant \eta$, $s_2\sigma - t_2\sigma \leqslant \eta$.

From the above choice of s_1, s_2, it follows that

$$P_\theta[T_n \in j_n(\theta; t_1, t_2)] \leqslant P_\theta[T_n \in j_n(\theta; s_1, s_2)]$$

and hence

$$\limsup P_\theta[T_n \in j_n(\theta; t_1, t_2)] \leqslant \lim P_\theta[T_n \in j_n(\theta; s_1, s_2)]. \quad (5.18)$$

By means of (5.5), the right-hand side of (5.18) is bounded by $B(\theta; s_1, s_2)$ a.e. $[l]$. Thus one has

$$\limsup P_\theta[T_n \in j_n(\theta; t_1, t_2)] \leqslant B(\theta; s_1, s_2) \text{ a.e. } [l]. \quad (5.19)$$

But $s_1\sigma - t_1\sigma = |(-t_1)\sigma - (-s_1)\sigma| \leqslant \eta$, $s_2\sigma - t_2\sigma \leqslant \eta$

imply

$$\left|\Phi(-t_1\sigma)-\Phi(-s_1\sigma)\right| \leqslant \epsilon, \quad \left|\Phi(s_2\sigma)-\Phi(t_2\sigma)\right| \leqslant \epsilon,$$

and hence

$$-\Phi(-s_1\sigma) \leqslant -\Phi(-t_1\sigma)+\epsilon, \quad \Phi(s_2\sigma) \leqslant \Phi(t_2\sigma)+\epsilon,$$

so that

$$\Phi(s_2\sigma)-\Phi(-s_1\sigma) \leqslant \Phi(t_2\sigma)-\Phi(-t_1\sigma)+2\epsilon.$$

By means of this last inequality, relation (5.19) becomes as follows

$$\limsup P_\theta[T_n \in j_n(\theta;t_1,t_2)] \leqslant B(\theta;t_1,t_2)+2\epsilon \quad \text{a.e.} \ [l].$$

Finally, letting $\epsilon \to 0$, we obtain the desired result. ∎

In Theorems 4.1, 4.2 and 5.1, it is a basic condition that $\{\mathscr{L}[(T_n-\theta)|P_\theta]\}$ converges (weakly) to a probability measure. However, this condition is a rather arbitrary restriction. Other criteria for defining a class of estimates could be used as well. Such an alternative criterion used by Pfanzagl [1] is *median-unbiasedness*. That is, the class $\mathscr{C}^*$, to be denoted here by $\mathscr{C}_{P'}$, is the following one.

$$\mathscr{C}_{P'} = \{\{T_n\}; P_\theta(T_n \geqslant \theta) \geqslant \tfrac{1}{2}, P_\theta(T_n \leqslant \theta) \geqslant \tfrac{1}{2}$$

$$\text{for all } \theta \in \Theta \text{ and all } n\}. \qquad (5.20)$$

Now we can prove the second main result in this section.

THEOREM 5.2 Let $\mathscr{C}_{P'}$, $J_n(\theta)$ and $B(\theta;t_1,t_2)$ be defined by (5.20), (5.6) and (4.10), respectively. Then one has

$$\limsup P_\theta[T_n \in j_n(\theta;t_1,t_2)] \leqslant B(\theta;t_1,t_2)$$

$$\text{for all } t_1,t_2 > 0 \text{ and all } \theta \in \Theta.$$

Proof The proof is based on the same ideas as those used in Theorem 5.1. That is, the utilization of (5.1) and (5.2) and the use of the Neyman–Pearson fundamental lemma. More precisely, let $\delta, h > 0$ be arbitrary and set $\theta_n = \theta + hn^{-\frac{1}{2}}$. Then the assumption of median-unbiasedness implies that for all n, one has $P_{\theta_n}(T_n \geqslant \theta_n) \geqslant \tfrac{1}{2}$, or, equivalently,

$$P_{\theta_n}(T_n < \theta_n) \leqslant \tfrac{1}{2}. \qquad (5.21)$$

Next, applying (5.2) with $x = (1+\delta)\, h\sigma$, one has

$$P_{\theta_n}\left[\frac{\Lambda_n + \tfrac{1}{2}h^2\sigma^2}{h\sigma} \leqslant (1+\delta)\, h\sigma\right] \to \Phi(\delta h\sigma) > \tfrac{1}{2}.$$

Therefore, for all sufficiently large n, $n \geqslant n_1$, say, we have

$$P_{\theta_n}\left[\frac{\Lambda_n + \tfrac{1}{2}h^2\sigma^2}{h\sigma} \leqslant (1+\delta)\, h\sigma\right] > \frac{1}{2}. \qquad (5.22)$$

Utilizing relations (5.21) and (5.22) and also the Neyman–Pearson fundamental lemma in the same way it was done in the course of the proof of Theorem 5.1, one has that for all sufficiently large n, $n \geqslant n_2$, say, and all $\theta \in \Theta$,

$$P_\theta(T_n < \theta_n) < P_\theta\left[\frac{\Lambda_n + \tfrac{1}{2}h^2\sigma^2}{h\sigma} \leqslant (1+\delta)\, h\sigma\right].$$

Letting first $n \to \infty$ and then $\delta \to 0$ and replacing h by t_2 in the resulting limits, one has

$$\limsup P[n^{\frac{1}{2}}(T_n - \theta) < t_2] \leqslant \Phi(t_2\sigma) \quad (t_2 > 0). \qquad (5.23)$$

Next taking $\theta_n = \theta - hn^{-\frac{1}{2}}$, $h > 0$, and $x = -\delta h\sigma$, $\delta > 0$, we obtain in a similar way the following result

$$\Phi(-t_1\sigma) \leqslant \liminf P_\theta[n^{\frac{1}{2}}(T_n - \theta) \leqslant -t_1] \quad (t_1 > 0). \qquad (5.24)$$

Therefore, by means of (5.23) and (5.24), one has

$$\limsup P_\theta[T_n \in j_n(\theta; t_1, t_2)] = \limsup P_\theta[-t_1 < n^{\frac{1}{2}}(T_n - \theta) < t_2]$$
$$\leqslant \limsup P_\theta[n^{\frac{1}{2}}(T_n - \theta) < t_2] - \liminf P_\theta[n^{\frac{1}{2}}(T_n - \theta) \leqslant -t_1]$$
$$= \Phi(t_2\sigma) - \Phi(-t_1\sigma) = B(\theta; t_1, t_2),$$

as was to be shown. ∎

This section is closed with a remark and two examples.

REMARK 5.1 It is to be pointed out that, as follows from Theorems 4.1, 4.2, 5.1 and 5.2, the variance $\sigma^2(\theta) = 4\mathscr{E}_\theta[\dot\phi_1(\theta)]^2$ plays the same role as Fisher's information number $I(\theta)$ under the standard conditions (of pointwise differentiability, etc.).

The following example shows that a sequence of estimates $\{V_n\}$ which is optimal with respect to $\mathscr{C}^*$ need not belong to $\mathscr{C}^*$.

EXAMPLE 5.1   Refer to Example 3.2 in Chapter 2. Then $\sigma^2(\theta) = \dfrac{1}{2\theta^2}$, as was seen in Example 1.2. Furthermore, with $S_n^2 = \dfrac{1}{n} \sum\limits_{j=1}^{n} (X_j - \mu)^2$, it follows that

$$\mathscr{L}[n^{\frac{1}{2}}(S_n^2 - \theta)|P_\theta] \Rightarrow N(0, 2\theta^2)$$

(see, e.g. Cramér [1], Sec. 28.3). Thus the upper bound $B(\theta; t_1, t_2)$ given by (4.10) is attained for $\{V_n\} = \{S_n^2\}$ in the sense of (1.5). Accordingly, $\{S_n^2\}$ is optimal with respect to all $\mathscr{C}_S$, $\mathscr{C}_P$ and $\mathscr{C}_{P'}$. However, $\{S_n^2\} \notin \mathscr{C}_{P'}$, because $P_\theta(S_n^2 \geq \theta) = P_\theta(\chi_n^2 \geq n)$ and therefore $P_\theta(S_n^2 \geq \theta) \geq \frac{1}{2}$ and $P_\theta(S_n^2 \leq \theta) \geq \frac{1}{2}$ cannot be simultaneously true. On the other hand, it is clear that $\{S_n^2\} \in \mathscr{C}_P$.

Furthermore, $\{S_n^2\} \in \mathscr{C}_S$ because

$$\frac{n^{\frac{1}{2}}(S_n^2 - \theta)}{\sqrt{2}} = \frac{\sum\limits_{j=1}^{n} Y_j}{\sigma_\theta\left(\sum\limits_{j=1}^{n} Y_j\right)},$$

where $Y_j = (X_j - \mu)^2 - \theta$ $(j = 1, ..., n)$. Next, $Y_1 = \theta(Z - 1)$ with Z being χ_1^2, so that $\mathscr{E}_\theta Y_1 = 0$, $\sigma_\theta^2 Y_1 = 2\theta^2$ and $\mathscr{E}_\theta Y_1^3 = 8\theta^3$ (since $\mathscr{E}Z = 1$, $\sigma^2 Z = 2$ and $\mathscr{E}_\theta Z^3 = 15$). Let F_n^* be the d.f. of

$$\left[\sum_{j=1}^{n} Y_j / \sigma_\theta\left(\sum_{j=1}^{n} Y_j\right)\right].$$

Then the Berry–Esseen theorem (see Loève [1], p. 288) gives

$$|F_n^*(x) - \Phi(x)| \leq \frac{cn8\theta^3}{[\theta\sqrt{(2n)}]^3} = \frac{4c}{\sqrt{(2n)}} \to 0 \quad \text{uniformly in } \theta,$$

where c is a numerical constant.

It follows that

$$\mathscr{L}\left[\frac{n^{\frac{1}{2}}(S_n^2 - \theta)}{\theta\sqrt{2}} \,\middle|\, P_\theta\right] \Rightarrow N(0, 1) \quad \text{uniformly in } \theta,$$

which implies that

$$\mathscr{L}[n^{\frac{1}{2}}(S_n^2 - \theta)|P_\theta] \Rightarrow N(0, 2\theta^2) \quad \text{uniformly in } \theta.$$

This result, together with Lemma 2.2 shows that this last convergence is continuous, as is required in the defining relation (4.8).

EXAMPLE 5.2 Refer to Example 3.3 in Chapter 2. Here $\sigma^2(\theta) = 1$, as was seen in Example 1.3. Furthermore, if V_n stands for the sample median, it was seen in Example 4.3 that

$$\mathscr{L}[n^{\frac{1}{2}}(V_n - \theta)|P_\theta] \Rightarrow N(0, 1).$$

It follows that the upper bound $B(\theta; t_1, t_2)$ is attained again for the sequence $\{V_n\}$. It is then immediate that $\{V_n\}$ is optimal within $\mathscr{C}_P$ and $\mathscr{C}_{P'}$ and also within $\mathscr{C}_S$ (see also the exercise at the end of this chapter).

The references Weiss and Wolfowitz [1], [2] and Hájek [3] are of related interest.

6 Asymptotic efficiency of estimates: the classical approach

The classical set-up for investigating asymptotic efficiency of estimates is the following. First the class $\mathscr{C}^*$ is specified by

$$\mathscr{C}^* = \{\{T_n\}; \mathscr{L}[n^{\frac{1}{2}}(T_n - \theta)|P_\theta] \Rightarrow N(0, \sigma_T^2(\theta))\} \quad \text{for all } \theta \in \Theta.$$

Next it is shown that, under suitable regularity conditions, one has for all $\{T_n\} \in \mathscr{C}^*$

$$\sigma_T^2(\theta) \geqslant I^{-1}(\theta) \quad \text{for all (a.a.}[l]) \quad (\theta \in \Theta). \tag{6.1}$$

Then any sequence of estimates $\{V_n\}$ in $\mathscr{C}^*$ for which $\sigma_V^2(\theta) = I^{-1}(\theta)$ for all $\theta \in \Theta$ is said to be *asymptotically efficient*. Under appropriate conditions, a sequence of MLEs enjoys this property.

In connection with the classical approach, the reader may wish to consult, e.g. Bahadur [1] and LeCam [2]. Another approach is presented in Rao [2]. In particular, for the Markov case, existence, consistency and asymptotic normality of an MLE is studied in Billingsley [1] and references cited there. A set of conditions for consistency of an MLE different from the ones used in the reference last cited, are presented in Roussas [3]. These conditions suffice when we restrict ourselves to compact subsets of Θ. Also of relevant interest is a paper by Gänssler [1]. Finally, conditions for asymptotic normality of an MLE weaker than those used in Billingsley are given in Roussas [4].

158 *Asymptotic efficiency of estimates*

What we intend to do in the present section is to show that the classical result (6.1) follows from theorems obtained in Sections 4 and 5 for two important specifications of the class $\mathscr{C}^*$. In this context, then, classical efficiency is a special case of W-efficiency.

Define two classes of sequences of estimates as follows.

$$\mathscr{C}_C = \{\{T_n\};\, P_\theta[n^{\frac{1}{2}}(T_n-\theta) \leqslant x] \to \Phi_T(x;\theta) \quad \text{for all}$$

$$x \in R \text{ and } \theta \in \Theta, \text{ where } \Phi_T(\cdot;\theta) \text{ is the d.f. of}$$

$$N(0, \sigma_T^2(\theta))\} \tag{6.2}$$

and

$$\mathscr{C}_{C'} = \{\{T_n\};\, P_\theta[n^{\frac{1}{2}}(T_n-\theta) \leqslant x] \to \Phi_T(x;\theta) \text{ } continuously$$

$$\text{in } \Theta \text{ for each } x \in R, \text{ where } \Phi_T(\cdot;\theta) \text{ is as above}\}. \tag{6.3}$$

Then the following theorem holds true.

THEOREM 6.1 With the classes $\mathscr{C}_C$ and $\mathscr{C}_{C'}$ defined by (6.2) and (6.3), respectively, one has

(i) $\sigma_T(\theta) \geqslant 1/\sigma(\theta)$ a.e. $[l]$, and
(ii) $\sigma_T(\theta) \geqslant 1/\sigma(\theta)$ for every $\theta \in \Theta$,

where $\sigma(\theta)$ is given by (5.3).

Proof (i) Consider the definition of the class $\mathscr{C}_P$ in (5.4). In the present case, $\Phi_T(0;\theta) = \frac{1}{2}$ for all $\theta \in \Theta$. Therefore $\mathscr{C}_C \subseteq \mathscr{C}_P$. Since the limiting d.f. $\Phi_T(\cdot;\theta)$ is also continuous, the conclusion of Theorem 5.1 holds with the lim sup replaced by lim. That is, by setting $\sigma = \sigma(\theta)$, one has

$$\lim P_\theta[-t_1 \leqslant n^{\frac{1}{2}}(T_n-\theta) < t_2]$$

$$\leqslant \Phi(t_2\sigma)-\Phi(-t_1\sigma) \quad \text{for all } t_1, t_2 > 0 \text{ and a.a. } [l] \quad (\theta \in \Theta). \tag{6.4}$$

Since $\Phi_T(x;\theta) = \Phi[x\sigma_T^{-1}(\theta)]$, relation (6.4) becomes

$$\Phi(t_2\sigma_T^{-1})-\Phi(-t_1\sigma_T^{-1})$$

$$\leqslant \Phi(t_2\sigma)-\Phi(-t_1\sigma) \quad \text{for all } t_1, t_2 > 0 \text{ and a.a. } [l] \quad (\theta \in \Theta), \tag{6.5}$$

where we set $\sigma_T = \sigma_T(\theta)$.

Letting $t_1 \to \infty$ in (6.5), we obtain the desired result in an obvious manner.

(ii) Consider the definition of the class $\mathscr{C}_S$ in (4.8). In the present case, the limiting d.f. is continuous in $x \in R$ for each fixed $\theta \in \Theta$. Therefore, by Polya's lemma and Lemma 2.2, the convergence in (6.3) is also continuous in R for each fixed $\theta \in \Theta$. Hence $\mathscr{C}_{C'} \subseteq \mathscr{C}_S$. Since also $l_T(\theta) = u_T(\theta) = 0$ for every $\theta \in \Theta$ and all $\{T_n\} \in \mathscr{C}_{C'}$, Theorem 4.1 implies that

$$\lim P_\theta[-t_1 < n^{\frac{1}{2}}(T_n - \theta) \leqslant t_2]$$

$$\leqslant \Phi(t_2\sigma) - \Phi(-t_1\sigma) \quad \text{for all } t_1, t_2 > 0 \text{ and every } \theta \in \Theta. \quad (6.6)$$

From (6.6), we obtain the desired result as in (i). ▌

The theorem just proved also provides us with a lower bound for the asymptotic squared error (see Bahadur [1], Proposition 3). Namely,

COROLLARY 6.1 One has
(i) $\liminf [n\mathscr{E}_\theta(T_n - \theta)^2] \geqslant 1/\sigma^2(\theta)$ a.e. $[l]$ if $\{T_n\} \in \mathscr{C}_C$,
and
(ii) $\liminf [n\mathscr{E}_\theta(T_n - \theta)^2] \geqslant 1/\sigma^2(\theta)$ for all $\theta \in \Theta$ if $\{T_n\} \in \mathscr{C}_{C'}$.

Proof Since $Y \geqslant 0$ implies $\mathscr{E}Y = \int_0^\infty [1 - F_Y(y)] \, \mathrm{d}y$, we have

$$n\mathscr{E}_\theta(T_n - \theta)^2 = n \int_0^\infty P_\theta[(T_n - \theta)^2 > x] \, \mathrm{d}x$$

$$= \int_0^\infty P_\theta[(T_n - \theta)^2 > x] \, \mathrm{d}(nx)$$

$$= \int_0^\infty P_\theta[(T_n - \theta)^2 > xn^{-1}] \, \mathrm{d}x$$

$$= \int_0^\infty P_\theta\{[n^{\frac{1}{2}}(T_n - \theta)]^2 > x\} \, \mathrm{d}x$$

$$= \int_0^\infty \{P_\theta[n^{\frac{1}{2}}(T_n - \theta) > \sqrt{x}]$$

$$+ P_\theta[n^{\frac{1}{2}}(T_n - \theta) < -\sqrt{x}]\} \, \mathrm{d}x.$$

Since the integrand in the last expression above is non-negative, Fatou's lemma applies and gives by means of (6.2) and (6.3)

$$\liminf [n\mathscr{E}_\theta(T_n-\theta)^2] \geqslant \int_0^\infty \left\{ \left[1 - \Phi\left(\frac{\sqrt{x}}{\sigma_T}\right) \right] + \Phi\left(-\frac{\sqrt{x}}{\sigma_T}\right) \right\} dx$$

$$= \int_0^\infty 2\Phi\left(-\frac{\sqrt{x}}{\sigma_T}\right) dx. \qquad (6.7)$$

But

$$\int_0^\infty \Phi\left(-\frac{\sqrt{x}}{\sigma_T}\right) dx = \int_0^{-\infty} \Phi(x)\, 2\sigma_T^2 x\, dx = -2\sigma_T^2 \int_{-\infty}^0 x\Phi(x)\, dx.$$

An integration by parts yields

$$\int_{-\infty}^0 x\Phi(x)\, dx = -\tfrac{1}{4},$$

so that

$$\int_0^\infty \Phi\left(-\frac{\sqrt{x}}{\sigma_T}\right) dx = \tfrac{1}{2}\sigma_T^2.$$

Thus (6.7) becomes

$$\liminf [n\mathscr{E}_\theta(T_n-\theta)^2] \geqslant \sigma_T^2.$$

Then the (i) and (ii) of the corollary follow from (i) and (ii), respectively, of the theorem. ∎

The present chapter is closed with a result referring to the multiparameter case and discussed in Bahadur [1] and Roussas [5].

7   Classical efficiency of estimates: the multiparameter case

In this section, we revert to the case that Θ is a k-dimensional subset of R^k. The class of estimates to be considered here is defined as follows, where inequalities are to be understood in the coordinatewise sense.

$$\mathscr{C}_{C''} = \{\{T_n\};\ P_\theta[n^{\frac{1}{2}}(T_n-\theta) \leqslant z] \to \Phi^{(k)}(z\,;C_T(\theta))$$

continuously in Θ for each $z \in R^k$, where $\Phi^{(k)}$

is the d.f. of $N(0, C_T(\theta))$ and $C_T(\theta)$ is assumed

to be positive definite for all $\theta \in \Theta\}.$ $\qquad (7.1)$

Then by recalling that $\Gamma(\theta) = 4\mathscr{E}_\theta[\phi_1(\theta)\,\phi_1'(\theta)]$, one has the following result.

THEOREM 7.1 Let $\mathscr{C}_{C''}$ be defined by (7.1). Then the following is true:

$$C_T(\theta) - \Gamma^{-1}(\theta) \quad \text{is positive semi-definite a.e. } [l^k]. \quad (7.2)$$

Clearly, the conclusion of the theorem is a direct generalization of the first conclusion in Theorem 6.1. The following important fact is also a consequence of (7.2).

COROLLARY 7.1 Let g be a real-valued function defined on R^k and suppose that $\dfrac{\partial}{\partial \theta_j} g(\theta), j = 1, \ldots k$ exist and are continuous in Θ. Let $\mathscr{C}_{C'', g}$ be a class of sequences of estimators $\{S_n\}$ of $g(\theta)$ defined as follows

$$\mathscr{C}_{C'', g} = \{\{S_n\} ; S_n = g(T_n) \quad \text{with} \quad \{T_n\} \in \mathscr{C}_{C''}\}.$$

Then one has
(i) $\mathscr{L}\{n^{\frac{1}{2}}[S_n - g(\theta)] | P_\theta\} \Rightarrow N(0, \sigma_S^2(\theta))$ for all $\theta \in \Theta$, provided $\sigma_S^2(\theta) > 0$, and
(ii) $\sigma_S^2(\theta) \geqslant \bar{g}' \Gamma^{-1}(\theta) \bar{g}$ a.e. $[l^k]$,
where

$$\sigma_S^2(\theta) = \bar{g}' C_T(\theta) \bar{g} \quad \text{and} \quad \bar{g} = \left(\frac{\partial}{\partial \theta_1} g(\theta), \ldots, \frac{\partial}{\partial \theta_k} g(\theta) \right)'.$$

Proof (i) This is a well-known result and its proof can be found, e.g. in Rao [3], p. 321, (ii).
(ii) It follows immediately from (7.2). ∎
For the proof of Theorem 7.1, the following auxiliary result is needed.

LEMMA 7.1 For each n, T_n with $\{T_n\} \in \mathscr{C}_{C''}$ and $v \in R^k$, define the function $f_n(\cdot\,; v)$ on Θ as follows:

$$f_n(\theta\,; v) = P_{n, \theta}(v' T_n \leqslant v'\theta). \quad (7.3)$$

Then $f_n(\cdot\,; v)$ is $\mathscr{C}$-measurable (recall that $\mathscr{C}$ is the σ-field of Borel subsets of Θ).

Proof For an arbitrary but fixed $\theta_0 \in \Theta$ and any $A \in \mathscr{A}_n$, one has

$$P_{n, \theta}(A) = \int_A g(X_0, \ldots, X_n; \theta_0, \theta)\, \mathrm{d}P_{n, \theta_0},$$

where
$$q(X_0, \ldots, X_n ; \theta_0, \theta) = q(X_0 ; \theta_0, \theta) \prod_{j=1}^{n} \phi_j^2(\theta_0, \theta)$$

(see Chapter 2, Section 1).

From (A 4) (ii), it follows that $P_{n,\theta}(A)$ is $\mathscr{C}$-measurable for each n and each fixed $A \in \mathscr{A}_n$. That is, for each n, $P_{n,\cdot}(\cdot)$ is a transition probability (see, e.g. Definition III, 2.1, p. 73 in Neveu [1]). Next, let $B = \{(\theta, \omega); v'T_n \leqslant v'\theta\}$. Then it follows that $B \in \mathscr{C} \times \mathscr{A}_n$. Therefore by setting $Z_\theta(\omega) = I_B(\theta, \omega)$, it follows by the Tulcea theorem (see, e.g. Neveu [1], p. 74) that

$$Y(\theta) = \int_\Omega P_{n,\theta}(\mathrm{d}\omega)\, Z_\theta(\omega)$$

is $\mathscr{C}$-measurable. Since, clearly,

$$\int_\Omega P_{n,\theta}(\mathrm{d}\omega)\, Z_\theta(\omega) = P_{n,\theta}(v'T_n \leqslant v'\theta),$$

the desired result follows from (7.3). ∎

Proof of Theorem 7.1 The convergence in (7.1) implies that $P_\theta[n^{\frac{1}{2}}v'(T_n - \theta) \leqslant 0] \to \frac{1}{2}$ and hence $f_n(\theta ; v) \to \frac{1}{2}$. For each n, define on R^k the function $f_n^*(\cdot ; v)$ as follows

$$f_n^*(\theta ; v) = \begin{cases} |f_n(\theta ; v) - \frac{1}{2}| & \text{if } \theta \in \Theta \\ 0 & \text{otherwise.} \end{cases} \qquad (7.4)$$

Then, clearly, $0 \leqslant f_n^*(\theta ; v) \leqslant \frac{1}{2}$ for all n and $\theta \in \Theta$, and also $f_n^*(\theta ; v) \to 0$, $\theta \in \Theta$. Thus the assumptions of Lemma 2.4 are satisfied. For $h \in R^k$, define the sequence $\{z_n(h)\}$ as follows: $z_n(h) = hn^{-\frac{1}{2}}$. Then, by Lemma 2.4, it follows that there exists a subsequence $\{m\} \subseteq \{n\}$ and an l^k-null set of θs (both of which may depend on $\{z_n(h)\}$ through h and also on v) such that

$$f_m^*(\theta + hm^{-\frac{1}{2}}; v) \to 0 \quad \text{for all } \theta \notin N(h, v), \qquad (7.5)$$

where $N(h, v)$ is the exceptional l^k-null set.

By means of (7.4), the convergence in (7.5) is equivalent to

$$f_m(\theta + hm^{-\frac{1}{2}}; v) \to \tfrac{1}{2} \quad \text{for all } \theta \notin N(h, v);$$

or, on account of (7.3) and by setting $\theta_m = \theta + hm^{-\frac{1}{2}}$, we have

$$P_{\theta_m}[m^{\frac{1}{2}}v'(T_m - \theta) > v'h] \to \tfrac{1}{2} \quad \text{for all } \theta \notin N(h, v).$$

Letting R_0^k be the set of rational numbers in R^k and setting $N_0 = \cup\,[N(h, v);\, h, v \in R_0^k]$, we have $l^k(N_0) = 0$ and

$$P_{\theta_m}[m^{\frac{1}{2}}v'(T_m - \theta) > v'h] \to \tfrac{1}{2} \quad \text{for every } \theta \notin N_0 \text{ and all } h, v \in R_0^k. \tag{7.6}$$

Use again the notation $\Lambda_m = \Lambda(\theta, \theta_m)$ and let

$$\tau^2 = \tau^2(h, \theta) = h'\Gamma(\theta)h.$$

The remaining part of the proof runs parallel to the lines of the corresponding part of the proof of Theorem 5.1 and also the proof of Lemma 5.1. An outline of it is as follows. In the first place we take h as above and $\neq 0$, so that $\tau \neq 0$. Then Theorems 4.3 and 4.5 in Chapter 2 imply in an obvious way that for $x \in R$,

$$P_\theta\left(\frac{\Lambda_m + \frac{1}{2}\tau^2}{\tau} \leqslant x\right) \to \Phi(x) \tag{7.7}$$

and $$P_{\theta_m}\left(\frac{\Lambda_m + \frac{1}{2}\tau^2}{\tau} \leqslant x\right) \to \Phi(x - \tau). \tag{7.8}$$

From (7.8) and for an arbitrary $\delta > 0$, one has

$$P_{\theta_m}\left[\frac{\Lambda_m + \frac{1}{2}\tau^2}{\tau} > (1+\delta)\tau\right] \to 1 - \Phi(\delta\tau), \tag{7.9}$$

so that $$1 - \Phi(\delta\tau) < \tfrac{1}{2}. \tag{7.10}$$

Working with $\theta \notin N_0$ and $h, v \in R_0^k$, relations (7.6), (7.9) and (7.10) imply that for all sufficiently large m,

$$P_{\theta_m}\left[\frac{\Lambda_m + \frac{1}{2}\tau^2}{\tau} > (1+\delta)\tau\right] < P_{\theta_m}[m^{\frac{1}{2}}v'(T_m - \theta) > v'h]. \tag{7.11}$$

At this point one employs the Neyman–Pearson fundamental lemma in a way analogous to the one used in the proof of Lemma 5.1 in order to conclude from (7.11) that

$$P_\theta\left[\frac{\Lambda_m + \frac{1}{2}\tau^2}{\tau} > (1+\delta)\tau\right] < P_\theta[m^{\frac{1}{2}}v'(T_m - \theta) > v'h]$$
$$\text{for all sufficiently large } m. \tag{7.12}$$

Letting $m \to \infty$ in (7.12) and utilizing (7.1) and (7.7), one has

$$1 - \Phi[(1+\delta)\tau] \leqslant 1 - \Phi[v'h(v'C_T(\theta)v)^{-\frac{1}{2}}],$$

or $$\Phi[(1+\delta)\,\tau] \geqslant \Phi[v'h(v'C_T(\theta)\,v)^{-\frac{1}{2}}]. \qquad (7.13)$$

Letting $\delta \to 0$ in (7.13) and replacing τ by $(h'\Gamma(\theta)\,h)^{\frac{1}{2}}$, it follows that
$$\Phi[(h'\Gamma(\theta)\,h)^{\frac{1}{2}}] \geqslant \Phi[v'h(v'C_T(\theta)\,v)^{-\frac{1}{2}}],$$

so that, finally

$(h'\Gamma(\theta)\,h)^{\frac{1}{2}} \geqslant v'h(v'C_T(\theta)\,v)^{-\frac{1}{2}}$ a.e. $[l^k]$ and for all $v, h \in R_0^k$.

Then Corollary 3.1 A gives

$C_T(\theta) - \Gamma^{-1}(\theta)$ is positive semi-definite a.e. $[l^k]$,

as was to be shown. ∎

For $\{T_n\}$ in $\mathscr{C}_{C''}$, one also has the following result.

PROPOSITION 7.1   Let $\mathscr{C}_{C''}$ be defined by (7.1). Then C_T is continuous in Θ (in any one of the usual norms for matrices).

Proof   Clearly, the continuity of C_T is equivalent to that of $h'C_T h$ for every $h \in R^k$. Fix an arbitrary $h \in R$ and set

$$\tau(\theta) = \tau(\theta, h) = (h'C_T(\theta)\,h)^{\frac{1}{2}}.$$

For this h and each $x \in R$, (7.1) gives

$$P_\theta[n^{\frac{1}{2}}h'(T_n - \theta) \leqslant x] \to \Phi[x/\tau(\theta)]   \text{continuously in } \theta \in \Theta.$$

Then Lemma 2.1 implies that $\Phi[x/\tau(\theta)]$ is continuous on Θ as a function of θ for each $x \in R$. (Here is where the continuous convergence in (7.1) is utilized.) In particular, this is true for $x = 1$. That is, $\Phi[1/\tau(\theta)]$ is a continuous function of θ on Θ. Also $z = \Phi(y)$ is continuous as a function of y and then so is its inverse $y = \Phi^{-1}(z)$.

Since $1/\tau(\theta) = \Phi^{-1}\{\Phi[1/\tau(\theta)]\}$, the continuity of $1/\tau(\theta)$, and hence that of $\tau(\theta)$ follows. ∎

By the proposition just established, Theorem 7.1 has the following corollary.

COROLLARY 7.2   Let $\mathscr{C}_{C''}$ be defined by (7.1) and suppose that $\Gamma(\theta)$ is continuous. Then

$C_T(\theta) - \Gamma^{-1}(\theta)$ is positive semi-definite for all $\theta \in \Theta$.

In particular, this is true if there exists $\{V_n\} \in \mathscr{C}_{C''}$ such that

$$\mathscr{L}[n^{\frac{1}{2}}(V_n - \theta)|P_\theta] \Rightarrow N(0, \Gamma^{-1}(\theta)). \qquad (7.14)$$

Proof The statement in (7.2) is equivalent to the inequality

$$h'[C_T(\theta) - \Gamma^{-1}(\theta)]h \geqslant 0 \quad \text{a.e. } [l^k] \text{ for every } h \in R^k. \quad (7.15)$$

Let $\Gamma(\theta)$ be continuous. Then so is $\Gamma^{-1}(\theta)$. Let $\theta \in N_0$, the exceptional l^k-null set. Then there exists $\{\theta_n\}$ with $\theta_n \in N_0^c$ for all n such that $\theta_n \to \theta$. Replacing θ by θ_n in (7.15) and letting $n \to \infty$, we obtain the desired result. If (7.14) holds true then $\Gamma(\theta)$ is continuous by Proposition 7.1, and hence the previous argument applies.]

Since $h'\Gamma(\theta)h = 4\mathscr{E}_\theta[h'\dot{\phi}_1(\theta)]^2$, it follows that $\Gamma(\theta)$ is continuous if for every $h \in R^k$, $\mathscr{E}_\theta[h'\dot{\phi}_1(\theta)]^2$ is continuous. The proposition to be established below also guarantees the continuity of $\Gamma(\theta)$ under (A1–4) and an additional assumption (A4′) to be formulated here. This assumption would be rather easy to check in practice.

ASSUMPTION (A4′) There exists $\delta > 0$ such that for each $\theta \in \Theta$ and each $h \in R^k$ there is a neighbourhood of θ, $v(\theta, h)$, with the property that

$$\mathscr{E}_{\theta*}|h'\dot{\phi}_1(\theta*)|^{2+\delta} \leqslant M(\theta, h) \, (<\infty) \quad \text{for all } \theta* \in v(\theta, h).$$

Then we have the following result.

PROPOSITION 7.2 Under the assumptions (A1–4) and (A4′), $\Gamma(\theta)$ is continuous on Θ.

Proof By Theorem 2.1A (with $r = 1$), assumption (A4) is equivalent to the convergence

$$q(X_0, X_1; \theta, \theta*) \to 1 \quad \text{in the first mean } [P_\theta] \text{ as } \theta* \to \theta,$$

and this in turn is equivalent to

$$\|P_{1,\theta*} - P_{1,\theta}\| \to 0 \quad \text{as} \quad \theta* \to \theta, \quad (7.16)$$

by Theorem 1.4A.

It is now claimed that

$$\mathscr{L}[h'\dot{\phi}_1(\theta*)|P_{\theta*}] \Rightarrow \mathscr{L}[h'\dot{\phi}_1(\theta)|P_\theta] \quad \text{as} \quad \theta* \to \theta. \quad (7.17)$$

In fact, let f be any real-valued, bounded and continuous function defined on R. One has then

$$\left| \int f[h'\phi_1(\theta^*)]\,\mathrm{d}P_{1,\theta^*} - \int f[h'\phi_1(\theta)]\,\mathrm{d}P_{1,\theta} \right|$$
$$\leq \int |f[h'\phi_1(\theta^*)] - f[h'\phi_1(\theta)]|\,\mathrm{d}P_{1,\theta}$$
$$+ \left| \int f[h'\phi_1(\theta^*)]\,\mathrm{d}(P_{1,\theta^*} - P_{1,\theta}) \right|$$
$$\leq \int |f[h'\phi_1(\theta^*)] - f[h'\phi_1(\theta)]|\,\mathrm{d}P_{1,\theta} + M\|P_{1,\theta^*} - P_{1,\theta}\|,$$

where $M(<\infty)$ is a bound for f. However, this last expression tends to zero as $\theta^* \to \theta$ because of (7.16), (A3)(i) and the dominated convergence theorem. Thus (7.17) holds true.

Next let F_{1,θ^*} and $F_{1,\theta}$ be the d.f.s of the r.v.s $h'\phi_1(\theta^*)$ and $h'\phi_1(\theta)$ under P_{1,θ^*} and $P_{1,\theta}$, respectively. Then by virtue of (A4'), one has

$$\mathscr{E}_{\theta^*}|h'\phi_1(\theta^*)|^{2+\delta}$$
$$= \int |x|^{2+\delta}\,\mathrm{d}F_{1,\theta^*} \leq M(\theta, h) \quad \text{for all } \theta^* \in v(\theta, h), \quad (7.18)$$

while (7.17) implies

$$F_{1,\theta^*} \Rightarrow F_{1,\theta} \quad \text{as} \quad \theta^* \to \theta. \quad (7.19)$$

By relations (7.17) and (7.18), the corollary on p. 184 in Loève [1] applies and gives

$$\int x^2\,\mathrm{d}F_{1,\theta^*} \to \int x^2\,\mathrm{d}F_{1,\theta} \quad \text{as} \quad \theta^* \to \theta,$$

which is equivalent to

$$h'\Gamma(\theta^*)h = \mathscr{E}_{\theta^*}[h'\phi_1(\theta^*)]^2 \to \mathscr{E}_\theta[h'\phi_1(\theta)]^2$$
$$= h'\Gamma(\theta)h \quad \text{as} \quad \theta^* \to \theta.$$

The proof is completed.∎

A brief account of most of the material discussed in this chapter can be found in Roussas [7].

Exercise

Refer to Example 5.2 and show that $\{V_n\} \in \mathscr{C}_{P'}$ and $\{V_n\} \in \mathscr{C}_S$, where $\mathscr{C}_{P'}$ and $\mathscr{C}_S$ are given by (5.20) and (4.8), respectively.

6. *Multiparameter asymtotically optimal tests*

1 Some notation and preliminary results

In the present chapter, we consider the problem of testing a simple hypothesis against a composite alternative in the framework described in Section 2 of Chapter 2 and under the assumption that the dimensionality of the parameter space Θ is greater than one. A sequence of tests having certain asymptotically optimal properties will be derived. In order to be able to formulate the problem and also state the relevant theorems, some preliminary results are needed. These are obtained in the present section and also in Sections 2 and 6.

As has been the practice in previous chapters, the dependence of various quantities on a fixed parameter point θ_0 will not be explicitly indicated. For instance, we shall write Δ_n, Δ_n^*, Γ, etc. rather than $\Delta_n(\theta_0)$, $\Delta_n^*(\theta_0)$, $\Gamma(\theta_0)$, etc., respectively.

The assumption has been made that the symmetric matrix Γ is positive definite. Therefore there exists a non-singular matrix M such that

$$M'M = \Gamma \tag{1.1}$$

(see Theorem 4.1 A). From (1.1), it immediately follows that

$$(M^{-1})' \, \Gamma M^{-1} = M\Gamma^{-1}M' = I, \tag{1.2}$$

where I is the $k \times k$ unit matrix.

Consider the following transformation of R^k onto itself

$$M(z - \theta_0) = v - \theta_0, \quad \text{where} \quad v = u + \theta_0 - M\theta_0 \quad \text{and} \quad u = Mz. \tag{1.3}$$

For $c > 0$, let $E(c)$ be the surface (of an ellipsoid) defined by

$$E(c) = \{z \in R^k; \ (z - \theta_0)' \, \Gamma(z - \theta_0) = c\}. \tag{1.4}$$

Then the transformation (1.3) sends the surface $E(c)$ onto the surface (of a sphere) $S(c)$, where

$$S(c) = \{z \in R^k;\ (z - \theta_0)'\,(z - \theta_0) = c\}. \tag{1.5}$$

For $z \in R^k$, with $z \neq \theta_0$, set $c(z) = (z - \theta_0)'\,\Gamma(z - \theta_0)$. Then this quantity is positive, since Γ is positive definite by assumption. By means of $E(c(z))$ and $S(c(z))$, define the function ξ as in Wald [2], p. 342. Namely, for any $\rho > 0$, define $\omega(z, \rho)$ by

$$\omega(z, \rho) = \{u \in E(c(z));\ \|u - z\| \leqslant \rho\}$$

and let $\omega'(z, \rho)$ be the image of the set $\omega(z, \rho)$ under the transformation (1.3). Also denote by $A(\omega(z, \rho))$ and $A(\omega'(z, \rho))$ the areas of the sets $\omega(z, \rho)$ and $\omega'(z, \rho)$, respectively. Then the function ξ is defined as follows

$$\xi(z) = \lim \frac{A(\omega'(z, \rho))}{A(\omega(z, \rho))} \quad \text{as} \quad \rho \to 0. \tag{1.6}$$

Thus one has the positive-valued function ξ defined on $R^k - \{\theta_0\}$ and it can be seen (see also Exercise 1) that its explicit form is

$$\xi(z) = \frac{[|\Gamma|\,(z - \theta_0)'\,\Gamma(z - \theta_0)]^{\frac{1}{2}}}{\|\Gamma(z - \theta_0)\|} \quad (z \neq \theta_0). \tag{1.7}$$

REMARK 1.1 The significance of the function ξ defined by (1.6) may be seen from the relation

$$\int_{E(c)} \xi(z)\,\mathrm{d}A = \text{area of } S(c), \quad \text{to be denoted by } A(c), \tag{1.8}$$

where $E(c)$ and $S(c)$ are defined by (1.4) and (1.5), respectively, and the integral in (1.8) is a surface integral (see also Exercise 2).

By setting

$$\zeta(c;\,z) = \frac{\xi(z)}{A(c)} \quad (z \in E(c)), \tag{1.9}$$

one obtains a weight function $\zeta(c;\,\cdot\,)$ (integrating to 1) over each one of the surfaces $E(c)$.

In the sequel, we will be interested in parameter points θ_n of the form
$$\theta_n = \theta_0 + hn^{-\frac{1}{2}} \quad (h \in R^k). \tag{1.10}$$

Also the hs, eventually, will be required to satisfy the condition $h'\Gamma h = c,\ c > 0$, so that $(\theta_n - \theta_0)'\,\Gamma(\theta_n - \theta_0) = cn^{-1}$. Set

$$E_{n,c} = E(cn^{-1}) = \{z \in R^k;\ (z - \theta_0)'\,\Gamma(z - \theta_0) = cn^{-1}), \tag{1.11}$$

and let
$$E^*(c) = \{z \in R^k; z' \Gamma z = c\}. \qquad (1.12)$$

The constant c appearing in (1.11) and (1.12) will eventually be chosen to lie in a compact subset K of $(0, \infty)$. Then it follows that for all sufficiently large $n, n \geqslant n(K)$, say, $E_{n,c} \subset \Theta$ for all $c \in K$. Thus, with θ_n being of the form (1.10), it follows that

$$\theta_n \in E_{n,c} \quad \text{if and only if} \quad h \in E^*(c)$$

and
$$E_{n,c} \subset \Theta, c \in K, n \geqslant n(K). \qquad (1.13)$$

REMARK 1.2   Let $\theta_n = \theta_0 + zn^{-\frac{1}{2}}$. Then from (1.7) it follows that $\xi(\theta_n) = \xi^*(z)$, where ξ^* is defined the same way as ξ except that θ_0 is replaced by zero. In particular, if θ_n is given by (1.10), then $\xi(\theta_n) = \xi^*(h)$, where $\theta_n \in E_{n,c}$, so that $h \in E^*(c)$ (see relations (1.11)–(1.13)).

Closing this section, we should like to mention that all results in the present chapter are derived under the basic assumptions (A 1–4) in Section 2, Chapter 2, although this will not always be stated explicitly.

2   Formulation of some of the main results

In the present chapter, the problem we are interested in is that of testing the hypothesis $H: \theta = \theta_0$, for some fixed parameter point θ_0, against the alternative $A: \theta \neq \theta_0$. To this end, let

$$\mathcal{L}_h = N(\Gamma h, \Gamma) \quad (h \in R^k), \qquad (2.1)$$

and define the set D as follows

$$D = \{z \in R^k; z' \Gamma^{-1} z \leqslant d\}, \quad \mathcal{L}_0(D) = 1 - \alpha \quad (0 < \alpha < 1). \qquad (2.2)$$

Also define the class $\bar{\mathscr{C}}$ by

$$\bar{\mathscr{C}} = \{C \in \mathscr{B}^k; C \text{ is closed and convex}\}, \qquad (2.3)$$

and set
$$\bar{\mathscr{F}} = \{\psi; \psi = I_{C^c}, C \in \bar{\mathscr{C}}\}. \qquad (2.4)$$

In particular, set
$$\phi = I_{D^c}. \qquad (2.5)$$

Then, clearly, $D \in \bar{\mathscr{C}}$, so that $\phi \in \bar{\mathscr{F}}$.

All tests herein will depend on the random vector Δ_n and for reasons to be explained in Section 3, we may confine ourselves to tests in $\bar{\mathscr{F}}$.

For $\psi \in \mathscr{F}$, that is $\psi = I_{C^c}$ for some $C \in \mathscr{C}$, set

$$\beta_n(\theta ; C^c) = \beta_n(\theta ; \psi) = \mathscr{E}_\theta \psi(\Delta_n). \tag{2.6}$$

REMARK 2.1 As will be seen in the theorems to be stated below, the sequence of tests $\{\phi(\Delta_n)\}$ possesses certain optimal asymptotic properties, where ϕ is defined by (2.5). Also with θ_n and $\mathscr{L}_h$ defined by (1.10) and (2.1), respectively, it will be shown later (see Lemma 4.1 (i)) that

$$\mathscr{E}_{\theta_0} \phi(\Delta_n) = P_{\theta_0}(\Delta_n \in D^c) \to \mathscr{L}_0(D^c).$$

Thus, if we decide to restrict attention to tests in the class $\mathscr{F}$ of asymptotic level of significance α – which we shall do – we must have $\mathscr{L}_0(D^c) = \alpha$. This fact provides the justification for the equation $\mathscr{L}_0(D) = 1 - \alpha$ employed in (2.2).

Unless otherwise explicitly specified, in all that follows and for each n, we shall consider only parameter points θ_n of the form $\theta_n = \theta_0 + hn^{-\frac{1}{2}}$ with $h \in E^*(c)$ or $h \in E_n^* = E^*(c_n)$, so that $\theta_n \in E_{n,c}$ or $\theta_n \in E_n = E_{n,c_n}$ (see (1.11) and (1.12) for the definition of $E_{n,c}$ and $E^*(c)$); here $\{c_n\}$ is a bounded sequence of positive numbers. Furthermore, although, θ_n – and also in certain instances h_n – would be a more appropriate notation, we shall simply write θ and h when no confusion is possible. It should be borne in mind, however, that $\theta = \theta_0 + hn^{-\frac{1}{2}}$, where $h \in E^*(c)$ or $h \in E_n^*$, so that $\theta \in E_{n,c}$ or $\theta \in E_n$, respectively. This will somewhat simplify an already cumbersome notation.

The first main result in this chapter is presented in the following theorem.

THEOREM 2.1 Let $\zeta_n = \zeta(cn^{-1} ; \cdot)$ and $E_{n,c}$ be defined by (1.9) and (1.11), respectively. Also let ϕ be given by (2.5) and let $\{\psi_n\}$ be any sequence of tests in $\mathscr{F}$ of asymptotic level of significance α. Then, under assumptions (A1–4), for testing H against A at symptotic level of significance α, one has

$$\liminf \left\{ \inf \left[\int_{E_{n,c}} \beta_n(\theta ; \phi) \, \zeta_n(\theta) \, dA \right. \right.$$

$$\left. \left. - \int_{E_{n,c}} \beta_n(\theta ; \psi_n) \, \zeta_n(\theta) \, dA \right] ; c \in K \right\} \geqslant 0, \tag{2.7}$$

where $\beta_n(\theta\,;\phi)$ and $\beta_n(\theta\,;\psi_n)$ are defined by (2.6) and K is an arbitrary compact subset of $(0,\infty)$.

(For a certain uniform (over the tests ψ) version of the result just stated and also its interpretation, the reader is referred to Theorem 4.1.)

The second main result herein is the following one.

THEOREM 2.2 Let $E_{n,c}, \phi, \{\psi_n\}, \beta_n(\theta\,;\phi)$ and $\beta_n(\theta\,;\psi_n)$ be as in Theorem 2.1. Then under assumptions (A 1–4), for testing H against A at asymptotic level of significance α, one has

(i) $\lim [\sup\{\sup[\beta_n(\theta\,;\phi);\ \theta\in E_{n,c}]$
$$- \inf[\beta_n(\theta\,;\phi);\ \theta\in E_{n,c}]\};\ c\in K] = 0,$$

where K is an arbitrary compact subset of $(0,\infty)$;

(ii) $\lim\inf[\inf\{\inf[\beta_n(\theta\,;\phi)-\beta_n(\theta\,;\psi_n);\ \theta\in E_{n,c}]\};\ c\in K] \geqslant 0$

for any tests ψ_n as described above and for which $\beta_n(\theta;\psi_n)$ satisfies (i).

The interpretation of the theorem is clear. Part (i) states that the power of the test ϕ on the surfaces $E_{n,c}$ is asymptotically constant. The second part asserts that, within the class of tests whose power on $E_{n,c}$ is asymptotically constant, the test ϕ is asymptotically most powerful.

The formulation (and proof) of the third main result in the present chapter is deferred to Section 6, since it requires substantial additional notation.

3 Restriction to the class of tests $\mathscr{F}$

Suppose that we are interested in testing the hypothesis $H\colon \theta = \theta_0$ against the alternative $A\colon \theta \neq \theta_0$ at asymptotic level of significance α. All tests are to be based on the random vector Δ_n and power is to be calculated under P_{θ_n}, where $\theta_n = \theta_0 + hn^{-\frac{1}{2}}$. This can be done without loss of generality by Theorem 5.1, Chapter 3. By Theorem 5.2, Chapter 3, one has that for any tests ψ_n (not necessarily in $\mathscr{F}$) and any bounded subset of R^k, B,

$$\sup\left[\left|\mathscr{E}_{\theta_n}\psi_n(\Delta_n) - \mathscr{E}_{\theta_n}\psi_n(\Delta_n^*)\right|;\ h\in B\right] \to 0,$$

where Δ_n^* is a suitable truncated version of Δ_n defined by (1.5) in Chapter 3. Therefore from an asymptotic point of view, it suffices to base any tests on the random vector Δ_n^* rather than Δ_n. This is so regardless of whether we are interested in pointwise power or average power (see Theorem 2.1 and also Exercise 3). Power is still to be calculated under P_{θ_n}. Next, by virtue of (1.6) and (1.7), Chapter 3, one has

$$[dR_{n,h}/dP_{\theta_0}] = \exp[-B_n(h) + h'\Delta_n^*], \tag{3.1}$$

where $\exp B_n(h) = \mathscr{E}_{\theta_0}(\exp h'\Delta_n^*) \quad (h \in R^k)$.

In the family of probability measure $R_{n,h}$, the parameter is h and its range is all of R^k. Since for any $\theta \in \Theta$, $\theta = \theta_0 + hn^{-\frac{1}{2}}$ for some $h \in R^k$, namely, $h = n^{\frac{1}{2}}(\theta - \theta_0)$, it follows that $\theta = \theta_0$ if and only if $h = 0$. For $h = 0$, it follows from (3.1) that $R_{n,0} = P_{\theta_0}$.

By Corollary 4.1, Chapter 3,

$$\sup(\|P_{\theta_n} - R_{n,h}\|; h \in B) \to 0,$$

where B is any bounded subset of R^k. Thus for any tests ψ_n, one has

$$\sup\{|\mathscr{E}_{\theta_n}\psi_n(\Delta_n^*) - \mathscr{E}[\psi_n(\Delta_n^*)|R_{n,h}]|; h \in B\}$$
$$\leqslant \sup(\|P_{\theta_n} - R_{n,h}\|; h \in B) \to 0.$$

Therefore it follows that, from asymptotic power viewpoint, power may be calculated under $R_{n,h}$ rather than P_{θ_n}. Furthermore, this is true regardless of whether our interest lies in pointwise power or average power in the sense of Theorem 2.1. (See Exercise 4.)

In order to summarize: the original testing hypothesis problem $H: \theta = \theta_0$ against $A: \theta \neq \theta_0$ at asymptotic level of significance α, where tests are to be based on the random vector Δ_n and power is to be calculated under P_{θ_n}, may be replaced by the equivalent testing hypothesis problem $H_0^*: h = 0$ against $A^*: h \neq 0$ at asymptotic level of significance α in connection with the family of probability measures defined by (3.1), where tests are to be based on the random vector Δ_n^* and power is to be calculated under $R_{n,h}$.

Now we introduce the random vector Z_n^*, where

$$Z_n^* = \Gamma^{-1}\Delta_n^*. \tag{3.2}$$

Also the following notation is needed:

$$\mathscr{L}^*_{n,\theta} = \mathscr{L}(\Delta^*_n | P_\theta), \theta \in \Theta, \quad \mathscr{L}^{**}_{n,h} = \mathscr{L}(\Delta^*_n | R_{n,h}) \quad (h \in R^k) \quad (3.3)$$

and $$L^*_{n,h} = \mathscr{L}(Z^*_n | R_{n,h}) \quad (h \in R^k). \quad (3.4)$$

The following result holds true.

LEMMA 3.1   (i) Let $\mathscr{L}^*_{n,\theta_0}$ and $\mathscr{L}^{**}_{n,h}$ be defined by (3.3). Then for every $h \in R^k$, one has

$$\mathscr{L}^{**}_{n,h} \ll \mathscr{L}^*_{n,\theta_0}$$

and $$[d\mathscr{L}^{**}_{n,h} / d\mathscr{L}^*_{n,\theta_0}] = \exp[-B_n(h) + h'z], z \in R^k.$$

(ii) Let $L^*_{n,h}$ be defined by (3.4). Then for every $h \in R^k$, one has

$$L^*_{n,h} \ll L^*_{n,0}$$

and $$[dL^*_{n,h} / dL^*_{n,0}] = \exp[-B_n(h) + h'\Gamma z], z \in R^k.$$

Proof   (i) For $A \in \mathscr{B}^k$ and by virtue of (3.3) and (3.1), one has

$$\mathscr{L}^{**}_{n,h}(A) = R_{n,h}(\Delta^*_n \in A) = \int_{(\Delta^*_n \in A)} \exp[-B_n(h) + h'\Delta^*_n] dP_{\theta_0}$$

$$= \int_A \exp[-B_n(h) + h'z] d\mathscr{L}(\Delta^*_n | P_{\theta_0})$$

$$= \int_A \exp[-B_n(h) + h'z] d\mathscr{L}^*_{n,\theta_0},$$

as was asserted.

(ii) With A as above and by virtue of (3.1), (3.2) and (3.3), one has

$$L^*_{n,h}(A) = R_{n,h}(Z^*_n \in A) = \int_{(Z^*_n \in A)} \exp[-B_n(h) + h'\Delta^*_n] dP_{\theta_0}$$

$$= \int_{(Z^*_n \in A)} \exp[-B_n(h) + h'\Gamma Z^*_n] dP_{\theta_0}$$

$$= \int_A \exp[-B_n(h) + h'\Gamma z] d\mathscr{L}(Z^*_n | P_{\theta_0})$$

$$= \int_A \exp[-B_n(h) + h'\Gamma z] d\mathscr{L}(Z^*_n | R_{n,0})$$

because $R_{n,0} = P_{\theta_0}$ by means of (3.1). Now, since

$$\mathscr{L}(Z^*_n | R_{n,0}) = L^*_{n,0},$$

we have
$$L_{n,h}^{*}(A) = \int_{A} \exp\left[-B_n(h) + h'\Gamma z\right] dL_{n,0}^{*},$$
as was asserted.]

From (3.2), it follows that, in testing the hypothesis last described, our tests may be based on the random vector Z_n^{*} rather than Δ_n^{*}.

Now for each n, the family of probability densities
$$[dL_{n,h}^{*}/dL_{n,0}^{*}] = \exp\left[-B_n(h) + h'\Gamma z\right], \quad z \in R^k$$
is of the form (4.21) in the Appendix with $T = R^k$,
$$C(t) = \exp\left[-B_n(h)\right] \quad \text{and} \quad \Sigma = \Gamma.$$

Therefore Corollary 4.2 A (see also Remark 4.1 A) applies and we conclude that an arbitrary test ψ_n' based on Z_n^{*} may be replaced by a test ψ_n based on Z_n^{*} of the form (4.10) in the Appendix. Thus for each n, we consider tests ψ_n based on Z_n^{*} such that

$$\psi_n(z) = \begin{cases} 1 & \text{if} \quad z \in C_n^c \\ 0 & \text{if} \quad z \in C_n^0 \text{ for some } C_n \in \bar{\mathscr{C}}, \end{cases} \tag{3.5}$$

where $\bar{\mathscr{C}}$ is given by (2.3); the test may be arbitrary (measurable) on ∂C_n, and $\mathscr{E}[\psi_n(Z_n^{*})|R_{n,0}] \to \alpha$.

Now it would be convenient to avoid arbitrariness of tests ψ_n on ∂C_n; for instance, it would be convenient to set $\psi_n(z) = 0$ for $z \in \partial C_n$, so that $\psi_n(z) = I_{C_n^c}$. In order for this modification to be valid, we would have to show that by changing the test ψ_n on ∂C_n in any arbitrary (measurable) way, both its asymptotic power and size remain intact. That this is, in fact, the case is the content of Lemma 3.3. In order to be able to prove the lemma, some additional notation and some preliminary results are needed. To this end, let

$$\mathscr{C}^{*} = \{C \in \mathscr{B}^k; \ C \text{ is convex}\}, \tag{3.6}$$

and also set
$$\mathscr{L}_{n,\theta} = \mathscr{L}(\Delta_n | P_n) \quad (\theta \in \Theta). \tag{3.7}$$

Then by Theorem 4.6, Chapter 2, for $\theta_n^{*} = \theta_0 + h_n n^{-\frac{1}{2}}$ with $h_n \to h \in R^k$, $\mathscr{L}_{n,\theta_n^{*}} \Rightarrow \mathscr{L}_h$, where $\mathscr{L}_h$ is defined by (2.1). Also $P_{\theta_n^{*}}(\Delta_n^{*} \neq \Delta_n) \to 0$ by Proposition 2.3 (ii), Chapter 3. Therefore $\mathscr{L}_{n,\theta_n^{*}}^{*} \Rightarrow \mathscr{L}_h$, where $\mathscr{L}_{n,\theta_n^{*}}^{*}$ and $\mathscr{L}_h$ are given by (3.3) and (2.1), respectively.

On the other hand, $\|P_{\theta_n^*} - R_{n,h_n}\| \to 0$ by Theorem 3.1, Chapter 3, so that Proposition 1.2, Chapter 1, applies and gives $\mathscr{L}_{n,h_n}^{**} \Rightarrow \mathscr{L}_h$, where $\mathscr{L}_{n,h_n}^{**}$ is given by (3.3). That is, we have

$$\mathscr{L}_{n,\theta_n^*}^* \Rightarrow \mathscr{L}_h \quad \text{and} \quad \mathscr{L}_{n,h_n}^{**} \Rightarrow \mathscr{L}_h. \tag{3.8}$$

The lemma below shows that these convergences are uniform over the class $\mathscr{C}^*$. More precisely, we have the following result.

LEMMA 3.2 Let θ_n, $\mathscr{L}_h$, $\mathscr{L}_{n,\theta_n}^*$, $\mathscr{L}_{n,h}^{**}$ and $\mathscr{C}^*$ be defined by (1.10), (2.1), (3.3) and (3.6), respectively. Then for any bounded subset B of R^k, one has

(i) $\sup\{\sup[|\mathscr{L}_{n,\theta_n}^*(C) - \mathscr{L}_h(C)|; C \in \mathscr{C}^*]; h \in B\} \to 0$,

and

(ii) $\sup\{\sup[|\mathscr{L}_{n,h}^{**}(C) - \mathscr{L}_h(C)|; C \in \mathscr{C}^*]; h \in B\} \to 0$.

Proof (i) The proof is by contradiction. Set

$$\delta_n(h) = \sup[|\mathscr{L}_{n,\theta_n}^*(C) - \mathscr{L}_h(C)|; C \in \mathscr{C}^*]$$

and suppose that $\sup[\delta_n(h); h \in B] \nrightarrow 0$. Then there is a subsequence $\{m\} \subseteq \{n\}$ and $h_m \in B$ such that $\delta_m(h_m) \to \delta$, for some $\delta > 0$. Equivalently,

$$\sup[|\mathscr{L}_{m,\theta_m^*}^*(C) - \mathscr{L}_{h_m}(C)|; C \in \mathscr{C}^*] \to \delta, \tag{3.9}$$

where $\theta_m^* = \theta_0 + h_m m^{-\frac{1}{2}}$.

Let $\{h_r\} \subseteq \{h_m\}$ be such that $h_r \to t \in R^k$. Then one has, by virtue of (3.8),

$$\mathscr{L}_{r,\theta_r^*}^* \Rightarrow \mathscr{L}_t.$$

Thus Theorem 4.3A applies and gives

$$\sup[|\mathscr{L}_{r,\theta_r^*}^*(C) - \mathscr{L}_t(C)|; C \in \mathscr{C}^*] \to 0. \tag{3.10}$$

On the other hand, we clearly have

$$\sup[|\mathscr{L}_{h_r}(C) - \mathscr{L}_t(C)|; C \in \mathscr{C}^*] \to 0. \tag{3.11}$$

Relations (3.10) and (3.11) then imply that

$$\sup[|\mathscr{L}_{r,\theta_r^*}^*(C) - \mathscr{L}_{h_r}(C)|; C \in \mathscr{C}^*] \to 0.$$

However, this contradicts (3.9) with m replaced by r.

(ii) We have

$$\|\mathscr{L}_{n,\theta_n}^* - \mathscr{L}_{n,h}^{**}\|$$
$$= 2\sup[|\mathscr{L}_{n,\theta_n}^*(A) - \mathscr{L}_{n,h}^{**}(A)|; \, A\in\mathscr{B}^k]$$
$$= 2\sup[|P_{\theta_n}(\Delta_n^*\in A) - R_{n,h}(\Delta_n^*\in A)|; \, A\in\mathscr{B}^k]$$
$$\leqslant 2\sup[|P_{\theta_n}(E) - R_{n,h}(E)|; \, E\in\mathscr{A}_n] = \|P_{\theta_n} - R_{n,h}\|.$$

But $\sup[\|P_{\theta_n} - R_{n,h}\|; \, h\in B] \to 0$ by Corollary 4.1, Chapter 3.
Therefore
$$\sup[\|\mathscr{L}_{n,\theta_n}^* - \mathscr{L}_{n,h}^{**}\|; \, h\in B] \to 0. \qquad (3.12)$$

Clearly, (3.12) implies that

$$\sup\{\sup[|\mathscr{L}_{n,\theta_n}^*(C) - \mathscr{L}_{n,h}^{**}(C)|; \, C\in\mathscr{C}^*]; \, h\in B\} \to 0.$$

This last convergence together with the first part of the lemma
yields the desired conclusion. ∎

The result just obtained is a strengthening of Lemma 2.1 in
Chibisov [1] in that taking the sup over h, we allow h to vary over
bounded rather than compact sets.

LEMMA 3.3   Let Z_n^*, $L_{n,h}^*$ and $\mathscr{C}^*$ be defined by (3.2), (3.4) and
(3.6), respectively. Then for any bounded subset B of R^k and any
sets $C_n\in\mathscr{C}^*$, one has

$$\sup[R_{n,h}(Z_n^*\in\partial C_n); \, h\in B] = \sup[L_{n,h}^*(\partial C_n); \, h\in B] \to 0.$$

Proof   For any $A\in\mathscr{B}^k$, one has $Z_n^*\in A$ if and only if $\Delta_n^*\in\hat{A}$,
where
$$\hat{A} = \{u\in R^k; \, u = \Gamma z, z\in A\}. \qquad (3.13)$$

Let A^0 and $\bar{A}$ denote the interior and the closure, respectively,
of the set A. We then have, by means of (3.3) and (2.1)

$$L_{n,h}^*(\partial C_n) = \mathscr{L}_{n,h}^{**}(\partial\hat{C}_n) = \mathscr{L}_{n,h}^{**}(\bar{\hat{C}}_n) - \mathscr{L}_{n,h}^{**}(\hat{C}_n^0)$$
$$= [\mathscr{L}_{n,h}^{**}(\bar{\hat{C}}_n) - \mathscr{L}_h(\bar{\hat{C}}_n)] - [\mathscr{L}_{n,h}^{**}(\hat{C}_n^0) - \mathscr{L}_h(\hat{C}_n^0)] \quad (3.14)$$

for any $C_n\in\mathscr{C}^*$, where $\hat{C}_n^0$ denotes the interior of the set $\hat{C}_n$. The
equality
$$\mathscr{L}_h(\bar{\hat{C}}_n) = \mathscr{L}_h(\hat{C}_n^0)$$

holds because $C_n\in\mathscr{C}^*$ if and only if $\hat{C}_n\in\mathscr{C}^*$ and the boundary
of any convex set in R^k has k-dimensional Lebesgue measure
zero and hence $\mathscr{L}_h$-measure zero. It is also well known that, both
the closure and the interior of a convex set are also convex.

Taking the sup of both sides of (3.14) as h varies in B, one obtains the desired result from Lemma 3.2 (ii). ▋

Returning now to the discussion following the definition of the test ψ_n by (3.5), we conclude that we may restrict ourselves to tests ψ_n based on Z_n^* and having the following form

$$\psi_n = I_{C_n^c} \quad \text{for some} \quad C_n = \bar{\mathscr{C}} \tag{3.15}$$

and

$$\mathscr{E}[\psi_n(Z_n^*)|R_{n,0}] \to \alpha. \tag{3.16}$$

4   Proof of the first main result

For the proof of the first theorem, we shall need some additional notation and also some preliminary results. Set

$$Z_n = \Gamma^{-1}\Delta_n \tag{4.1}$$

and

$$L_{n,\theta} = \mathscr{L}(Z_n|P_\theta) \quad (\theta \in \Theta). \tag{4.2}$$

Then $Z_n \in A$ if and only if $\Delta_n \in \hat{A}$, where $\hat{A}$ is given by (3.13). Also set

$$L_h = N(h, \Gamma^{-1}) \quad (h \in R^k). \tag{4.3}$$

One then has the following result.

LEMMA 4.1   (i) Let $\mathscr{L}_h$, $\mathscr{C}^*$ and $\mathscr{L}_{n,\theta}$ be defined by (2.1), (3.6) and (3.7), respectively. Then

$$\sup\{\sup[|\mathscr{L}_{n,\theta_n}(C) - \mathscr{L}_h(C)|; C \in \mathscr{C}^*]; h \in B\} \to 0,$$

where θ_n is given by (1.10) and B is any bounded subset of R^k.

(ii) Let θ_n, $\mathscr{C}^*$ and B be as above and let $L_{n,\theta}$ and L_h be defined by (4.2) and (4.3), respectively. Then

$$\sup\{\sup[|L_{n,\theta_n}(C) - L_h(C)|; C \in \mathscr{C}^*]; h \in B\} \to 0.$$

Proof (i) The proof is similar to that of Lemma 3.2 (i) and the details are left as an exercise (see Exercise 5).

(ii) For $A \in \mathscr{B}^k$, we have $L_{n,\theta_n}(A) = \mathscr{L}_{n,\theta_n}(\hat{A})$, where $\hat{A}$ is given by (3.13) and $A \in \mathscr{C}^*(A \in \bar{\mathscr{C}})$ if and only if $\hat{A} \in \mathscr{C}^*(\hat{A} \in \bar{\mathscr{C}})$. It is also readily seen (see Exercise 6) that

$$L_h(A) = \mathscr{L}_h(\hat{A}). \tag{4.4}$$

Therefore $\sup\left[\,|L_{n,\,\theta_n}(C) - L_h(C)|\,;\;C \in \mathscr{C}^*\right]$

$\qquad = \sup\left[\,|\mathscr{L}_{n,\,\theta_n}(\hat{C}) - \mathscr{L}_h(\hat{C})|\,;\;\hat{C} \in \mathscr{C}^*\right]$

$\qquad = \sup\left[\,|\mathscr{L}_{n,\,\theta_n}(C) - \mathscr{L}_h(C)|\,;\;C \in \mathscr{C}^*\right].$

Then taking the sup of both sides of this last relation as h varies in B and utilizing the first part of the lemma, we obtain the desired result.

From (4.1), it follows that $\Delta_n \in A$ if and only if $Z_n \in \tilde{A}$, where

$$\tilde{A} = \{u \in R^k;\; u = \Gamma^{-1}z,\, z \in A\}. \tag{4.5}$$

Therefore by setting

$$\beta_n(\theta;A) = P_\theta(\Delta_n \in A) \quad \text{and} \quad \tilde{\beta}_n(\theta;\tilde{A}) = P_\theta(Z_n \in \tilde{A}) \quad (A \in \mathscr{B}^k),$$
$$\tag{4.6}$$

we have $\beta_n(\theta;A) = \tilde{\beta}_n(\theta;\tilde{A}) \quad (\theta \in \Theta,\, A \in \mathscr{B}^k). \tag{4.7}$

We may now proceed with the proof of the first main result.

Proof of Theorem 2.1   Throughout the proof of the theorem, we always assume that $n > n(K)$ in addition to any other conditions which may be imposed on n, so that $E_{n,c} \subset \Theta$ for all $c \in K$. We now proceed with the proof which is by contradiction. To this end, suppose that (2.7) is not true. Then there exists a subsequence $\{m\} \subseteq \{n\}$ and $\{c_m\} \subset K$ such that

$$\int_{E_m} \beta_m(\theta;\phi)\,\zeta_m(\theta)\,\mathrm{d}A$$
$$- \int_{E_m} \beta_m(\theta;\psi_m)\,\zeta_m(\theta)\,\mathrm{d}A \to \delta, \quad \text{for some} \quad \delta < 0, \tag{4.8}$$

where, we recall, that E_m stands for E_{m,c_m} (see (1.11)). By employing the notation in (2.6), this is rewritten as follows

$$\int_{E_m} \beta_m(\theta;D^c)\,\zeta_m(\theta)\,\mathrm{d}A - \int_{E_m} \beta_m(\theta;C_m^c)\,\zeta_m(\theta)\,\mathrm{d}A \to \delta,$$

or $\displaystyle\int_{E_m} \beta_m(\theta;C_m)\,\zeta_m(\theta)\,\mathrm{d}A - \int_{E_m} \beta_m(\theta;D)\,\zeta_m(\theta)\,\mathrm{d}A \to \delta.$

By virtue of (4.7), this becomes

$$\int_{E_m} \tilde{\beta}_m(\theta;\tilde{C}_m)\,\zeta_m(\theta)\,\mathrm{d}A - \int_{E_m} \tilde{\beta}_m(\theta;\tilde{D})\,\zeta_m(\theta)\,\mathrm{d}A \to \delta. \tag{4.9}$$

Now set $\qquad \tilde{\beta}(h; A) = L_h(A), h \in R^k \quad (A \in \mathscr{B}^k),$ (4.10)

where L_h is given by (4.3).

Then on account of (4.6) and (4.10), Lemma 4.1 (ii) implies that for arbitrary sets $D_m \in \mathscr{C}^*$

$$\sup [|\tilde{\beta}_m(\theta_m; D_m) - \tilde{\beta}(h; D_m)|; h \in B] \to 0$$

for any bounded subset B of R^k. In particular,

$$\sup [|\tilde{\beta}_m(\theta_m; D_m) - \tilde{\beta}(h; D_m)|; h \in E_m^*] \to 0, \qquad (4.11)$$

where $E_m^* = E^*(c_m)$, and $E^*(c_m)$ is given by (1.12).

At this point, we set

$$\tilde{\beta}(h; A) = \tilde{\beta}(m^{\frac{1}{2}}(\theta_m - \theta_0); A) = \beta^*(\theta_m; A) \quad (A \in \mathscr{B}^k) \quad (4.12)$$

and we recall that, by (1.13), $h \in E_m^*$ if and only if $\theta_m \in E_m$. The convergence in (4.11) then becomes

$$\sup [|\tilde{\beta}_m(\theta_m; D_m) - \beta^*(\theta_m; D_m)|; \theta_m \in E_m] \to 0,$$

or $\qquad \sup [|\tilde{\beta}_m(\theta; D_m) - \beta^*(\theta; D_m)|; \theta \in E_m] \to 0.$ (4.13)

Utilizing (4.13) with D_m replaced by C_m and D successively, we obtain

$$\int_{E_m} \tilde{\beta}_m(\theta; \tilde{C}_m)\, \zeta_m(\theta)\, \mathrm{d}A - \int_{E_m} \beta^*(\theta; \tilde{C}_m)\, \zeta_m(\theta)\, \mathrm{d}A \to 0 \quad (4.14)$$

and $\qquad \displaystyle\int_{E_m} \tilde{\beta}_m(\theta; \tilde{D})\, \zeta_m(\theta)\, \mathrm{d}A - \int_{E_m} \beta^*(\theta; \tilde{D})\, \zeta_m(\theta)\, \mathrm{d}A \to 0.$ (4.15)

From (4.9), (4.14) and (4.15), we obtain

$$\int_{E_m} \beta^*(\theta; \tilde{C}_m)\, \zeta_m(\theta)\, \mathrm{d}A - \int_{E_m} \beta^*(\theta; \tilde{D})\, \zeta_m(\theta)\, \mathrm{d}A \to \delta,$$

or equivalently,

$$\int_{E_m} \beta^*(\theta; \tilde{D}^c)\, \zeta_m(\theta)\, \mathrm{d}A - \int_{E_m} \beta^*(\theta; \tilde{C}_m^c)\, \zeta_m(\theta)\, \mathrm{d}A \to \delta,$$

where $\tilde{D}^c$ denotes the complement of the set $\tilde{D}$. Thus for all sufficiently large m, $m \geqslant m_1$, say, we have

$$\int_{E_m} \beta^*(\theta; \tilde{D}^c)\, \zeta_m(\theta)\, \mathrm{d}A$$

$$< \int_{E_m} \beta^*(\theta; \tilde{C}_m^c)\, \zeta_m(\theta)\, \mathrm{d}A + \tfrac{1}{2}\delta \quad \text{(recall that } \delta < 0\text{),}$$

or by means of (4.12),

$$\int_{E_m} \beta(m^{\frac{1}{2}}(\theta - \theta_0); \tilde{D}^c) \, \zeta_m(\theta) \, \mathrm{d}A$$

$$< \int_{E_m} \beta(m^{\frac{1}{2}}(\theta - \theta_0); \tilde{C}_m^c) \, \zeta_m(\theta) \, \mathrm{d}A + \tfrac{1}{2}\delta \quad \text{for all} \quad m \geqslant m_1.$$
(4.16)

Let $A_m = \int_{E_m} \xi(z) \, \mathrm{d}A$. (See also (1.8).) Then, on account of (1.9), (4.16) becomes

$$\int_{E_m} \beta(m^{\frac{1}{2}}(\theta - \theta_0); \tilde{D}^c) \, \xi(\theta) \, \mathrm{d}A$$

$$< \int_{E_m} \beta(m^{\frac{1}{2}}(\theta - \theta_0); \tilde{C}_m^c) \, \xi(\theta) \, \mathrm{d}A + \tfrac{1}{2}\delta A_m \quad \text{for all} \quad m \geqslant m_1.$$
(4.17)

Set $\quad m^{\frac{1}{2}}(\theta - \theta_0) = h,\quad$ so that $\quad \theta = \theta_0 + hm^{-\frac{1}{2}}$

and $\quad\quad h \in E_m^*,\quad$ given by (1.12). $\quad\quad\quad\quad$ (4.18)

Then by virtue of (4.12) and Remark 1.2, the inequality in (4.17) becomes

$$\int_{E_m^*} \beta(h; \tilde{D}^c) \, \xi^*(h) \, \mathrm{d}A$$

$$< \int_{E_m^*} \beta(h; \tilde{C}_m^c) \, \xi^*(h) \, \mathrm{d}A + \frac{\delta A_m}{2|J_m|} \quad \text{for all} \quad m \geqslant m_1, \quad (4.19)$$

where the scaling factor J_m results from the transformation in (4.18). It is not hard to show (see Exercise 7) that $|J_m| = m^{-\frac{1}{2}(k-1)}$, whereas A_m, which is the surface area of the sphere with radius $(c_m m^{-1})^{\frac{1}{2}}$ corresponding to $E_m = E_{m, c_m}$ in (1.11), is equal to

$$\frac{2\pi^{\frac{1}{2}k}}{\Gamma(\frac{1}{2}k)} \frac{c_m^{\frac{1}{2}(k-1)}}{m^{\frac{1}{2}(k-1)}},$$

as is well known. Therefore

$$\frac{A_m}{2|J_m|} = \frac{\pi^{\frac{1}{2}k}}{\Gamma(\frac{1}{2}k)} c_m^{\frac{1}{2}(k-1)}.$$

Now, since $c_m \in K$, a compact subset of $(0, \infty)$, and by passing to a subsequence if necessary, we may assume that $c_m \to c \geq 0$. First consider the case that $c > 0$. Then for all sufficiently large m, $m \geq m_2$, say, we have $c_m \geq \frac{1}{2}c$, so that $\dfrac{A_m}{2|J_m|} \geq \delta_1$, where

$$\delta_1 = \frac{\pi^{\frac{1}{2}k}}{\Gamma(\frac{1}{2}k)} (\tfrac{1}{2}c)^{\frac{1}{2}(k-1)}.$$

Hence for $m \geq m_3 = \max(m_1, m_2)$, (4.19) becomes

$$\int_{E_m^{\bullet}} \beta(h; \tilde{D}^c)\, \xi^*(h)\, \mathrm{d}A < \int_{E_m^{\bullet}} \beta(h; \tilde{C}_m^c)\, \xi^*(h)\, \mathrm{d}A + \delta_2, \quad (4.20)$$

where $$\delta_2 = \delta\delta_1 (< 0). \quad (4.21)$$

At this point, we recall that $\beta(h; \tilde{C}_m^c) = L_h(\tilde{C}_m^c)$ by (4.10). On the other hand, it is clear from (4.5) that $\tilde{A}^c = \widetilde{A}^c$, whereas

$$\hat{\tilde{A}} = A,$$

as follows in an obvious manner from (3.13) and (4.5). Therefore one has $L_h(\tilde{C}_m^c) = L_h(\widetilde{C}_m^c)$ and, by (4.4), this is equal to

$$\mathscr{L}_h(\widehat{\widetilde{C}_m^c}) = \mathscr{L}_h(C_m^c).$$

In order to summarize, one has

$$\beta(h; \tilde{C}_m^c) = L_n(\tilde{C}_m^c) = \mathscr{L}_h(C_m^c). \quad (4.22)$$

By (3.7), $\qquad \mathscr{L}_{m,\theta_m}(C_m^c) = P_{\theta_m}(\Delta_m \in C_m^c),$

so that $\qquad \mathscr{L}_{m,\theta_0}(C_m^c) = P_{\theta_0}(\Delta_m \in C_m^c)$

and this converges to α. Then Lemma 4.1 (i) gives that

$$\mathscr{L}_0(C_m^c) \to \alpha.$$

But $\mathscr{L}_0(C_m^c) = L_0(\tilde{C}_m^c)$, by means of (4.22), and this is also equal to $L_0(\widetilde{C}_m^c)$, since $\tilde{C}_m^c = \widetilde{C}_m^c$ as was remarked before. Thus

$$L_0(\widetilde{C}_m^c) \to \alpha.$$

Now let $C \in \mathscr{F}$ be such that $L_0(C^c) = \alpha$. Then

$$L_0(\widetilde{C}_m^c) - L_0(C^c) \to 0$$

and from this it also follows (see Exercise 8) that

$$\sup\left[\,|L_h(\widetilde{C}_m^c) - L_h(C^c)|\,;\, h \in B\right] \to 0, \quad (4.23)$$

for any bounded subset B of R^k.

Since E_m^* remains bounded as $m \to \infty$, it follows that for all sufficiently large m, $m \geqslant m_4$, say, one has $L_h(\tilde{C}_m^c) \leqslant L_h(C^c) + \epsilon$. This, together with (4.10), gives

$$\int_{E_m^*} \beta(h; \tilde{C}_m^c)\, \xi^*(h)\, \mathrm{d}A \leqslant \int_{E_m^*} \beta(h; C^c)\, \xi^*(h)\, \mathrm{d}A + \epsilon A_m^* \quad (m \geqslant m_4),$$

where $A_m^* = \int_{E_m^*} \xi^*(h)\, \mathrm{d}A$ is the surface area of the sphere with radius $c_m^{\frac{1}{2}}$ corresponding to $E_m^* = E^*(c_m)$ in (1.12). According to the formulae given above,

$$A_m^* = \frac{2\pi^{\frac{1}{2}k}}{\Gamma(\frac{1}{2}k)} c_m^{\frac{1}{2}(k-1)}.$$

Combining this inequality with (4.20), we obtain that for $m \geqslant m_5 = \max(m_3, m_4)$,

$$\int_{E_m^*} \beta(h; \tilde{D}^c)\, \xi^*(h)\, \mathrm{d}A < \int_{E_m^*} \beta(h; C^c)\, \xi^*(h)\, \mathrm{d}A + \epsilon A_m^* + \delta_2. \quad (4.24)$$

From this expression of A_m^* given above, it follows that $\{A_m^*\}$ stays bounded. This result, together with (4.21), implies that for $m \geqslant m_6$, some m_6, and some $\delta_3 < 0$,

$$\int_{E_m^*} \beta(h; \tilde{D}^c)\, \xi^*(h)\, \mathrm{d}A < \int_{E_m^*} \beta(h; \tilde{C}^c)\, \xi^*(h)\, \mathrm{d}A + \delta_3, \quad (4.25)$$

for sufficiently small ϵ. By (4.10), $\beta(h; A)$ is the power of the test $\psi = I_A$, based on the random vector Z whose distribution, under h, is $L_h = N(h, \Gamma^{-1})$ (see (4.3)). On account of (4.5), the set $\tilde{D}$ is given by
$$\tilde{D} = \{u \in R^k;\, u = \Gamma^{-1}z,\, z \in D\}.$$

By taking into consideration (2.2), one has

$$\tilde{D} = \{u \in R^k;\, u'\Gamma u \leqslant d\}.$$

Applying (4.4) with $A = \tilde{D}$ and $h = 0$ and also utilizing (2.2) and the fact that
$$\hat{\tilde{D}} = D,$$

we obtain $L_0(\tilde{D}) = 1 - \alpha$. That is,

$$\tilde{D} = \{u \in R^k;\, u'\Gamma u \leqslant d\}, \quad L_0(\tilde{D}) = 1 - \alpha$$

and also $\qquad\qquad\qquad L_0(C) = 1 - \alpha.$

Now for all sufficiently large m, $m \geqslant m_7$, say, look at $\tilde{D}$ given above and also $E_m^* = E^*(c_m)$ (see (1.12)), and consider the problem of testing the hypothesis H': $h = 0$ in connection with the distribution $L_h = N(h, \Gamma^{-1})$. Then we have that both $\tilde{D}$ and E_m^* are of the type required by Theorem 4.4A. Hence relation (4.25) cannot hold true. The desired result is then established.

In order to complete the proof of the theorem, we have to show that its conclusion is true if $c = 0$; that is, if $c_m \to 0$. In this case, by setting h_m instead of h in $h'\Gamma h = c_m$, we have that $h_m \to 0$, or equivalently $m^{\frac{1}{2}}(\theta_m - \theta_0) \to 0$. Then repeating the arguments which led us to (3.17) in Chapter 4, we obtain $\|P_{\theta_m} - P_{\theta_0}\| \to 0$. From this and by a simple contradiction argument, we also get $\sup(\|P_{\theta_m} - P_{\theta_0}\|; \theta_m \in E_m) \to 0$. Therefore uniformly in ψ_m in $\mathcal{F}$ and $\theta_m \in E_m$, one has

$$\beta_m(\theta_m; \psi_m) - \beta_m(\theta_0; \psi_m) \to 0.$$

Hence $\beta_m(\theta_m; \psi_m) \to \alpha$ uniformly in $\psi_m \in \mathcal{F}$ and $\theta_m \in E_m$, since $\beta_m(\theta_0; \psi_m) \to \alpha$. Applying this result for $\psi_m = \phi$, we obtain

$$\beta_m(\theta_m; \phi) - \beta_m(\theta_m; \psi_m) \to 0$$

uniformly in $\psi_m \in \mathcal{F}$ and $\theta_m \in E_m$, so that

$$\int_{E_m} \beta_m(\theta; \phi)\, \zeta_m(\theta)\, \mathrm{d}A - \int_{E_m} \beta_m(\theta; \psi_m)\, \zeta_m(\theta)\, \mathrm{d}A \to 0.$$

Thus the left-hand side of (2.7) (with lim inf replaced by lim) is equal to zero. The proof is completed.]

From Lemma 4.1 (i), it follows that from asymptotic power viewpoint (both in pointwise and the average power sense), there is no loss by restricting ourselves, for each n, to tests lying in the class $\mathcal{F}_0$ defined by (6.2) below. In this case, the proof of Theorem 2.1 is considerably simpler in that one may deduce the desired contradiction from (4.19). This is so because $\hat{\beta}(0; C_m^c) = \alpha$. Also, in this case, one may formulate and prove a uniform version of Theorem 2.1. More precisely, one has the following result.

THEOREM 4.1 Refer to the testing hypothesis problem considered in Theorem 2.1. Then with the same notation as that employed in Theorem 2.1 and under the same assumptions, one has

$$\liminf \left[\inf \left\{ \inf \left[\int_{E_{n,\,c}} \beta_n(\theta\,;\phi)\,\zeta_n(\theta)\,\mathrm{d}A \right. \right.\right.$$

$$\left.\left.\left. - \int_{E_{n,\,c}} \beta_n(\theta\,;\psi)\,\zeta_n(\theta)\,\mathrm{d}A\,;\psi \in \mathscr{F}_0 \right];\, c \in K \right\} \right] = 0, \quad (4.26)$$

where $\mathscr{F}_0$ is given by (6.2) below.

Thus, according to the conclusion of the theorem, for every $\epsilon > 0$ and all sufficiently large n, one has

$$\int_{E_{n,\,c}} \beta_n(\theta\,;\phi)\,\zeta_n(\theta)\,\mathrm{d}A > \int_{E_{n,\,c}} \beta_n(\theta\,;\psi)\,\zeta_n(\theta)\,\mathrm{d}A - \epsilon$$

simultaneously for all $\psi \in \mathscr{F}_0$ and all $c \in K$.

Proof of Theorem 4.1 In the first place, the left-hand side of (4.26) cannot be positive since $\phi \in \mathscr{F}_0$. This follows immediately from the definition of ϕ by (2.5) and (6.2). Next, suppose that (4.26) is not true and let the left-hand side of it be equal to some $\delta < 0$. Then there is a subsequence $\{m\} \subseteq \{n\}$ and $\{c_m\} \subset K$ such that

$$\inf \left[\int_{E_m} \beta_m(\theta\,;\phi)\,\zeta_m(\theta)\,\mathrm{d}A - \int_{E_m} \beta_m(\theta\,;\psi)\,\zeta_m(\theta)\,\mathrm{d}A\,;\psi \in \mathscr{F}_0 \right] \to \delta,$$

where $E_m = E_{m,\,c_m}$ (see (1.11)). From this it follows that there exists a sequence $\{\psi_m\}$ of tests in $\mathscr{F}_0$ such that

$$\int_{E_m} \beta_m(\theta\,;\phi)\,\zeta_m(\theta)\,\mathrm{d}A - \int_{E_m} \beta_m(\theta\,;\psi_m)\,\zeta_m(\theta)\,\mathrm{d}A \to \delta.$$

This is the same as relation (4.8) and a repetition of the arguments used in the proof of Theorem 2.1, leads us to (4.19). The desired contradiction then follows as indicated above. ∎

5 Proof of the second main result

The following inequalities will be useful in the proof of the second theorem.

Let $\{\alpha_j, j \in I\}$ and $\{\beta_j, j \in I\}$ be any collections of bounded real numbers and let I be any index set. Then the following inequalities hold (see Exercise 9).

$$\left|\sup\left(\alpha_j;j\in I\right)-\sup\left(\beta_j;j\in I\right)\right|\leqslant\sup\left(|\alpha_j-\beta_j|;j\in I\right)\quad(5.1)$$

and

$$\left|\inf\left(\alpha_j;j\in I\right)-\inf\left(\beta_j;j\in I\right)\right|\leqslant\sup\left(|\alpha_j-\beta_j|;j\in I\right).\quad(5.2)$$

A few more facts will be needed before we proceed with the proof of the theorem.

We recall from Chapter 3 that Δ stands for the identity mapping in R^k. Also $L_h=N(h,\Gamma^{-1})$ by (4.3). Therefore

$$\mathscr{L}(\Delta|L_h)=N(h,\Gamma^{-1})$$

and hence (see, e.g. Rao [3], p. 152)

$$\mathscr{L}(\Delta'\Gamma\Delta|L_h)=\chi^2_{k;\delta(h)},\quad\text{where}\quad\delta(h)=h'\Gamma h.\quad(5.3)$$

From the definition of $\tilde{A}$ by (4.5), it is immediate that $\tilde{A}^c=\widetilde{A^c}$. On the other hand, as was mentioned in the proof of Theorem 2.1,

$$\hat{\tilde{A}}=A,$$

where $\hat{A}$ is given by (3.13). Utilizing these facts with

$$A=\widetilde{D^c}(=\tilde{D}^c),$$

where $\tilde{D}$ is defined by (4.20), relation (4.4) becomes

$$L_h(\widetilde{D^c})=\mathscr{L}_h(D^c).\quad(5.4)$$

The quantities $\mathscr{L}_h$ and D are defined by (2.1) and (2.2), respectively. From (5.3) and (5.4), one obtains

$$\mathscr{L}_h(D^c)=1-P[\chi^2_{k;\delta(h)}\leqslant d]=\text{constant on each }E_n^*,\quad(5.5)$$

where $E_n^*=E^*(c_n)$ (see (1.12)) and $\{c_n\}$ is a bounded sequence of positive numbers.

Finally, let $h\in E_n^*$ and let $\theta_n=\theta_0+hn^{-\frac{1}{2}}$, so that $\theta_n\in E_n$, where $E_n=E_{n,c_n}$ (see (1.11)). Then by virtue of (4.10) and (4.12), we have $L_h(A)=\beta^*(\theta_n;A)$. Taking $A=\widetilde{D^c}$ and employing (5.4), one obtains $\beta^*(\theta_n;\widetilde{D^c})=\mathscr{L}_h(D^c)$. This, together with (5.5), implies that

$$\beta^*(\theta;\widetilde{D^c})=\text{constant on each }E_n.\quad(5.6)$$

We may now start with the proof of the result.

Proof of Theorem 2.2   By a familiar contradiction argument, it suffices to show that

(i′) $\lim \{\sup [\beta_n(\theta; \phi); \theta \in E_n] - \inf [\beta_n(\theta; \phi); \theta \in E_n]\} = 0$,

(ii′) $\liminf \{\inf [\beta_n(\theta; \phi) - \beta_n(\theta; \psi_n); \theta \in E_n]\} \geqslant 0$,

where $E_n = E_{n, c_n}$ (see (1.11)) and $\{c_n\}$ is any sequence with elements in K.

(i′) By employing (5.6), we have

$$|\sup [\beta_n(\theta; \phi); \theta \in E_n] - \inf [\beta_n(\theta; \phi); \theta \in E_n]|$$
$$\leqslant |\sup [\beta_n(\theta; \phi); \theta \in E_n] - \sup [\beta^*(\theta; \widetilde{D^c}); \theta \in E_n]|$$
$$+ |\inf [\beta_n(\theta; \phi); \theta \in E_n] - \inf [\beta^*(\theta; \widetilde{D^c}); \theta \in E_n]|,$$

and by means of (5.1) and (5.2), the right-hand side above is bounded above by

$$2 \sup [|\beta_n(\theta; \phi) - \beta^*(\theta; \widetilde{D^c})|; \theta \in E_n]$$
$$= 2 \sup [|\beta_n(\theta; D^c) - \beta^*(\theta; \widetilde{D^c})|; \theta \in E_n]$$
$$= 2 \sup [|\beta_n(\theta; D) - \beta^*(\theta; \widetilde{D})|; \theta \in E_n].$$

Set $\theta_n = \theta_0 + hn^{-\frac{1}{2}}$ with $h \in E_n^* = E^*(c_n)$ (see (1.12)). Then on account of (4.6) and (3.7), one has

$$\beta_n(\theta_n; D) = P_{\theta_n}(\Delta_n \in D) = \mathscr{L}_{n, \theta_n}(D).$$

From (3.13) and (4.5), it follows that

$$\hat{D} = D.$$

Therefore, by virtue of (4.12), (4.10) and (4.4), we have

$$\beta^*(\theta_n; \widetilde{D}) = \hat{\beta}(h; \widetilde{D}) = L_h(\widetilde{D}) = \mathscr{L}_h(D).$$

Hence $$2 \sup [|\beta_n(\theta; D) - \beta^*(\theta; \widetilde{D})|; \theta \in E_n]$$
$$= 2 \sup [|\mathscr{L}_{n, \theta_n}(D) - \mathscr{L}_h(D)|; h \in E_n^*].$$

Thus $$|\sup [\beta_n(\theta; \phi); \theta \in E_n] - \inf [\beta_n(\theta; \phi); \theta \in E_n]|$$
$$\leqslant 2 \sup [|\mathscr{L}_{n, \theta_n}(D) - \mathscr{L}_h(D)|; h \in E_n^*].$$

Since the expression on the right-hand side above converges to zero by Lemma 4.1 (i), the proof of part (i′) is completed.

(ii′) We first show that

$$\liminf \{\sup [\beta_n(\theta; \phi) - \beta_n(\theta; \psi_n); \theta \in E_n]\} \geqslant 0. \qquad (5.7)$$

The proof is by contradiction. Suppose that (5.7) is not true and let the left-hand side be equal to some $\delta < 0$. Then there is a subsequence $\{m\} \subseteq \{n\}$ such that for all sufficiently large m, $m \geqslant m_1$, say, one has

$$\sup\left[\beta_m(\theta; \phi) - \beta_m(\theta; \psi_m); \theta \in E_m\right] < \tfrac{1}{2}\delta.$$

This is equivalent to

$$\beta_m(\theta; \phi) - \beta_m(\theta; \psi_m) < \tfrac{1}{2}\delta \quad \text{for all} \quad \theta \in E_m \quad \text{and all} \quad m \geqslant m_1,$$

or

$$\left[\beta_m(\theta; \phi) - \beta_m(\theta; \psi_m)\right] \zeta_m(\theta) < \tfrac{1}{2}\delta\zeta_m(\theta) \quad \text{for all} \quad \theta \in E_m$$

and all $\qquad\qquad m \geqslant m_1$.

Hence

$$\liminf\left[\int_{E_m} \beta_m(\theta; \phi)\, \zeta_m\, \mathrm{d}A - \int_{E_m} \beta_m(\theta; \psi_m)\, \zeta_m(\theta)\, \mathrm{d}A\right] \leqslant \tfrac{1}{2}\delta,$$

since $\displaystyle\int_{E_m} \zeta_m(\theta)\, \mathrm{d}A = 1$, and this contradicts (2.7).

We now continue as follows

$$\begin{aligned}
\inf[\beta_n(\theta; \phi) &- \beta_n(\theta; \psi_n); \theta \in E_n] \\
&= \inf\{\beta_n(\theta; \phi) + [-\beta_n(\theta; \psi_n)]; \theta \in E_n\} \\
&\geqslant \inf[\beta_n(\theta; \phi); \theta \in E_n] + \inf[-\beta_n(\theta; \psi_n); \theta \in E_n] \\
&= \inf[\beta_n(\theta; \phi); \theta \in E_n] - \sup[\beta_n(\theta; \psi_n); \theta \in E_n].
\end{aligned}$$

Adding and subtracting appropriate quantities, the last expression on the right-hand side above becomes equal to

$$\begin{aligned}
&-\{\sup[\beta_n(\theta; \phi); \theta \in E_n] - \inf[\beta_n(\theta; \phi); \theta \in E_n]\} \\
&-\{\sup[\beta_n(\theta; \psi_n); \theta \in E_n] - \inf[\beta_n(\theta; \psi_n); \theta \in E_n]\} \\
&+\{\sup[\beta_n(\theta; \phi); \theta \in E_n] - \inf[\beta_n(\theta; \psi_n); \theta \in E_n]\}.
\end{aligned}$$

The third of these terms is further written as

$$\sup[\beta_n(\theta; \phi); \theta \in E_n] + \sup[-\beta_n(\theta; \psi_n); \theta \in E_n]$$

and this is bounded below by

$$\sup[\beta_n(\theta; \phi) - \beta_n(\theta; \psi_n); \theta \in E_n].$$

Combining these results, we obtain

$$\inf[\beta_n(\theta;\phi)-\beta_n(\theta;\psi_n);\ \theta\in E_n]$$

$$\geqslant -\{\sup[\beta_n(\theta;\phi);\ \theta\in E_n]-\inf[\beta_n(\theta;\phi);\ \theta\in E_n]\}$$
$$-\{\sup[\beta_n(\theta;\psi_n);\ \theta\in E_n]-\inf[\beta_n(\theta;\psi_n);\ \theta\in E_n]\}$$
$$+\sup[\beta_n(\theta;\phi)-\beta_n(\theta;\psi_n);\ \theta\in E_n]. \qquad (5.8)$$

Letting now $n\to\infty$ in (5.8), we have that the limit of the first term on the right-hand side is equal to zero by the first part of the theorem, the limit of the second term on the same side is equal to zero, by assumption, and the lim inf of the third term on the same side is $\geqslant 0$ by (5.7). This establishes (ii') and hence the theorem itself. ∎

6   Formulation and proof of the third main result

Up to this point we have dealt with tests ψ_n in $\bar{\mathscr{F}}$ defined by (2.4), depending on the random vector Δ_n and having asymptotic level of significance α. In this section, we are going to further restrict the class $\bar{\mathscr{F}}$ of tests by introducing another class contained in $\bar{\mathscr{F}}$ and denoted by $\mathscr{F}_0$. The reasons for this restriction are implicit in the definition of the envelope power functions by (6.10) and (6.14) and also Lemma 6.2 below which is needed in the proof of Theorem 6.1. However, in order for the restriction under question to be legitimate, we must show that nothing is lost, asymptotically, in the process neither in terms of power nor in terms of asymptotic level of significance.

Set

$$\mathscr{C}_0 = \{C\in\mathscr{B}^k;\ C \text{ is closed, convex and } \mathscr{L}_0(C)=1-\alpha\}, \quad (6.1)$$

where $\mathscr{L}_0$ is given by (2.1), and define the class $\mathscr{F}_0$ by

$$\mathscr{F}_0 = \{\psi;\ \psi = I_{C^c},\ C\in\mathscr{C}_0\}. \qquad (6.2)$$

Since $\mathscr{L}(\Delta_n|P_{\theta_0}) = \mathscr{L}_{n,\theta_0} \Rightarrow N(0,\Gamma) = \mathscr{L}_0$

by Theorem 4.2, Chapter 2, we have that $\mathscr{L}_{n,\theta_0}(C_n)-\mathscr{L}_0(C_n)\to 0$ for any sets $C_n\in\mathscr{C}^*$; this is so by Theorem 4.3A. Thus if $\omega_n = I_{C_n^c}$ is in $\mathscr{F}_0$, so that $\mathscr{L}_0(C_n) = 1-\alpha$, the last convergence above gives

$\mathscr{L}_{n,\theta_0}(C_n^c) \to \alpha$. That is, tests in $\mathscr{F}_0$ are of asymptotic level of significance α. Then it suffices for us to show that every test ψ_n in $\mathscr{F}$ which is of asymptotic level of significance α, can be replaced, from asymptotic power point of view, by tests ω_n in $\mathscr{F}_0$. More precisely, it suffices to establish the following result.

LEMMA 6.1   For any sequence of tests $\{\psi_n\}$ in $\mathscr{F}$ for which $\mathscr{E}_{\theta_0}\psi_n(\Delta_n) \to \alpha$, there is a sequence of tests ω_n in $\mathscr{F}_0$ such that

$$\sup\left[\left|\mathscr{E}_{\theta_n}\psi_n(\Delta_n) - \mathscr{E}_{\theta_n}\omega_n(\Delta_n)\right|; h \in B\right] \to 0, \qquad (6.3)$$

where $\theta_n = \theta_0 + hn^{-\frac{1}{2}}$ and B is any bounded subset of R^k.

Proof   Since $\psi_n \in \mathscr{F}$ and $\mathscr{E}_{\theta_0}\psi_n(\Delta_n) \to \alpha$, we have $\psi_n = I_{C_n^c}$ for some $C_n \in \mathscr{F}$ and $\mathscr{L}_{n,\theta_0}(C_n) \to 1-\alpha$. Setting $h = 0$ in Lemma 4.1 (i), we obtain $\mathscr{L}_{n,\theta_0}(C_n) - \mathscr{L}_0(C_n) \to 0$, so that

$$\mathscr{L}_0(C_n) \to 1-\alpha. \qquad (6.4)$$

Thus for all sufficiently large n, $n \geqslant n_1$, say, $\mathscr{L}_0(C_n) > 0$. This implies that for $n \geqslant n_1$, the sets C_n are k-dimensional because if their dimensionality were less than k, then their k-dimensional Lebesgue measure would be zero and hence their $\mathscr{L}_0$-measure zero, a contradiction.

Then for each $n \geqslant n_1$, consider the following modification of the set C_n. If $\mathscr{L}_0(C_n) < 1-\alpha$, enlarge C_n, so that it remains closed and convex and $\mathscr{L}_0(C_n) = 1-\alpha$. If $\mathscr{L}_0(C_n) > 1-\alpha$, shrink the set C_n until $\mathscr{L}_0(C_n) = 1-\alpha$. If $\mathscr{L}_0(C_n) = 1-\alpha$, the set C_n is left intact. Denote the resulting set by $C_{n,0}$. Then $C_{n,0}$ is closed and convex and

$$\mathscr{L}_0(C_{n,0}) = 1-\alpha. \qquad (6.5)$$

Thus setting $\qquad \omega_n = I_{C_{n,0}^c},$

we have $\omega_n \in \mathscr{F}_0$. From (6.4) and (6.5), it follows that

$$\mathscr{L}_0(C_n) - \mathscr{L}_0(C_{n,0}) \to 0. \qquad (6.6)$$

From the process of arriving at $C_{n,0}$, it follows that $C_{n,0} \subseteq C_n$ or $C_{n,0} \supseteq C_n$. Therefore for each $h \in R^k$, we have

$$\mathscr{L}_h(C_n \Delta C_{n,0}) = \mathscr{L}_h(C_n - C_{n,0})$$
$$= \mathscr{L}_h(C_n) - \mathscr{L}_h(C_{n,0}) \quad \text{if} \quad C_{n,0} \subseteq C_n,$$

and $\quad \mathscr{L}_h(C_n \Delta C_{n,0}) = \mathscr{L}_h(C_{n,0} - C_n)$

$$= \mathscr{L}_h(C_{n,0}) - \mathscr{L}_h(C_n) \quad \text{if} \quad C_{n,0} \supseteq C_n.$$

Thus $\qquad \mathscr{L}_h(C_n \Delta C_{n,0}) = |\mathscr{L}_h(C_n) - \mathscr{L}_h(C_{n,0})|.$ $\qquad$ (6.7)

For $h = 0$, $\mathscr{L}_h(C_n \Delta C_{n,0}) \to 0$ by (6.6). On the other hand, $\mathscr{L}_h \ll \mathscr{L}_0$ for every $h \in R^k$. Thus $\mathscr{L}_h(C_n \Delta C_{n,0}) \to 0$. It can be further seen that for any bounded subset B of R^k, one has (see Exercise 10)

$$\sup \left[\mathscr{L}_h(C_n \Delta C_{n,0}); h \in B \right] \to 0. \qquad (6.8)$$

Therefore, (6.7) and (6.8) give

$$\sup \left[|\mathscr{L}_h(C_n) - \mathscr{L}_h(C_{n,0})|; h \in B \right] \to 0. \qquad (6.9)$$

Now, for any B as described above, one has

$$\sup \left[|\mathscr{E}_{\theta_n} \psi_n(\Delta_n) - \mathscr{E}_{\theta_n} \omega_n(\Delta_n)|; h \in B \right]$$

$$= \sup \left[|\mathscr{L}_{n,\theta_n}(C_n) - \mathscr{L}_{n,\theta_n}(C_{n,0})|; h \in B \right]$$

$$\leqslant \sup \left[|\mathscr{L}_{n,\theta_n}(C_n) - \mathscr{L}_h(C_n)|; h \in B \right]$$

$$+ \sup \left[|\mathscr{L}_{n,\theta_n}(C_{n,0}) - \mathscr{L}_h(C_{n,0})|; h \in B \right]$$

$$+ \sup \left[|\mathscr{L}_h(C_n) - \mathscr{L}_h(C_{n,0})|; h \in B \right],$$

and each one of the terms on the right-hand side above converges to zero on account of Lemma 4.1 (i) and (6.9). The proof of (6.3) is completed and we have the justification for confining ourselves to the class of tests, $\mathscr{F}_0$. $\blacksquare$

Before we are able to formulate the third main result, we shall have to introduce a further piece of notation. To this end, let $\beta_n(\theta; \psi) = \mathscr{E}_\theta \psi(\Delta_n)$, as given in (2.6), and suppose that the test ψ lies in

$$\mathscr{F}_0 = \{\psi; \psi = I_{C^c}, C \in \mathscr{C}_0\},$$

where

$$\mathscr{C}_0 = \{C \in \mathscr{B}^k; C \text{ is closed, convex and } \mathscr{L}_0(C) = 1 - \alpha\};$$

we also recall that $\mathscr{L}_0 = N(0, \Gamma)$. Next, define the *envelope power function* $\beta_n(\theta; \alpha)$ by

$$\beta_n(\theta; \alpha) = \sup \{\beta_n(\theta; \psi); \psi \in \mathscr{F}_0\}. \qquad (6.10)$$

Then the third main result in this chapter is as follows.

THEOREM 6.1 Let $E_{n,c}$ and ϕ be defined by (1.11) and (2.5), respectively, and let ψ_n be any tests in $\mathcal{F}_0$, where $\mathcal{F}_0$ is given in (6.2). Also let $\beta_n(\theta; \phi)$ and $\beta_n(\theta; \psi_n)$ be defined by (2.6) and let $\beta_n(\theta; \alpha)$ be given by (6.10). Then under assumptions (A 1–4), for testing H against A at asymptotic level of significance α, one has

$$\limsup \left[\sup \{ \sup [\beta_n(\theta; \alpha) - \beta_n(\theta; \phi); \theta \in E_{n,c}] \right.$$
$$\left. - \sup [\beta_n(\theta; \alpha) - \beta_n(\theta; \psi_n); \theta \in E_{n,c}]; c \in K \} \right] \leqslant 0, \quad (6.11)$$

where K is an arbitrary compact subset of $(0, \infty)$.

The interpretation of the theorem is that within the class $\mathcal{F}_0$, the test ϕ is *asymptotically most stringent* on $E_{n,c}$. We recall that in the present framework and for each n, the test ϕ_n would be said to be *most stringent* on $E_{n,c}$ within the class $\mathcal{F}_0$, if the quantity $\sup [\beta_n(\theta; \alpha) - \beta_n(\theta; \psi_n); \theta \in E_{n,c}]$ were minimized for $\psi_n = \phi_n$.

For the proof of Theorem 6.1, a couple of auxiliary results will be needed. For their formulation, let us recall once again that $\mathcal{L}_h = N(\Gamma h, \Gamma)$ and set

$$\bar{\beta}(h; A) = \mathcal{L}_h(A) \quad (h \in R^k, A \in \mathcal{B}^k). \quad (6.12)$$

For each n and a sequence of positive numbers $\{c_n\}$, let

$$h \in E_n^* = E^*(c_n)$$

(see (1.12)) and transform h to $\theta \in E_n = E_{n, c_n}$ (see (1.11)) through the transformation $\theta = \theta_0 + hn^{-\frac{1}{2}}$. Also set

$$\bar{\beta}(h; A) = \bar{\beta}(n^{\frac{1}{2}}(\theta - \theta_0); A) = \beta'(\theta; A). \quad (6.13)$$

Next, by means of $\beta'(\theta; A)$, define the *envelope power function* $\beta'(\theta; \alpha)$ as follows

$$\beta'(\theta; \alpha) = \sup [\beta'(\theta; C^c); C \in \mathcal{C}_0]. \quad (6.14)$$

The first auxiliary result is given in the following lemma.

LEMMA 6.2 The function $\beta'(\theta; \alpha)$ defined by (6.14) remains constant on each E_n, where $E_n = E_{n, c_n}$ (see (1.11)) and $\{c_n\}$ is any bounded sequence of positive numbers.

Proof Clearly, $\beta'(\theta; \alpha) = \bar{\beta}(h; \alpha)$, where

$$\bar{\beta}(h; \alpha) = \sup [\bar{\beta}(h; C^c); C \in \mathcal{C}_0] \quad \text{and} \quad h = n^{\frac{1}{2}}(\theta - \theta_0).$$

Thus it suffices to show that $\bar{\beta}(h;\alpha)$ stays constant on each E_n^*, where $E_n^* = E^*(c_n)$ (see (1.12)). With M defined by (1.1), consider the transformation $t = Mh$. Then, by (1.2), E_n^* is transformed into $S_n^* = \{t \in R^k; t't = \|t\|^2 = c_n\}$. Also the class of sets $\mathscr{C}_0$ is transformed into the class of sets $\mathscr{C}_*$, where

$$\mathscr{C}_* = \{C \in \mathscr{B}^k; C \text{ is closed, convex and } N_0(C) = 1-\alpha\}, \quad (6.15)$$

and $$N_t = N(MM't, M\Gamma M') \quad (t \in R^k) \quad (6.16)$$

(see Exercise 11). Therefore, by setting

$$\left.\begin{array}{l} \beta^0(t;A) = N_t(A), \quad t \in R^k, \quad A \in \mathscr{B}^k \\ \beta^0(t;\alpha) = \sup[\beta^0(t;C^c); C \in \mathscr{C}_*], \end{array}\right\} \quad (6.17)$$

it suffices to show that $\beta^0(t;\alpha)$ is constant on each S_n^*. For contradiction, suppose that this is not so. Then there exist $t_1, t_2 \in S_n^*$ for which $\beta^0(t_1;\alpha) \neq \beta^0(t_2;\alpha)$ and let

$$\beta^0(t_1;\alpha) < \beta^0(t_2;\alpha). \quad (6.18)$$

From the definition of $\beta^0(t;\alpha)$, there exists a set $C \in \mathscr{C}_*$ such that

$$\beta^0(t_1;\alpha) < \beta^0(t_2;C^c). \quad (6.19)$$

Now, clearly, $\beta^0(t_2;C^c) = N_{t_2}(C^c)$ is equal to the $N(t_2, I)$-measure of the set $(MM')^{-1}C^c$, and by symmetry, this is equal to the $N(t_1, I)$-measure of D, where D is the symmetric image of $(MM')^{-1}C^c$ with respect to the hyperplane through the origin that is perpendicular to the line segment connecting the points t_1 and t_2. But the $N(t_1, I)$-measure of D is equal to $N_{t_1}((MM')D)$, and, clearly, $(MM')D$ is the complement of a closed, convex set, C_0^c, say. Then, by symmetry, one, clearly, has $N_0(C_0) = 1-\alpha$, so that $C_0 \in \mathscr{C}_*$, and also $\beta^0(t_2;C^c) = \beta^0(t_1;C_0^c)$. Then (6.19) gives $\beta^0(t_1;\alpha) < \beta^0(t_1;C_0^c)$. However, this contradicts the definition of $\beta^0(t_1;\alpha)$ by (6.17). We reach the same conclusion if the inequality in (6.18) is reversed. Thus the proof of the lemma is completed. ∎

The second auxiliary result referred to above is the following lemma. This lemma, as well as the one just established, is of some interest in their own right.

LEMMA 6.3   Let $\beta_n(\theta;\alpha)$ and $\beta'(\theta;\alpha)$ be defined by (6.10) and (6.13), respectively. Then for each n, one has

(i) $$\sup \left[|\beta_n(\theta; \alpha) - \beta'(\theta; \alpha)| ; \theta \in E_n \right] \to 0;$$

and

(ii) $$\sup \left[\beta_n(\theta; \alpha); \theta \in E_n \right] - \inf \left[\beta_n(\theta; \alpha); \theta \in E_n \right] \to 0,$$

where $E_n = E_{n, c_n}$ (see (1.11)) and $\{c_n\}$ is any bounded sequence of positive numbers.

Proof (i) In the first place, relation (5.1) justifies the inequality below

$$
\begin{aligned}
|\beta_n(\theta; \alpha) - \beta'(\theta; \alpha)| \\
&= \left| \sup [\beta_n(\theta; \psi); \psi \in \mathcal{F}_0] - \sup [\beta'(\theta; C^c); C \in \mathcal{C}_0] \right| \\
&= \left| \sup [\beta_n(\theta; C^c); C \in \mathcal{C}_0] - \sup [\beta'(\theta; C^c); C \in \mathcal{C}_0] \right| \\
&\leqslant \sup \left[|\beta_n(\theta; C^c) - \beta'(\theta; C^c)| ; C \in \mathcal{C}_0 \right] \\
&= \sup \left[|\beta_n(\theta; C) - \beta'(\theta; C)| ; C \in \mathcal{C}_0 \right],
\end{aligned}
$$

and this last expression is equal to

$$\sup \left[|\mathcal{L}_{n, \theta}(C) - \mathcal{L}_h(C)| ; C \in \mathcal{C}_0 \right], \quad \text{where} \quad h = n^{\frac{1}{2}}(\theta - \theta_0),$$

since $$\beta_n(\theta; C) = P_\theta(\Delta_n \in C) = \mathcal{L}_{n, \theta}(C)$$

and $$\beta'(\theta; C) = \bar{\beta}(h; C) = \mathcal{L}_h(C).$$

That is, with $h = n^{\frac{1}{2}}(\theta - \theta_0)$,

$$|\beta_n(\theta; \alpha) - \beta'(\theta; \alpha)| \leqslant \sup \left[|\mathcal{L}_{n, \theta}(C) - \mathcal{L}_h(C)| ; C \in \mathcal{C}_0 \right].$$

Hence, with $E_n^* = E^*(c_n)$ (see (1.12)), one has

$$
\begin{aligned}
\sup \left[|\beta_n(\theta; \alpha) - \beta'(\theta; \alpha)| ; \theta \in E_n \right] \\
= \sup \left\{ \sup \left[|\mathcal{L}_{n, \theta}(C) - \mathcal{L}_h(C)| ; C \in \mathcal{C}_0 \right]; h \in E_n^* \right\},
\end{aligned}
$$

and the expression on the right-hand side converges to zero by Lemma 4.1 (i).

(ii) Letting $\theta \in E_n$ and utilizing Lemma 6.2 and inequalities (5.1) and (5.2), one has

$$
\begin{aligned}
|\sup [\beta_n(\theta; \alpha); \theta \in E_n] &- \inf [\beta_n(\theta; \alpha); \theta \in E_n]| \\
&\leqslant |\sup [\beta_n(\theta; \alpha); \theta \in E_n] - \beta'(\theta; \alpha)| \\
&\quad + |\inf [\beta_n(\theta; \alpha); \theta \in E_n] - \beta'(\theta; \alpha)| \\
&= |\sup [\beta_n(\theta; \alpha); \theta \in E_n] - \sup [\beta'(\theta; \alpha); \theta \in E_n]| \\
&\quad + |\inf [\beta_n(\theta; \alpha); \theta \in E_n] - \inf [\beta'(\theta; \alpha); \theta \in E_n]| \\
&\leqslant 2 \sup [|\beta_n(\theta; \alpha) - \beta'(\theta; \alpha)| ; \theta \in E_n],
\end{aligned}
$$

and this last expression tends to zero by part (i). This establishes the lemma. ▌

We may now proceed with the proof of the third main result herein.

Proof of Theorem 6.1   The proof is by contradiction. Suppose that the theorem is not true and let the left-hand side of (6.11) be equal to some $4\delta > 0$. Then there exists a subsequence $\{m\} \subseteq \{n\}$ for which

$$\sup [\beta_m(\theta; \alpha) - \beta_m(\theta; \phi); \theta \in E_m]$$
$$> \sup [\beta_m(\theta; \alpha) - \beta_m(\theta; \psi_m); \theta \in E_m] + 3\delta \qquad (6.20)$$

for all sufficiently large m, $m \geqslant m_1$, say.

The left-hand side of the inequality above is bounded from above by

$$\sup [\beta_m(\theta; \alpha); \theta \in E_m] - \inf [\beta_m(\theta; \phi); \theta \in E_m] \qquad (6.21)$$

and its right-hand side is bounded from below by

$$\inf [\beta_m(\theta; \alpha); \theta \in E_m] - \inf [\beta_m(\theta; \psi_m); \theta \in E_m] + 3\delta \qquad (6.22)$$

by virtue of (5.2).

By means of (6.21) and (6.22), relation (6.20) gives that, for all $m \geqslant m_1$,

$$\sup [\beta_m(\theta; \alpha); \theta \in E_m] - \inf [\beta_m(\theta; \phi); \theta \in E_m]$$
$$> \inf [\beta_m(\theta; \alpha); \theta \in E_m] - \inf [\beta_m(\theta; \psi_m); \theta \in E_m] + 3\delta,$$

or

$$\sup [\beta_m(\theta; \alpha); \theta \in E_m] - \inf [\beta_m(\theta; \alpha); \theta \in E_m]$$
$$> \inf [\beta_m(\theta; \phi); \theta \in E_m] - \inf [\beta_m(\theta; \psi_m); \theta \in E_m] + 3\delta. \qquad (6.23)$$

By Lemma 6.3 (ii), the left-hand side of (6.23) tends to zero and hence it remains less than δ for all sufficiently large m, $m \geqslant m_2$, say. On the other hand, Theorem 2.2 (i) yields

$$\inf [\beta_m(\theta; \phi); \theta \in E_m] > \sup [\beta_m(\theta; \phi); \theta \in E_m] - \delta$$

for all sufficiently large m, $m \geqslant m_3$, say. On account of these facts, inequality (6.23) becomes then

$$\delta > \sup [\beta_m(\theta; \phi); \theta \in E_m] - \inf [\beta_m(\theta; \psi_m); \theta \in E_m] + 2\delta,$$

or

$$\inf [\beta_m(\theta; \psi_m); \theta \in E_m] - \delta = \inf [\beta_m(\theta; \psi_m) - \delta; \theta \in E_m]$$
$$> \sup [\beta_m(\theta; \phi); \theta \in E_m]$$

for all $m \geqslant m_4 = \max (m_1, m_2, m_3)$. Hence

$$\beta_m(\theta; \psi_m) - \delta > \beta_m(\theta; \phi)$$

for all $m \geqslant m_4$ and every $\theta \in E_m$. Therefore

$$\int_{E_m} \beta_m(\theta; \phi) \zeta_m(\theta) \, dA - \int_{E_m} \beta_m(\theta; \psi_m) \zeta_m(\theta) \, dA < -\delta$$

for all $m \geqslant m_4$, and this implies that

$$\liminf \left[\int_{E_m} \beta_m(\theta; \phi) \zeta_m(\theta) \, dA - \int_{E_m} \beta_m(\theta; \psi_m) \zeta_m(\theta) \, dA \right] < -\delta.$$

However, this contradicts Theorem 2.1. The desired result follows. ▋

The following uniform (over the tests ψ) version of Theorem 6.1 is also true.

THEOREM 6.2   Refer to the testing hypothesis problem considered in Theorem 6.1. Then with the same notation as that employed in Theorem 6.1 and under the same assumptions, one has

$$\limsup \{\sup [\sup [\beta_n(\theta; \alpha) - \beta_n(\theta; \phi); \theta \in E_{n,c}]$$
$$- \inf \{\sup [\beta_n(\theta; \alpha) - \beta_n(\theta; \psi); \theta \in E_{n,c}]; \psi \in \mathscr{F}_0\}; c \in K\} \leqslant 0,$$
$$(6.24)$$

where $\mathscr{F}_0$ is given by (6.2).

Thus, according to the conclusion of the theorem, if $\delta(\leqslant 0)$ is equal to the left-hand side of (6.24), then for every $\epsilon > 0$ and all sufficiently large n, one has

$$\sup [\beta_n(\theta; \alpha) - \beta_n(\theta; \phi); \theta \in E_{n,c}]$$
$$< \sup [\beta_n(\theta; \alpha) - \beta_n(\theta; \psi); \theta \in E_{n,c}] + \delta + \epsilon$$

simultaneously for all $\psi \in \mathscr{F}_0$ and all $c \in K$.

Proof of Theorem 6.2   Suppose that the theorem is not true and let the left-hand side of (6.24) be equal to some $\delta > 0$. Then there exists a subsequence $\{m\} \subseteq \{n\}$ and a sequence $\{c_m\}$ with elements in K, such that, if $E_m = E_{m,c_m}$ (see (1.11)), then

$$\sup[\beta_m(\theta; \alpha) - \beta_m(\theta; \phi); \theta \in E_m]$$
$$- \inf\{\sup[\beta_m(\theta; \alpha) - \beta_m(\theta; \psi); \theta \in E_m]; \psi \in \mathscr{F}_0\} \to \delta.$$

Thus for $\epsilon > 0$ and all $m \geqslant m_5$, say, one has

$$\sup[\beta_m(\theta; \alpha) - \beta_m(\theta; \phi); \theta \in E_m] - \delta - \epsilon$$
$$< \inf\{\sup[\beta_m(\theta; \alpha) - \beta_m(\theta; \psi); \theta \in E_m]; \psi \in \mathscr{F}_0\}$$
$$< \sup[\beta_m(\theta; \alpha) - \beta_m(\theta; \phi); \theta \in E_m] - \delta + \epsilon.$$

Therefore, for each $m \geqslant m_5$, there exists a test $\psi_m \in \mathscr{F}_0$ such that

$$\sup[\beta_m(\theta; \alpha) - \beta_m(\theta; \phi); \theta \in E_m] - \delta - \epsilon$$
$$< \sup[\beta_m(\theta; \alpha) - \beta_m(\theta; \psi_m); \theta \in E_m]$$
$$< \sup[\beta_m(\theta; \alpha) - \beta_m(\theta; \phi); \theta \in E_m] - \delta + \epsilon,$$

or equivalently,

$$\delta - \epsilon < \sup[\beta_m(\theta; \alpha) - \beta_m(\theta; \phi); \theta \in E_m]$$
$$- \sup[\beta_m(\theta; \alpha) - \beta_m(\theta; \psi_m); \theta \in E_m] < \delta + \epsilon,$$

provided $m \geqslant m_5$.

It follows that

$$\sup[\beta_m(\theta; \alpha) - \beta_m(\theta; \phi); \theta \in E_m]$$
$$- \sup[\beta_m(\theta; \alpha) - \beta_m(\theta; \psi_m); \theta \in E_m] \to \delta(> 0).$$

However, this result contradicts (6.11). The proof of the theorem is completed. ∎

7 Behaviour of the power under non-local alternatives

Recall that $\phi = I_{D^c}$, where D is given by (2.2). Also recall that the power of the test ϕ, based on $\Delta_n = \Delta_n(\theta_0)$, has been denoted by $\beta_n(\theta; \phi)$, $\theta \in \Theta$. Then the theorems formulated and proved in the previous sections, provide us with some optimal properties of the test ϕ. However, these properties are local in character, since the alternatives are required to lie close to the hypothesis being tested; actually, they are required to converge to θ_0 and at a specified rate.

The underlying basic assumptions (A 1–4) employed throughout this chapter, do not suffice for establishing optimal properties of the power function at alternatives removed from θ_0 or not converging to it at the specified rate. This can be done, however, under the following additional condition (A 5).

ASSUMPTION (A 5) Consider a sequence $\{\theta_n\}$ with $\theta_n \in \Theta$ for all n. Then $\|\Delta_n(\theta_0)\| \to \infty$ in P_{n,θ_n}-probability whenever

$$\|n^{\frac{1}{2}}(\theta_n - \theta_0)\| \to \infty.$$

The following result can now be established.

THEOREM 7.1 Under assumptions (A 1–5), for testing the hypothesis $H: \theta = \theta_0$ against the alternative $A: \theta \neq \theta_0$ at asymptotic level of significance α, the test defined by (2.5) possesses the optimal properties mentioned in Theorems 2.1, 2.2, 4.1, 6.1, 6.2 and also has the property that its power converges to 1, i.e. $\beta_n(\theta_n; \phi) \to 1$, whenever $\|n^{\frac{1}{2}}(\theta_n - \theta_0)\| \to \infty$.

Proof The proof is immediate. Since Γ is positive definite, so is Γ^{-1}. Thus there exists a positive number p such that $z'\Gamma^{-1}z \geqslant p\|z\|^2$ for all $z \in R^k$. Therefore

$$\beta_n(\theta_n; \phi) = \mathscr{E}_{\theta_n}\phi(\Delta_n) = P_{\theta_n}(\Delta_n \in D^c) = P_{\theta_n}(\Delta_n'\Gamma^{-1}\Delta_n > d).$$

However, this last quantity is greater than or equal to

$$P_{\theta_n}(p\|\Delta_n\|^2 > d)$$

which converges to 1 by (A 5). Thus $\beta_n(\theta_n; \phi) \to 1$, as was to be seen. ∎

Exercises

1. Verify (1.7).
2. Verify (1.8).
3. Show that if (2.7) in Theorem 2.1 is true, then it is also true if all tests involved are based on the random vector Δ_n^* rather than Δ_n and conversely.
4. For each n and a bounded sequence $\{c_n\}$ of positive numbers, let $h \in R^k$ and set $\theta_n = \theta_0 + hn^{-\frac{1}{2}}$. Then $h \in E_n^*$ if and only if $\theta \in E_n$, where $E_n = E_{n,c_n}$ and $E_n^* = E^*(c_n)$ are defined

by (1.11) and (1.12), respectively. For $h \in E_n^*$ and θ as above, set $R_{n,h} = R_{n,(\theta_n - \theta_0)n^{\frac{1}{2}}} = R_{n,\theta_n}^*$. Then show that if (2.7) in Theorem 2.1 holds true, it is also true if all tests involved are based on the random vector Δ_n^* rather than Δ_n and power is calculated under R_{n,θ_n}^* rather than P_{θ_n}.

   5. Provide a detailed proof of Lemma 4.1 (i).
   6. Verify (4.4).
   7. Justify the passage from (4.17) to (4.19) by means of transformation (4.18) and also the fact that the Jacobian of the transformation is given by $|J_m| = J_m = m^{-\frac{1}{2}(k-1)}$.
   8. Verify (4.23)
   9. Establish inequalities (5.1) and (5.2).
   10. Verify (6.8).
   11. Verify (6.16).

Appendix

In this appendix, we gather together some of the basic theorems used in the body of the monograph. Some of the proofs are supplied and for the omitted ones the reader is referred to appropriate sources. In formulating the theorems herein, we do not strive to present them in their most general form. Simplicity and the needs of the present monograph have been the criteria employed rather than obtaining the most general results possible.

1 Some theorems employed in Chapter 1

Let S be a metric space and let $\mathscr{S}$ be the topological Borel σ-field in S. Then for a probability measure $\mathscr{L}$ defined on $\mathscr{S}$, a set $A \in \mathscr{S}$ is said to be an $\mathscr{L}$-*continuity* set if $\mathscr{L}(\partial A) = 0$, where ∂A is the boundary of A.

THEOREM 1.1A   Let S be a metric space and let $\mathscr{S}$ be the topological Borel σ-field in S. For $n = 1, 2, ...,$ let $\mathscr{L}_n$, $\mathscr{L}$ be probability measures defined on $\mathscr{S}$. Then $\mathscr{L}_n \Rightarrow \mathscr{L}$ if and only if $\mathscr{L}_n(A) \to \mathscr{L}(A)$ for all $\mathscr{L}$-continuity sets A in $\mathscr{S}$.

This theorem is contained in Theorem 2.1 in Billingsley [4], pp. 11–14.

THEOREM 1.2A   Suppose $(S, \mathscr{S}) = (R^m, B^m)$, and for

$$n = 1, 2, ...,$$

let $\mathscr{L}_n$, $\mathscr{L}$ be probability measures defined on $\mathscr{S}$. Then $\mathscr{L}_n \Rightarrow \mathscr{L}$ if and only if $\int f d\mathscr{L}_n \to \int f d\mathscr{L}$ for every real-valued continuous (and bounded) function f defined on S and vanishing outside a compact set.

Proof   One direction of the theorem is immediate from the definition of weak convergence. Thus it suffices to prove the other one. To this end, let g be a real-valued, bounded and continuous function defined on S. We wish to show that

$$\int g\,\mathrm{d}\mathscr{L}_n \to \int g\,\mathrm{d}\mathscr{L}. \tag{1.1}$$

For $j = 1, 2, \ldots$, let S_j be the (open) sphere of radius j centred at the origin. Then $S = \bigcup\limits_{j=1}^{\infty} S_j$, and $\bar{S}_j$ (the closure of S_j) is a compact subset of S_{j+1}. Let d be the metric in S and for $A \subseteq S$, define $d(x, A)$ by

$$d(x, A) = \inf\{d(x, y); y \in A\}. \tag{1.2}$$

Let also the function ϕ be defined by

$$\phi(t) = \begin{cases} 1 & \text{if } t \leqslant 0, \\ 1 - t & \text{if } 0 < t < 1, \\ 0 & \text{if } t \geqslant 1. \end{cases} \tag{1.3}$$

Then for some $0 < \delta < 1$, set

$$f_j(x) = \phi\left[\frac{1}{\delta} d(x, \bar{S}_j)\right]. \tag{1.4}$$

Then f_j is continuous and vanishes outside the compact set $K_j = \bar{S}_{j+\delta}$. It follows that

$$gf_j = g \quad \text{on} \quad \bar{S}_j \quad \text{and} \quad gf_j = 0 \quad \text{on} \quad K_j^c = \bar{S}_{j+\delta}^c. \tag{1.5}$$

Next $|gf_j| \leqslant |g|$, integrable with respect to any probability measure, and $gf_j \to g$ pointwise, since $f_j \to 1$ pointwise. Therefore the dominated convergence theorem applies and gives that

$$\int gf_j\,\mathrm{d}\mathscr{L}_n \to \int g\,\mathrm{d}\mathscr{L}_n \quad \text{as} \quad j \to \infty, \quad \text{for} \quad n \geqslant 1 \tag{1.6}$$

and

$$\int gf_j\,\mathrm{d}\mathscr{L} \to \int g\,\mathrm{d}\mathscr{L} \quad \text{as} \quad j \to \infty. \tag{1.7}$$

Hence by means of our assumption, (1.5), (1.6) and (1.7), one has that

$$\lim_{n\to\infty} \int g\,\mathrm{d}\mathscr{L}_n = \lim_{n\to\infty}\left(\lim_{j\to\infty} \int gf_j\,\mathrm{d}\mathscr{L}_n\right) = \lim_{j\to\infty}\left(\lim_{n\to\infty} \int gf_j\,\mathrm{d}\mathscr{L}_n\right)$$

$$= \lim_{j\to\infty} \int gf_j\,\mathrm{d}\mathscr{L} = \int g\,\mathrm{d}\mathscr{L},$$

provided

$$\lim_{n \to \infty} \left(\lim_{j \to \infty} \int g f_j \, \mathrm{d}\mathscr{L}_n \right) = \lim_{j \to \infty} \left(\lim_{n \to \infty} \int g f_j \, \mathrm{d}\mathscr{L}_n \right). \qquad (1.8)$$

For (1.8) to be true, it suffices to show that

$$\int g f_j \, \mathrm{d}\mathscr{L}_n \underset{j \to \infty}{\to} \int g \, \mathrm{d}\mathscr{L}_n \quad \text{uniformly in } n \geqslant 1. \qquad (1.9)$$

Let M be a bound of $|g|$. We have then, on account of (1.5) and (1.4),

$$\left| \int g f_j \, \mathrm{d}\mathscr{L}_n - \int g \, \mathrm{d}\mathscr{L}_n \right| = \left| \int_{K_j - \bar{S}_j} g f_j \, \mathrm{d}\mathscr{L}_n - \int_{\bar{S}_j^c} g \, \mathrm{d}\mathscr{L}_n \right|$$

$$\leqslant M[\mathscr{L}_n(K_j - \bar{S}_j) + \mathscr{L}_n(\bar{S}_j^c)]. \qquad (1.10)$$

For $\epsilon > 0$, set $\epsilon' = \epsilon/2M$. Then there exists a compact set $K = K(\epsilon')$ such that
$$\mathscr{L}(K) > 1 - \epsilon'. \qquad (1.11)$$

Let j_1 be the smallest j for which $K \subset S_{j_1}$. Then, clearly,

$$K \subset \bar{S}_j \subset K_j \subset \bar{S}_{j+1} \quad (j \geqslant j_1). \qquad (1.12)$$

By means of this and (1.4), we have then that

$$I_K \leqslant f_{j_1} \leqslant I_{\bar{S}_{j_1+1}},$$

so that
$$\int f_{j_1} \, \mathrm{d}\mathscr{L}_n \leqslant \mathscr{L}_n(\bar{S}_{j+1}) \quad (n \geqslant 1), \qquad (1.13)$$

and
$$\mathscr{L}(K) \leqslant \int f_{j_1} \, \mathrm{d}\mathscr{L}. \qquad (1.14)$$

Taking the limits in (1.13) as $n \to \infty$, utilizing the fact that $\int f_{j_1} \, \mathrm{d}\mathscr{L}_n \to \int f_{j_1} \, \mathrm{d}\mathscr{L}$ and also taking into consideration (1.11) and (1.14), we obtain

$$1 - \epsilon' < \mathscr{L}(K) \leqslant \int f_{j_1} \, \mathrm{d}\mathscr{L} \leqslant \liminf \mathscr{L}_n(\bar{S}_{j_1+1}).$$

Hence there exists n_0 such that $\mathscr{L}_n(\bar{S}_j) > 1 - \epsilon', j \geqslant j_1 + 1, n \geqslant n_0$; equivalently,
$$\mathscr{L}_n(\bar{S}_j^c) < \epsilon' \quad (j \geqslant j_1 + 1, n \geqslant n_0). \qquad (1.15)$$

From (1.12) and (1.15), one has then

$$\mathcal{L}_n(K - \bar{S}_j)$$
$$= \mathcal{L}_n(K_j) - \mathcal{L}_n(\bar{S}_j) < 1 - (1 - \epsilon') = \epsilon' \quad (j \geqslant j_1 + 1, n \geqslant n_0).$$
$$(1.16)$$

On account of (1.15) and (1.16), relation (1.10) gives

$$\left| \int gf_j \, d\mathcal{L}_n - \int g \, d\mathcal{L}_n \right| \leqslant M2\epsilon' = \epsilon \quad (j \geqslant j_1 + 1, n \geqslant n_0). \quad (1.17)$$

Finally, applying (1.6) for $n = 1, ..., n_0 - 1$, one can find j_2 such that

$$\left| \int gf_j \, d\mathcal{L}_n - \int g \, d\mathcal{L}_n \right| \leqslant \epsilon \quad (j \geqslant j_2, n = 1, ..., n_0 - 1).$$

From this result and (1.17), it follows then that there exists j_0 such that

$$\left| \int gf_j \, d\mathcal{L}_n - \int g \, d\mathcal{L}_n \right| \leqslant \epsilon \, (j \geqslant j_0, n \geqslant 1),$$

which is what (1.9) asserts. ∎

REMARK 1.1A   This theorem may be established under more general conditions regarding the space S than the ones employed herein.

THEOREM 1.3A   Let S be a metric space and let $\mathcal{S}$ be the topological Borel σ-field in S. For $n = 1, 2, ...$, let $\mathcal{L}_n, \mathcal{L}$ be probability measures defined on $\mathcal{S}$. Then if $\{\mathcal{L}_n\}$ is tight, it is relatively compact. If S is separable and complete, then the converse is also true.

For the proof of this theorem, the reader is referred to Billingsley [4], Theorems 6.1 and 6.2, p. 37.

THEOREM 1.4A   Let P, Q be probability measures defined on the σ-field $\mathcal{A}$ of subsets of Ω and let f, g be densities of P, Q, respectively, relative to a dominating σ-finite measure μ (e.g. $\mu = P + Q$). Then

$$\|P - Q\| = \int |f - g| \, d\mu,$$

where $\|P - Q\| = 2 \sup [|P(A) - Q(A)|; A \in \mathcal{A}].$

Proof For any $A \in \mathscr{A}$, one has

$$P(A) - Q(A) = \int_A f \, \mathrm{d}\mu - \int_A g \, \mathrm{d}\mu = \int_A (f-g) \, \mathrm{d}\mu$$

$$= \int_{A_1} (f-g) \, \mathrm{d}\mu + \int_{A_2} (f-g) \, \mathrm{d}\mu$$

$$= \int_{A_1} (f-g) \, \mathrm{d}\mu - \int_{A_2} (g-f) \, \mathrm{d}\mu,$$

where $A_1 = A \cap (f-g > 0)$, $A_2 = A \cap (f-g < 0)$. Thus

$$|P(A) - Q(A)| = \left| \int_{A_1} (f-g) \, \mathrm{d}\mu - \int_{A_2} (g-f) \, \mathrm{d}\mu \right|. \quad (1.18)$$

Since $f-g > 0$ on A_1 and $g-f > 0$ on A_2, (1.18) is maximised by maximising $\displaystyle\int_{A_1} (f-g) \, \mathrm{d}\mu$ and by minimising $\displaystyle\int_{A_2} (g-f) \, \mathrm{d}\mu$. This is achieved by taking $A = B = (f-g > 0)$, so that $A_1 = B$, $A_2 = \varnothing$. Thus

$$\max [|P(A) - Q(A)|; A \in \mathscr{A}] = \int_B (f-g) \, \mathrm{d}\mu. \quad (1.19)$$

In an entirely similar way, one obtains

$$\max [|P(A) - Q(A)|; A \in \mathscr{A}] = \int_C (g-f) \, \mathrm{d}\mu, \quad (1.20)$$

where $C = (g-f > 0)$. From (1.19) and (1.20), one has then that

$$\int |f-g| \, \mathrm{d}\mu = \int_B (f-g) \, \mathrm{d}\mu + \int_C (g-f) \, \mathrm{d}\mu$$

$$= 2 \max [|P(A) - Q(A)|; A \in \mathscr{A}],$$

as was to be shown. ∎

THEOREM 1.5A Let S be a metric space and let $\mathscr{S}$ be the topological Borel σ-field in S. Then if P, Q are probability measures on $\mathscr{S}$ such that $\int f \, \mathrm{d}P = \int f \, \mathrm{d}Q$ for every real-valued, bounded and continuous function f on S, it follows that P, Q are identical.

This is Theorem 1.3 in Billingsley [4], p. 9.

2   Some theorems employed in Chapter 2

In this section, we prove Vitali's theorem which is instrumental in establishing differentiability in quadratic mean, and also the central limit theorem for martingales which is of basic importance in deriving asymptotic distributions.

THEOREM 2.1 A (*Vitali's theorem*)   For $n = 1, 2, \ldots$, let X_n, X be r.v.s defined on the probability space $(\Omega, \mathscr{A}, P)$. Then $X_n \overset{r}{\to} X$ if and only if $X_n \overset{P}{\to} X$ and $\mathscr{E}|X_n|^r \to \mathscr{E}|X|^r$ finite $(r > 0)$.

For the proof of this theorem, we shall utilize the following lemma due to Scheffé.

LEMMA 2.1 A   For $n = 1, 2, \ldots$, let Y_n, Y be non-negative r.v.s defined on the probability space $(\Omega, \mathscr{A}, P)$. Then $Y_n \overset{P}{\to} Y$ and $\mathscr{E}Y_n \to \mathscr{E}Y$ finite if and only if $Y_n \overset{1}{\to} Y$.

Proof   One direction of the lemma is immediate. Thus it suffices to prove the other one. Now $Y_n \overset{P}{\to} Y$ implies that $(Y - Y_n)^+ \to 0$. Furthermore, the fact that $Y_n \geqslant 0$ also implies that $(Y - Y_n)^+ \leqslant Y$. Then the dominated convergence theorem applies and gives that $\mathscr{E}(Y - Y_n)^+ \to 0$. By assumption, we also have that $\mathscr{E}(Y - Y_n) \to 0$. Hence $\mathscr{E}(Y - Y_n)^- \to 0$. This result, along with $\mathscr{E}(Y - Y_n)^+ \to 0$, implies then that $\mathscr{E}|Y - Y_n| \to 0$, as was to be seen. ▮

Proof of Theorem 2.1 A   Since one direction of the theorem is immediate, we concentrate in establishing the other one. To this end, set $Y_n = |X_n|^r$, $Y = |X|^r$. Then Lemma 2.1 A applies and gives that $Y_n \overset{1}{\to} Y$, or equivalently, $|X_n|^r \overset{1}{\to} |X|^r$. Hence $\int |X_n|^r \mathrm{d}P$ are uniformly continuous by the L_r-convergence theorem (see Loève [1], p. 163). This result, along with the fact that $X_n \overset{P}{\to} X$, implies that $X_n \overset{r}{\to} X$ again by the theorem just cited. The proof of the theorem is completed. ▮

The theorem to be proved below is known as the central limit theorem for martingales for reasons to be explained shortly. A proof of this result is presented in Billingsley [1], Theorem 9.1, pp. 52–61 under the assumption that the underlying r.v.s have a finite moment of order $2 + \delta \, (\delta > 0)$. Subsequently, this assumption was weakened by the same author (see Billingsley [2]), so that δ may be taken to be equal to zero. However, none of these proofs is particularly easy to follow. An alternative proof of the same theorem is also discussed in Billingsley [4], Theorem 23.1, pp. 206–8, but this proof is based on previous results on weak convergence of probability measures. Finally, an exposition of the proof of the theorem, based only on the standard martingale and ergodic theory theorems, is given by Ibragimov [1]. The proof to be presented herein is based on this last paper.

THEOREM 2.2A *(central limit theorem for martingales)* Let $\{X_n\}, n = 1, 2, \ldots$ be a stationary and ergodic stochastic process defined on the probability space $(\Omega, \mathscr{A}, P)$ and such that

$$\mathscr{E}X_n^2 = \sigma^2, \; 0 < \sigma^2 < \infty, \; \mathscr{E}(X_n | \mathscr{A}_{n-1}) = 0 \text{ a.s.} \quad (n \geqslant 1), \quad (2.1)$$

where $\mathscr{A}_n = \mathscr{B}(X_1, \ldots, X_n), n \geqslant 1, \mathscr{A}_0 = \{\varnothing, \Omega\}$. Then

$$\mathscr{L}\left(\frac{1}{\sigma\sqrt{n}} \sum_{j=1}^{n} X_j \Big| P\right) \Rightarrow N(0, 1). \quad (2.2)$$

REMARK 2.1A

(i) Clearly, $\mathscr{E}X_n = 0, \; n \geqslant 1$, so that $\sigma^2 = \mathscr{E}X_n^2, \; n \geqslant 1$.

(ii) For all $n \geqslant 1$ and $k \geqslant 1$, $\mathscr{E}(X_n X_{n+k})$ is finite, by the Hölder inequality, and

$$\mathscr{E}(X_n X_{n+k}) = \mathscr{E}[\mathscr{E}(X_n X_{n+k} | \mathscr{A}_{n+k-1})]$$
$$= \mathscr{E}[X_n \mathscr{E}(X_{n+k} | \mathscr{A}_{n+k-1})] = 0$$

by means of (2.1); i.e. $\mathscr{E}(X_n X_{n+k}) = 0, \; n, k \geqslant 1$, and hence

$$\sigma^2\left(\sum_{j=1}^{n} X_j\right) = \sum_{j=1}^{n} \sigma^2(X_j) = n\sigma^2.$$

Therefore (2.2) is the conclusion of the central limit theorem in its ordinary form. Furthermore, the partial sums

$$\left\{\sum_{j=1}^{n} X_j\right\} \quad (n \geqslant 1),$$

form a martingale because, by means of (2.1), one has

$$
\mathscr{E}\left(\sum_{j=1}^{n} X_j \,\middle|\, \sum_{j=1}^{i} X_j, i = 1, \dots, n-1\right)
$$
$$
\overset{\text{a.s.}}{=} \mathscr{E}\left[\mathscr{E}\left(\sum_{j=1}^{n} X_j \,\middle|\, \sum_{j=1}^{i} X_j, i = 1, \dots, n-1\right)\middle|\mathscr{A}_{n-1}\right]
$$
$$
\overset{\text{a.s.}}{=} \mathscr{E}\left[\mathscr{E}\left(\sum_{j=1}^{n} X_j \middle| \mathscr{A}_{n-1}\right)\middle|\sum_{j=1}^{i} X_j, i = 1, \dots, n-1\right]
$$
$$
\overset{\text{a.s.}}{=} \mathscr{E}\left[\sum_{j=1}^{n-1} X_j + \mathscr{E}(X_n|\mathscr{A}_{n-1})\middle|\sum_{j=1}^{i} X_j, i = 1, \dots, n-1\right]
$$
$$
\overset{\text{a.s.}}{=} \mathscr{E}\left(\sum_{j=1}^{n-1} X_j \,\middle|\, \sum_{j=1}^{i} X_j, i = 1, \dots, n-1\right) \overset{\text{a.s.}}{=} \sum_{j=1}^{n-1} X_j.
$$

Hence the qualification of the theorem as central limit theorem for martingales.

(iii) Let $Y_n = X_n/\sigma$. Then $\{Y_n\}$, $n \geqslant 1$, is also stationary and ergodic (see Proposition 6.6 and Proposition 6.31 in Breiman [1], pp. 105, 119) and

$$
\mathscr{E}Y_n^2 = 1, \mathscr{E}(Y_n|Y_1, \dots, Y_{n-1}) = 0 \text{ a.s.} \quad (n \geqslant 1).
$$

So in the original process, we may assume that $\sigma = 1$ and we shall do so.

(iv) Let $\{\hat{X}_n\}$, $n \geqslant 1$, be the coordinate process and let $\hat{P}$ be the distribution of $(X_1, X_2, \dots)$ under P. Then the distribution of $(X_1, X_2, \dots)$ under P is the same as the distribution of $(\hat{X}_1, \hat{X}_2, \dots)$ under $\hat{P}$, so that

$$
\mathscr{L}\left(n^{-\frac{1}{2}} \sum_{j=1}^{n} X_j\middle|P\right) = \mathscr{L}\left(n^{-\frac{1}{2}} \sum_{j=1}^{n} \hat{X}_j\middle|\hat{P}\right).
$$

Thus we may assume that $(\Omega, \mathscr{A}, P) = (R^\infty, \mathscr{B}^\infty, \hat{P})$ and that the original process is the coordinate process, and we shall do so.

(v) The stationary and ergodic coordinate process $\{X_n\}$, $n \geqslant 1$, may be extended to the process $\{X_n\}, n = 0, \pm 1, \dots$ which is also stationary and ergodic and has the following property

$$
\mathscr{E}(X_n|X_{n-1}, X_{n-2}, \dots) = 0 \text{ a.s.} \quad (n = 0, \pm 1, \dots) \qquad (2.3)
$$

(see also exercise at the end of this section).

That (2.3) holds is seen as follows. Set

$$
\mathscr{F}_n = \mathscr{B}(X_n, X_{n-1}, \dots) \quad (n = 0, \pm 1, \dots) \qquad (2.4)
$$

and $$\mathscr{F}_{n,\,n-k} = \mathscr{B}(X_n, ..., X_{n-k}) \quad (k \geqslant 1).$$

Then for each n,

$$\mathscr{E}(X_n | \mathscr{F}_{n-1}) = \lim_{k \to \infty} \mathscr{E}(X_n | \mathscr{F}_{n-1,\,n-k}) \quad \text{a.s.} \tag{2.5}$$

(see, e.g. Theorem 5.21 in Breiman [1], p. 92). Now for every $A \in \mathscr{F}_{n-1,\,n-k}$, there exists $B \in \mathscr{B}^k$ such that

$$A = (X_{n-k}, ..., X_{n-1})^{-1} B$$

and, by stationarity,

$$\int_A X_n \, d\hat{P} = \int X_n I_{(X_{n-k}, ..., X_{n-1})^{-1}B} (X_{n-k}, ..., X_{n-1}) \, d\hat{P}$$

$$= \int X_{k+1} I_{(X_1, ... \, X_k)^{-1}B} (X_1, ..., X_k) \, d\hat{P} = \int_{A'} X_{k+1} \, d\hat{P},$$

where $A' = (X_1, ..., X_k)^{-1} B$. But by means of (2.1),

$$\int_{A'} X_{k+1} \, d\hat{P} = \int_{A'} \mathscr{E}(X_{k+1} | \mathscr{A}_k) \, d\hat{P} = 0,$$

so that $$\int_A X_n \, d\hat{P} = 0 \quad \text{for every} \quad A \in \mathscr{F}_{n-1,\,n-k}.$$

Finally, $$\int_A X_n \, d\hat{P} = \int_A \mathscr{E}(X_n | \mathscr{F}_{n-1\,\,n-k}) \, d\hat{P},$$

so that

$$\int_A \mathscr{E}(X_n | \mathscr{F}_{n-1,\,n-k}) \, d\hat{P} = 0 \quad \text{for every} \quad A \in \mathscr{F}_{n-1\,\,n-k}.$$

It follows that $\mathscr{E}(X_n | \mathscr{F}_{n-1,\,n-k}) = 0$ a.s. and therefore from (2.5) one concludes (2.3).

Proof of Theorem 2.2A Set

$$\left. \begin{aligned} S_k &= \sum_{j=1}^{k} X_j & (k = 1, ..., n, S_0 = 0), \\ Z_k &= Z_{k,\,n} = n^{-\frac{1}{2}} S_k & (k = 0, 1, ..., n; n \geqslant 1), \\ f_n(t) &= \mathscr{E}[\exp{(itZ_n)}] = \mathscr{E}[\exp{(itS_n n^{-\frac{1}{2}})}]. \end{aligned} \right\} \tag{2.6}$$

Since f_n defined by (2.6) is the characteristic function of $S_n n^{-\frac{1}{2}}$, by the continuity theorem, it suffices to show that

$$f_n(t) - e^{-\frac{1}{2}t^2} \to 0 \quad (t \in R). \tag{2.7}$$

Set

$$\phi_k(t) = \phi_{k,n}(t) = \exp\left(-\frac{n-k}{n}\frac{t^2}{2}\right) \quad (t \in R, k = 0, 1, \ldots, n, n \geqslant 1), \Bigg\}$$
$$\psi_k(t) = \psi_{k,n}(t) = \phi_k(t)\mathcal{E}[\exp(itZ_k)] \quad (t \in R, k = 0, 1, \ldots, n, n \geqslant 1). \Bigg\}$$
(2.8)

From (2.6) and (2.8), we clearly have that

$$\psi_0(t) = \phi_0(t) = e^{-\frac{1}{2}t^2}, \ \phi_n(t) = 1, \ \psi_n(t) = f_n(t) \quad (t \in R, n \geqslant 1).$$
(2.9)

On the other hand, by means of (2.9), one has

$$\sum_{k=0}^{n-1} [\psi_{k+1}(t) - \psi_k(t)] = \psi_n(t) - \psi_0(t) = f_n(t) - e^{-\frac{1}{2}t^2}.$$

Therefore, on account of (2.7), in order to establish the theorem it suffices to show that

$$\sum_{k=0}^{n-1} [\psi_{k+1}(t) - \psi_k(t)] \to 0 \quad (t \in R).$$
(2.10)

Since the proof of (2.10) is rather long, we present the intermediate auxiliary results as lemmas.

LEMMA 2.2A   For $n \geqslant 1$ and $k = 0, 1, \ldots, n$, let $\mathscr{F}_k$ be defined by (2.4) and let ϕ_k and ψ_k be defined by (2.8). Then for every $t \in R$, one has that

$$\left|\sum_{k=0}^{n-1} [\psi_{k+1}(t) - \psi_k(t)]\right| \leqslant \left[\frac{|t|^3}{n^{\frac{1}{5}}} + \frac{t^4}{n} + t^2 \int_{(|X_1| > n^{1/10})} X_1^2 \, d\hat{P}\right]$$
$$+ \frac{t^2}{n}\left|\sum_{k=0}^{n-1} \phi_{k+1}(t)\,\mathcal{E}\{\exp(itZ_k)\,[1 - \mathcal{E}(X_{k+1}^2|\mathscr{F}_k)]\}\right|. \quad (2.11)$$

Proof   From (2.8) and (2.6), one has

$$\psi_{k+1}(t) - \psi_k(t) = \exp\left(-\frac{n-(k+1)}{n}\frac{t^2}{2}\right)\mathcal{E}[\exp(itZ_{k+1})]$$

$$-\exp\left(-\frac{n-k}{n}\frac{t^2}{2}\right)\mathcal{E}[\exp(itZ_k)]$$

$$= \exp\left(-\frac{n-(k+1)}{n}\frac{t^2}{2}\right)\Big\{\mathcal{E}[\exp(itZ_{k+1})]$$

$$- \exp\left(-\frac{t^2}{2n}\right)\mathscr{E}\left[\exp\left(itZ_k\right)\right]\Big\}$$

$$= \phi_{k+1}(t)\,\mathscr{E}\left[\exp\left(itZ_k + itX_{k+1}n^{-\frac{1}{2}}\right)\right]$$

$$- \exp\left(-\frac{t^2}{2n}\right)\mathscr{E}\left[\exp\left(itZ_k\right)\right];$$

i.e. $\quad \psi_{k+1}(t) - \psi_k(t) = \phi_{k+1}(t)\Big\{\mathscr{E}\left[\exp\left(itZ_k\right)\exp\left(itX_{k+1}n^{-\frac{1}{2}}\right)\right]$

$$- \exp\left(-\frac{t^2}{2n}\right)\mathscr{E}\left[\exp\left(itZ_k\right)\right]\Big\}. \qquad (2.12)$$

Now consider the expansion

$$e^{ix} = 1 + \frac{ix}{1!} + \frac{(ix)^2}{2!} + \dots + \frac{(ix)^m}{m!} + R(m, x), \quad |R(m, x)| \leqslant \left|\frac{(ix)^{m+1}}{(m+1)!}\right|$$

and apply it for $m = 1$ and $m = 2$. We get then

$$e^{ix} = 1 + ix + R(1, x), \quad |R(1, x)| \leqslant \tfrac{1}{2}x^2$$

and $\quad e^{ix} = 1 + ix - \tfrac{1}{2}x^2 + R(2, x), \quad |R(2, x)| \leqslant \dfrac{|x|^3}{3!} \leqslant |x|^3.$

Also $\quad |R(2, x)| = \left|e^{ix} - (1 + ix) + \tfrac{1}{2}x^2\right|$

$$= \left|R(1, x) + \tfrac{1}{2}x^2\right| \leqslant \tfrac{1}{2}x^2 + \tfrac{1}{2}x^2 = x^2.$$

That is, one has

$$\left.\begin{array}{l} e^{ix} = 1 + ix - \tfrac{1}{2}x^2 + R(2, x), \\[2mm] \text{where} \quad |R(2, x)| \leqslant |x|^3 \quad \text{and also} \quad |R(2, x)| \leqslant x^2. \end{array}\right\} \qquad (2.13)$$

Replacing x by $tX_{k+1}n^{-\frac{1}{2}}$ in (2.13), we get

$$\left.\begin{array}{l} \exp\left(itX_{k+1}n^{-\frac{1}{2}}\right) = 1 + \dfrac{itX_{k+1}}{\sqrt{n}} - \dfrac{t^2X_{k+1}^2}{2n} + R_n(t;\,X_{k+1}) \\[4mm] \text{with} \\[2mm] |R_n(t;\,X_{k+1})| \leqslant \dfrac{|tX_{k+1}|^3}{n\sqrt{n}} \quad \text{and also} \quad |R_n(t;\,X_{k+1})| \leqslant \dfrac{t^2X_{k+1}^2}{n}. \end{array}\right\}$$
$$(2.14)$$

Next $\quad \mathscr{E}\left[\exp\left(itZ_k\right)\exp\left(itX_{k+1}n^{-\frac{1}{2}}\right)|\mathscr{F}_k\right]$

$$\overset{\text{a.s.}}{=} \exp\left(itZ_k\right)\mathscr{E}\left[\exp\left(itX_{k+1}n^{-\frac{1}{2}}\right)|\mathscr{F}_k\right]$$

since Z_k is $\mathscr{F}_k$-measurable. On the other hand, (2.14) and (2.3) imply that

$$\mathscr{E}[\exp{(itX_{k+1}n^{-\frac{1}{2}})}|\mathscr{F}_k]$$
$$\overset{\text{a.s.}}{=} 1 - \frac{t^2}{2n}\mathscr{E}(X_{k+1}^2|\mathscr{F}_k) + \mathscr{E}[R_n(t; X_{k+1})|\mathscr{F}_k].$$

Therefore

$$\mathscr{E}[\exp{(itZ_k)}\exp{(itX_{k+1}n^{-\frac{1}{2}})}]$$
$$= \mathscr{E}\{\mathscr{E}[\exp{(itZ_k)}\exp{(itX_{k+1}n^{-\frac{1}{2}})}]|\mathscr{F}_k\}$$
$$= \mathscr{E}\{\exp{(itZ_k)}\,\mathscr{E}[\exp{(itX_{k+1}n^{-\frac{1}{2}})}|\mathscr{F}_k]\}$$
$$= \mathscr{E}\left[\exp{(itZ_k)}\left\{1 - \frac{t^2}{2n}\mathscr{E}(X_{k+1}^2|\mathscr{F}_k) + \mathscr{E}[R_n(t; X_{k+1})|\mathscr{F}_k]\right\}\right].$$

By means of this, (2.12) becomes as follows

$$\psi_{k+1}(t) - \psi_k(t)$$
$$= \phi_{k+1}(t)\left\{\mathscr{E}[\exp{(itZ_k)}] - \frac{t^2}{2n}\mathscr{E}[\exp{(itZ_k)}\,\mathscr{E}(X_{k+1}^2|\mathscr{F}_k)]\right\}$$
$$+ \mathscr{E}\{\exp{(itZ_k)}\,\mathscr{E}[R_n(t; X_{k+1})|\mathscr{F}_k]\} - \exp\left(-\frac{t^2}{2n}\right)\mathscr{E}[\exp{(itZ_k)}].$$

$$(2.15)$$

But
$$|\mathscr{E}\{\exp{(itZ_k)}\,\mathscr{E}[R_n(t; X_{k+1})|\mathscr{F}_k]\}|$$
$$\leqslant \mathscr{E}\,|\exp{(itZ_k)}\,\mathscr{E}[R_n(t; X_{k+1})|\mathscr{F}_k]|$$
$$\leqslant \mathscr{E}\,|\mathscr{E}[R_n(t; X_{k+1})|\mathscr{F}_k]|$$
$$\leqslant \mathscr{E}\,|R_n(t; X_{k+1})|,\qquad\qquad (2.16)$$

whereas by means of (2.14) and stationarity (see, e.g. Proposition 6.6 in Breiman [1], p. 105), one has

$$\mathscr{E}\,|R_n(t; X_{k+1})|$$
$$= \int_{(|X_{k+1}|\leqslant n^{1/10})} |R_n(t; X_{k+1})|\,\mathrm{d}\hat{P}$$
$$\quad + \int_{(|X_{k+1}|>n^{1/10})} |R_n(t; X_{k+1})|\,\mathrm{d}\hat{P}$$
$$\leqslant \frac{|t|^3}{n\sqrt{n}}\int_{(|X_{k+1}|\leqslant n^{1/10})} |X_{k+1}|^3\,\mathrm{d}\hat{P} + \frac{t^2}{n}\int_{(|X_{k+1}|>n^{1/10})} X_{k+1}^2\,\mathrm{d}\hat{P}$$
$$\leqslant \frac{|t|^3}{n^{1+\frac{1}{5}}} + \frac{t^2}{n}\int_{(|X_1|>n^{1/10})} X_1^2\,\mathrm{d}\hat{P}.$$

On account of this, (2.16) becomes as follows

$$|\mathscr{E}\{\exp(itZ_k)\,\mathscr{E}[R_n(t;X_{k+1})|\mathscr{F}_k]\}| \leqslant \frac{|t|^3}{n^{1+\frac{1}{5}}} + \frac{t^2}{n}\int_{(|X_1|>n^{1/10})} X_1^2\,d\hat{P}. \tag{2.17}$$

In the expansion

$$e^x = 1 + x + \tfrac{1}{2}x^2\,e^{x\theta(x)}, \quad 0 < \theta(x) < 1,$$

replace x by $-\dfrac{t^2}{2n}$. Then we obtain

$$\exp\left(-\frac{t^2}{2n}\right) = 1 - \frac{t^2}{2n} + \frac{t^4}{8n^2}\theta_n^*(t),$$

$$0 < \theta_n^*(t) = \exp\left(-\frac{t^2\theta_n(t)}{2n}\right) < 1.$$

By means of this and the fact that $|\phi_k(t)| \leqslant 1$, one has that

$$\left|\sum_{k=0}^{n-1}\phi_{k+1}(t)\Big\{\mathscr{E}[\exp(itZ_k)] - \frac{t^2}{2n}\mathscr{E}[\exp(itZ_k)\,\mathscr{E}(X_{k+1}^2|\mathscr{F}_k)]\right.$$

$$\left. - \exp\left(-\frac{t^2}{2n}\right)\mathscr{E}[\exp(itZ_k)]\Big\}\right|$$

$$= \left|\sum_{k=0}^{n-1}\phi_{k+1}(t)\Big\{\mathscr{E}[\exp(itZ_k)] - \frac{t^2}{2n}\mathscr{E}[\exp(itZ_k)\,\mathscr{E}(X_{k+1}^2|\mathscr{F}_k)]\right.$$

$$\left. - \mathscr{E}[\exp(itZ_k)] + \frac{t^2}{2n}\mathscr{E}[\exp(itZ_k)] - \frac{t^4}{8n^2}\theta_n^*(t)\,\mathscr{E}[\exp(itZ_k)]\Big\}\right|$$

$$\leqslant \frac{t^4}{n} + \frac{t^2}{n}\left|\sum_{k=0}^{n-1}\phi_{k+1}(t)\,\mathscr{E}\{\exp(itZ_k)\,[1 - \mathscr{E}(X_{k+1}^2|\mathscr{F}_k)]\}\right|. \tag{2.18}$$

Then summing over k in (2.15) and utilizing (2.17) and (2.18), one obtains (2.11). The proof of the lemma is completed. ∎

It has been stated that in order to establish the theorem it suffices to show (2.10). By means of Lemma 2.2A, this will be true if the right-hand side of (2.11) converges to zero (as $n \to \infty$). Since the first summand does so by the fact that $\mathscr{E}X_1^2 < \infty$, it suffices to show that

$$\frac{t^2}{n}\left|\sum_{k=0}^{n-1}\phi_{k+1}(t)\,\mathscr{E}\{\exp(itZ_k)\,[1 - \mathscr{E}(X_{k+1}^2|\mathscr{F}_k)]\}\right| \to 0,$$

or by setting

$$Y_k = \mathscr{E}(X_{k+1}^2|\mathscr{F}_k) \quad (k = 0, 1, \ldots, n), \tag{2.19}$$

it suffices to show that

$$\frac{t^2}{n}\left|\sum_{k=0}^{n-1}\phi_{k+1}(t)\,\mathscr{E}[\exp{(\mathrm{i}tZ_k)}\,(1-Y_k)]\right| \to 0. \qquad (2.20)$$

For $n \geqslant 1$ and $k = 0, 1, ..., n$, define the r.v.s V_{kn} by

$$V_{kn} = \mathscr{E}(X_{k+1}^2 | \mathscr{F}_{k-n}), \qquad (2.21)$$

where $\mathscr{F}_{k-n}$ is given by (2.4).

For each fixed k and as $n \to \infty$, (2.4) gives that

$$\mathscr{F}_{n-k} \downarrow \mathscr{C}_k = \bigcap_{n=1}^{\infty} \mathscr{F}_{k-n}. \qquad (2.22)$$

Therefore Theorem 5.24 in Breiman [1], p. 93 applies and gives that

$$V_{kn} \underset{(1)}{\overset{\text{a.s.}}{\to}} V_k = \mathscr{E}(X_{k+1}^2 | \mathscr{C}_k). \qquad (2.23)$$

Furthermore,

$$\mathscr{E}V_k = \mathscr{E}[\mathscr{E}(X_{k+1}^2 | \mathscr{C}_k)] = \mathscr{E}X_{k+1}^2 = 1. \qquad (2.24)$$

Now, in order to facilitate the proof of (2.20), we establish first the following result.

LEMMA 2.3A For each $k \geqslant 0$, let Y_k and V_k be defined by (2.19) and (2.23), respectively. Then one has that

$$\frac{t^2}{n}\left|\sum_{k=0}^{n-1}\phi_{k+1}(t)\,\mathscr{E}[\exp{(\mathrm{i}tZ_k)}\,(Y_k-1)]\right.$$
$$\left. - \sum_{k=0}^{n-1}\phi_{k+1}(t)\,\mathscr{E}[\exp{(\mathrm{i}tZ_k)}\,(V_k-1)]\right| \to 0. \qquad (2.25)$$

Proof From (2.23), one has, in particular, that $V_{0n} \overset{(1)}{\to} V_0$. Therefore for $\epsilon > 0$ there exists an integer $m = m(\epsilon) \, (> 0)$ such that
$$\mathscr{E}|V_{0m} - V_0| < \epsilon. \qquad (2.26)$$

Next for $k > m$, and stationarity, one has that

$$\mathscr{E}[\exp{(\mathrm{i}tZ_k)}\,(Y_k - V_k)] = \mathscr{E}\left[\exp\left(\mathrm{i}tn^{-\frac{1}{2}}\sum_{j=1}^{k}X_j\right)(Y_k - V_k)\right]$$

$$= \mathscr{E}\left[\exp\left(\mathrm{i}tn^{-\frac{1}{2}}\sum_{j=-k+1}^{0}X_j\right)(Y_0 - V_0)\right]$$

$$= \mathscr{E}\left\{\exp\left(\mathrm{i}tn^{-\frac{1}{2}}\sum_{j=-k+1}^{-m}X_j\right)(Y_0 - V_0) + \exp\left(\mathrm{i}tn^{-\frac{1}{2}}\sum_{j=-k+1}^{-m}X_j\right)\right.$$

$$\left. \times\left[\exp\left(\mathrm{i}tn^{-\frac{1}{2}}\sum_{j=-m+1}^{0}X_j\right) - 1\right](Y_0 - V_0)\right\}. \qquad (2.27)$$

From (2.23) and (2.22), it follows that V_0 is $\mathscr{F}_{-n}$-measurable for all $n \geqslant 1$. Also, from (2.19), Y_0 is $\mathscr{F}_0$-measurable. Therefore

$$\mathscr{E}\left[\exp\left(itn^{-\frac{1}{2}}\sum_{j=-k+1}^{-m}X_j\right)(Y_0-V_0)\right]$$

$$= \mathscr{E}\left\{\mathscr{E}\left[\exp\left(itn^{-\frac{1}{2}}\sum_{j=-k+1}^{-m}X_j\right)(Y_0-V_0)|\mathscr{F}_{-m}\right]\right\}$$

$$= \mathscr{E}\left\{\exp\left(itn^{-\frac{1}{2}}\sum_{j=-k+1}^{-m}X_j\right)[\mathscr{E}(Y_0|\mathscr{F}_{-m})-\mathscr{E}(V_0|\mathscr{F}_{-m})]\right\}$$

$$= \mathscr{E}\left\{\exp\left(itn^{-\frac{1}{2}}\sum_{!j=-k+1}^{-m}X_j\right)[\mathscr{E}(Y_0|\mathscr{F}_{-m})-V_0]\right\}$$

$$= \mathscr{E}\left[\exp\left(itn^{-\frac{1}{2}}\sum_{j=-k+1}^{-m}X_j\right)(V_{0m}-V_0)\right],$$

the last equality being so because, on account of (2.19) and (2.21),

$$\mathscr{E}(Y_0|\mathscr{F}_{-m}) = \mathscr{E}[\mathscr{E}(X_1^2|\mathscr{F}_0)|\mathscr{F}_{-m}] \overset{\text{a.s.}}{=} \mathscr{E}(X_1^2|\mathscr{F}_{-m}) = V_{0m}.$$

Therefore

$$\left|\mathscr{E}\left[\exp\left(itn^{-\frac{1}{2}}\sum_{j=-k+1}^{-m}X_j\right)(Y_0-V_0)\right]\right| \leqslant \mathscr{E}|V_{0m}-V_0| < \epsilon, \quad (2.28)$$

the last inequality being true because of (2.26).

Next, by employing the inequality $|e^{iz}-1| \leqslant |z|$, we obtain

$$\left|\mathscr{E}\left\{\exp\left(itn^{-\frac{1}{2}}\sum_{j=-k+1}^{-m}X_j\right)\left[\exp\left(itn^{-\frac{1}{2}}\sum_{j=-m+1}^{0}X_j\right)-1\right](Y_0-V_0)\right\}\right|$$

$$\leqslant \mathscr{E}\left|\left[\exp\left(itn^{-\frac{1}{2}}\sum_{j=-m+1}^{0}X_j\right)-1\right](Y_0-V_0)\right|$$

$$= \int_{\left(\left|\sum_{j=-m+1}^{0}X_j\right|>\epsilon\sqrt{n}\right)}\left\{\left|\left[\exp\left(itn^{-\frac{1}{2}}\sum_{j=-m+1}^{0}X_j\right)-1\right](Y_0-V_0)\right|\right\}\mathrm{d}\hat{P}$$

$$+ \int_{\left(\left|\sum_{j=-m+1}^{0}X_j\right|\leqslant\epsilon\sqrt{n}\right)}\left\{\left|\left[\exp\left(itn^{-\frac{1}{2}}\sum_{j=-m+1}^{0}X_j\right)-1\right](Y_0-V_0)\right|\right\}\mathrm{d}\hat{P}$$

$$= 2 \int_{\left(\left|\sum\limits_{j=-m+1}^{0} X_j\right| > \epsilon\sqrt{n}\right)} |Y_0 - V_0| \, \mathrm{d}\hat{P} + \int_{\left(\left|\sum\limits_{j=-m+1}^{0} X_j\right| \leqslant \epsilon\sqrt{n}\right)}$$

$$\times \left(|t| \, n^{-\frac{1}{2}} \left|\sum_{j=-m+1}^{0} X_j\right| |Y_0 - V_0|\right) \mathrm{d}\hat{P}$$

$$\leqslant 2 \int_{\left(\left|\sum\limits_{j=-m+1}^{0} X_j\right| > \epsilon\sqrt{n}\right)} |Y_0 - V_0| \, \mathrm{d}\hat{P} + \epsilon |t| \int |Y_0 - V_0| \, \mathrm{d}\hat{P}$$

and this last expression can be made less than a multiple of ϵ since the first summand goes to zero (as $n \to \infty$) because of the finiteness of the integral $\int |Y_0 - V_0| \, \mathrm{d}\hat{P}$. By means of this result and (2.28), (2.25) follows in an obvious manner.]

Now by means of (2.20) and Lemma 2.3 A, in order to prove the theorem it suffices to show that

$$\frac{t^2}{n} \left|\sum_{k=0}^{n-1} \phi_{k+1}(t) \, \mathscr{E}[\exp(\mathrm{i}tZ_k)(V_k - 1)]\right| \to 0,$$

or by setting $\qquad W_k = V_k - 1,$ (2.29)

it suffices to show that

$$\frac{t^2}{n} \left|\sum_{k=0}^{n-1} \phi_{k+1}(t) \, \mathscr{E}[\exp(\mathrm{i}tZ_k) W_k]\right| \to 0. \qquad (2.30)$$

This will be shown by appropriately truncating W_k. More precisely, for $M > 0$, set

$$W_k(M) = \begin{cases} W_k & \text{if } |W_k| \leqslant M \\ 0 & \text{if } |W_k| > M \end{cases}; \quad \overline{W}_k(M) = \begin{cases} 0 & \text{if } |W_k| \leqslant M \\ W_k & \text{if } |W_k| > M. \end{cases}$$
$$(2.31)$$

From (2.31), it follows that $W_k = W_k(M) + \overline{W}_k(M)$, so that

$$\frac{t^2}{n} \left|\sum_{k=0}^{n-1} \phi_{k+1}(t) \, \mathscr{E}[\exp(\mathrm{i}tZ_k) W_k]\right|$$

$$\leqslant \frac{t^2}{n} \left|\sum_{k=0}^{n-1} \phi_{k+1}(t) \, \mathscr{E}[\exp(\mathrm{i}tZ_k) W_k(M)]\right|$$

$$+ \frac{t^2}{n} \left|\sum_{k=0}^{n-1} \phi_{k+1}(t) \, \mathscr{E}[\exp(\mathrm{i}tZ_k) \overline{W}_k(M)]\right|$$

$$\leqslant \frac{t^2}{n} \left|\sum_{k=0}^{n-1} \phi_{k+1}(t) \, \mathscr{E}[\exp(\mathrm{i}tZ_k) W_k(M)]\right| + \frac{t^2}{n} \sum_{k=0}^{n-1} \mathscr{E} |\overline{W}_k(M)|. \quad (2.32)$$

By (2.31), stationarity and the fact that $\mathscr{E}\,|W_0| < \infty$ (by (2.29) and (2.24)), one has that

$$\mathscr{E}\,|\overline{W}_k(M)| = \int_{(|W_k|>M)} |W_k|\,\mathrm{d}\hat{P} = \int_{(|W_0|>M)} |W_0|\,\mathrm{d}\hat{P} < \epsilon$$

for sufficiently large M. For such an M,

$$\frac{t^2}{n}\sum_{k=0}^{n-1} \mathscr{E}\,|\overline{W}_k(M)| \leqslant \epsilon t^2$$

and therefore from (2.32) it follows that in order to establish (2.30), and hence the theorem, it suffices to show that

$$\Lambda_n = \frac{t^2}{n}\sum_{k=0}^{n-1} \phi_{k+1}(t)\,\mathscr{E}[\exp{(\mathrm{i}tZ_k)}\,W_k(M)] \to 0. \qquad (2.33)$$

The proof of it will be facilitated by the following auxiliary result.

LEMMA 2.4A With Λ_n given by (2.33), one has that

$$\begin{aligned}
\Lambda_n = {}& \frac{t^4}{4n^2}\sum_{k=1}^{n-1} \phi_{k+1}(t)\,\mathscr{E}\left\{Y_{k-1}\exp{(\mathrm{i}tZ_{k+1})}\left[\sum_{j=0}^{k-1} W_j(M)\right]\right\} \\
& - \frac{t^4}{4n^2}\theta_n(t)\sum_{k=1}^{n-1} \phi_{k+1}(t)\,\mathscr{E}\left\{\exp{(\mathrm{i}tZ_{k-1})}\left[\sum_{j=0}^{k-1} W_j(M)\right]\right\} \\
& - \frac{t^2}{2n}\sum_{k=1}^{n-1} \phi_{k+1}(t)\,\mathscr{E}\left\{\exp{(\mathrm{i}tZ_{k-1})}\left[\sum_{j=0}^{k-1} W_j(M)\right]R_n(t, X_k)\right\} \\
& + \frac{t^2}{2n}\phi_n(t)\,\mathscr{E}\left\{\exp{(\mathrm{i}tZ_{n-1})}\left[\sum_{j=0}^{n-1} W_j(M)\right]\right\}, \qquad (2.34)
\end{aligned}$$

where $0 < \theta_n(t) < 1$ and $|R_n(t, X_k)| \leqslant \dfrac{t^2 X_k^2}{n}$.

Proof We have

$$\sum_{k=0}^{n-1} \phi_{k+1}(t)\exp{(\mathrm{i}tZ_k)}\,W_k(M) = \sum_{k=1}^{n} \phi_k(t)\exp{(\mathrm{i}tZ_{k-1})}\,W_{k-1}(M). \qquad (2.35)$$

Now for α_k, β_k, $k = 1, \ldots, n$ and $s_k = \sum_{k=1}^{k}\beta_j$, consider the following Abel's identity

$$\sum_{k=1}^{n}\alpha_k\beta_k = \sum_{k=1}^{n-1} s_k(\alpha_k - \alpha_{k+1}) + s_n\alpha_n = \sum_{k=1}^{n-1}\left(\sum_{j=1}^{k}\beta_j\right)(\alpha_k - \alpha_{k+1}) + s_n\alpha_n.$$

Applying this identity with

$$\alpha_k = \phi_k(t) \exp(\mathrm{it}Z_{k-1}) \quad \text{and} \quad \beta_k = W_{k-1}(M),$$

(2.35) becomes as follows

$$\sum_{k=1}^{n} \phi_k(t) \exp(\mathrm{it}Z_{k-1}) W_{k-1}(M)$$

$$= \sum_{k=1}^{n-1} \left[\sum_{j=1}^{k} W_{j-1}(M) \right] [\phi_k(t) \exp(\mathrm{it}Z_{k-1}) - \phi_{k+1}(t) \exp(\mathrm{it}Z_k)]$$

$$+ \left[\sum_{j=1}^{n} W_{j-1}(M) \right] \phi_n(t) \exp(\mathrm{it}Z_{n-1})$$

$$= \sum_{k=1}^{n-1} \left[\sum_{j=0}^{k-1} W_j(M) \right] [\phi_k(t) \exp(\mathrm{it}Z_{k-1}) - \phi_{k+1}(t) \exp(\mathrm{it}Z_k)]$$

$$+ \left[\sum_{j=0}^{n-1} W_j(M) \right] \phi_n(t) \exp(\mathrm{it}Z_{n-1}). \tag{2.36}$$

On account of (2.8), one has

$$\phi_k(t) \exp(\mathrm{it}Z_{k-1}) - \phi_{k+1}(t) \exp(\mathrm{it}Z_k)$$

$$= \exp\left(-\frac{n-k}{n}\frac{t^2}{2}\right) \exp(\mathrm{it}Z_{k-1}) - \exp\left(-\frac{n-(k+1)}{n}\frac{t^2}{2}\right) \exp(\mathrm{it}Z_k)$$

$$= \exp\left(-\frac{n-(k+1)}{n}\frac{t^2}{2}\right) \exp(\mathrm{it}Z_{k-1}) \left[\exp\left(-\frac{t^2}{2n}\right) - \exp(\mathrm{it}X_k n^{-\frac{1}{2}})\right]$$

$$= \phi_{k+1}(t) \exp(\mathrm{it}Z_{k-1}) \left[\exp\left(-\frac{t^2}{2n}\right) - \exp(\mathrm{it}X_k n^{-\frac{1}{2}})\right],$$

since $Z_k - Z_{k-1} = X_n n^{-\frac{1}{2}}$ as follows from (2.6). By means of this result, (2.36) becomes as follows

$$\sum_{k=1}^{n} \phi_k(t) \exp(\mathrm{it}Z_{k-1}) W_{k-1}(M)$$

$$= \sum_{k=1}^{n-1} \left[\sum_{j=0}^{k-1} W_j(M) \right] \phi_{k+1}(t) \exp(\mathrm{it}Z_{k-1})$$

$$\times \left[\exp\left(-\frac{t^2}{2n}\right) - \exp(\mathrm{it}X_k n^{-\frac{1}{2}})\right]$$

$$+ \left[\sum_{j=0}^{n-1} W_j(M) \right] \phi_n(t) \exp(\mathrm{it}Z_{n-1}). \tag{2.37}$$

Multiplying both sides of (2.37) by t^2/n, taking the expectations and utilizing the definition of Λ_n by (2.33), as well as (2.35), we obtain

$$\Lambda_n = \frac{t^2}{n}\left[\sum_{k=1}^{n-1}\phi_{k+1}(t)\,\mathscr{E}\left\{\exp\left(itZ_{k-1}\right)\left[\sum_{j=0}^{k-1}W_j(M)\right]\right]\left[\exp\left(-\frac{t^2}{2n}\right)\right.\right.$$

$$\left.\left. - \exp\left(itX_k n^{-\frac{1}{2}}\right)\right]\right\} + \phi_n(t)\,\mathscr{E}\left\{\exp\left(itZ_{k-1}\right)\left[\sum_{j=0}^{n-1}W_j(M)\right]\right\}\right].$$

(2.38)

Now in the expansion

$$e^x = 1 + x\theta(x), \quad 0 < \theta(x) < 1,$$

we replace x by $-t^2/2n$ and we get

$$\exp\left(-\frac{t^2}{2n}\right) = 1 - \frac{t^2}{2n}\theta_n(t), \quad 0 < \theta_n(t) < 1. \tag{2.39}$$

Also in the expansion

$$e^{ix} = 1 + ix - \tfrac{1}{2}x^2 + R(x), \quad |R(x)| \leqslant x^2,$$

we replace x by $tX_k n^{-\frac{1}{2}}$ and obtain

$$\exp\left(itX_k n^{-\frac{1}{2}}\right) = 1 + \frac{itX_k}{\sqrt{n}} - \frac{t^2X_k^2}{2n} + R_n(t, X_k), \quad |R_n(t, X_k)| \leqslant \frac{t^2X_k^2}{n}.$$

(2.40)

From (2.39) and (2.40), we get

$$\exp\left(-\frac{t^2}{2n}\right) - \exp\left(itX_k n^{-\frac{1}{2}}\right) = -\frac{t^2}{2n}\theta_n(t) - \frac{itX_k}{\sqrt{n}} + \frac{t^2X_k^2}{2n} - R_n(t, X_k),$$

so that

$$\mathscr{E}\left\{\exp\left(itZ_{k-1}\right)\left[\sum_{j=0}^{k-1}W_j(M)\right]\left[\exp\left(-\frac{t^2}{2n}\right) - \exp\left(itX_k n^{-\frac{1}{2}}\right)\right]\right\}$$

$$= \frac{t^2}{2n}\mathscr{E}\left\{X_k^2\exp\left(itZ_{k-1}\right)\left[\sum_{j=0}^{k-1}W_j(M)\right]\right\}$$

$$- \frac{t^2}{2n}\theta_n(t)\,\mathscr{E}\left\{\exp\left(itZ_{k-1}\right)\left[\sum_{j=0}^{k-1}W_j(M)\right]\right\}$$

$$- \mathscr{E}\left\{\exp\left(itZ_{k-1}\right)\left[\sum_{j=0}^{k-1}W_j(M)\right]R_n(t, X_k)\right\}$$

$$- itn^{-\frac{1}{2}}\mathscr{E}\left\{X_k\exp\left(itZ_{k-1}\right)\left[\sum_{j=0}^{k-1}W_j(M)\right]\right\}. \tag{2.41}$$

Next

$$\mathscr{E}\left\{X_k \exp\left(\mathrm{i}tZ_{k-1}\right)\left[\sum_{j=0}^{k-1} W_j(M)\right]\right\}$$

$$= \mathscr{E}\left[\mathscr{E}\left\{X_k \exp\left(\mathrm{i}tZ_{k-1}\right)\left[\sum_{j=0}^{k-1} W_j(M)\right]\Big|\mathscr{F}_{k-1}\right\}\right]$$

$$= \mathscr{E}\left\{\exp\left(\mathrm{i}tZ_{k-1}\right)\left[\sum_{j=0}^{k-1} W_j(M)\right]\mathscr{E}(X_k|\mathscr{F}_{k-1})\right\} = 0 \quad (2.42)$$

by means of (2.3), whereas on account of (2.19), one has

$$\mathscr{E}\left\{X_k^2 \exp\left(\mathrm{i}tZ_{k-1}\right)\left[\sum_{j=0}^{k-1} W_j(M)\right]\right\}$$

$$= \mathscr{E}\left[\mathscr{E}\left\{X_k^2 \exp\left(\mathrm{i}tZ_{k-1}\right)\left[\sum_{j=0}^{k-1} W_j(M)\right]\Big|\mathscr{F}_{k-1}\right\}\right]$$

$$= \mathscr{E}\left\{\exp\left(\mathrm{i}tZ_{k-1}\right)\left[\sum_{j=0}^{k-1} W_j(M)\right]\mathscr{E}(X_k^2|\mathscr{F}_{k-1})\right\}$$

$$= \mathscr{E}\left\{Y_{k-1} \exp\left(\mathrm{i}tZ_{k-1}\right)\left[\sum_{j=0}^{k-1} W_j(M)\right]\right\}. \quad (2.43)$$

By means of (2.22) and (2.43), (2.41) becomes then as follows

$$\mathscr{E}\left\{\exp\left(\mathrm{i}tZ_{k-1}\right)\left[\sum_{j=0}^{k-1} W_j(M)\right]\left[\exp\left(-\frac{t^2}{2n}\right) - \exp\left(\mathrm{i}tX_k n^{-\frac{1}{2}}\right)\right]\right\}$$

$$= \frac{t^2}{2n}\mathscr{E}\left\{Y_{k-1} \exp\left(\mathrm{i}tZ_{k-1}\right)\left[\sum_{j=0}^{k-1} W_j(M)\right]\right\}$$

$$- \frac{t^2}{2n}\theta_n(t)\,\mathscr{E}\left\{\exp\left(\mathrm{i}tZ_{k-1}\right)\left[\sum_{j=0}^{k-1} W_j(M)\right]\right\}$$

$$- \mathscr{E}\left\{\exp\left(\mathrm{i}tZ_{k-1}\right)\left[\sum_{j=0}^{k-1} W_j(M)\right]R_n(t, X_k)\right\}. \quad (2.44)$$

Finally, multiplying both sides of (2.44) by $\dfrac{t^2}{2n}\phi_{k+1}(t)$ and summing over k, as k goes from 1 to $n-1$, and also utilizing (2.28), we obtain (2.34). The proof of the lemma is completed. ∎

Thus on the basis of (2.33), in order to prove the theorem it suffices to show that each one of the four terms on the right-hand side of (2.34) converges to zero (as $n \to \infty$). This is done in the following four lemmas.

LEMMA 2.5A   Referring to (2.34), one has

$$\frac{t^4}{n^2}\sum_{k=1}^{n-1}\phi_{k+1}(t)\,\mathscr{E}\left\{Y_{k-1}\exp\left(itZ_{k-1}\right)\left[\sum_{j=0}^{k-1}W_j(M)\right]\right\}\to 0. \quad (2.45)$$

Proof   From (2.29) and (2.24), it follows that $\mathscr{E}W_k = 0$ and $\mathscr{E}|W_k| \leqslant 2$. On the other hand, from (2.31) and for each k, one has

$$W_k(M)\to W_k \quad \text{as} \quad M\to\infty \quad \text{and} \quad |W_k(M)|\leqslant|W_k|.$$

Then the dominated convergence theorem implies that

$$\mathscr{E}W_k(M)\to 0 \quad \text{as} \quad M\to\infty. \quad (2.46)$$

Next, by stationarity, one has that

$$\mathscr{E}\left\{Y_{k-1}\exp\left(itZ_{k-1}\right)\left[\sum_{j=0}^{k-1}W_j(M)\right]\right\}$$

$$= \mathscr{E}\left\{Y_0\exp\left(itZ_0\right)\left[\sum_{j=-(k-1)}^{0}W_j(M)\right]\right\}$$

$$= \mathscr{E}\left\{Y_0\exp\left(itZ_0\right)\left[\sum_{j=-(k-1)}^{0}W_j(M)-k\mathscr{E}W_0(M)\right]\right\}$$

$$\qquad\qquad\qquad + k\mathscr{E}[Y_0\exp\left(itZ_0\right)\mathscr{E}W_0(M)]$$

$$= k\mathscr{E}\left\{Y_0\exp\left(itZ_0\right)\left[\frac{1}{k}\sum_{j=-(k-1)}^{0}W_j(M)-\mathscr{E}W_0(M)\right]\right\}$$

$$\qquad\qquad\qquad + k\mathscr{E}[Y_0\exp\left(itZ_0\right)\mathscr{E}W_0(M)]$$

$$= k\mathscr{E}[Y_0\exp\left(itZ_0\right)\gamma_k]+k\mathscr{E}[Y_0\exp\left(itZ_0\right)\mathscr{E}W_0(M)], \quad (2.47)$$

where $$\gamma_k = \frac{1}{k}\sum_{j=-(k-1)}^{0}W_j(M)-\mathscr{E}W_0(M). \quad (2.48)$$

Therefore multiplying both sides of (2.47) by $\dfrac{t^4}{n^2}\phi_{k+1}(t)$ and summing over k, as k goes from 1 to $n-1$, one has

$$\left|\frac{t^4}{n^2}\sum_{k=1}^{n-1}\phi_{k+1}(t)\,\mathscr{E}\left\{Y_{k-1}\exp\left(itZ_{k-1}\right)\left[\sum_{j=0}^{k-1}W_j(M)\right]\right\}\right|$$

$$\leqslant \frac{t^4}{n^2}\sum_{k=1}^{n-1}\mathscr{E}|Y_0\gamma_k|+\frac{t^4 n^2}{n^2}(\mathscr{E}Y_0)|\mathscr{E}W_0(M)|$$

$$= \frac{t^4}{n}\sum_{k=1}^{n-1}\mathscr{E}|Y_0\gamma_k|+t^4(\mathscr{E}Y_0)|\mathscr{E}W_0(M)|. \quad (2.49)$$

Now on account of (2.19) and (2.46), $\mathscr{E}Y_0 = 1$ and $\mathscr{E}W_0(M) \to 0$ as $M \to \infty$. Therefore for all sufficiently large M, one has that $t^4(\mathscr{E}Y_0)|\mathscr{E}W_0(M)| < \epsilon$. Thus, in order to show (2.45), it suffices to show that the first summand at the right-hand side of (2.49) converges to zero; i.e.

$$\frac{1}{n}\sum_{k=1}^{n-1} \mathscr{E}|Y_0\gamma_k| \to 0. \qquad (2.50)$$

By stationarity and ergodicity, one has that

$$\mathscr{E}|\gamma_k| \to 0 \quad \text{as} \quad k \to \infty, \qquad (2.51)$$

where γ_k is given by (2.48)

Thus for all sufficiently large k, $k \geqslant k_0 = k(\epsilon)$, say, one has that

$$\mathscr{E}|\gamma_k| < \epsilon. \qquad (2.52)$$

Set

$$\xi_n = \frac{1}{n-1}\sum_{k=1}^{n-1}|\gamma_k|. \qquad (2.53)$$

Then for $n > k_0 + 1$ and by means of (2.52), we have

$$\mathscr{E}\xi_n = \frac{1}{n-1}\sum_{k=1}^{k_0-1}\mathscr{E}|\gamma_k| + \frac{1}{n-1}\sum_{k=k_0}^{n-1}\mathscr{E}|\gamma_k|$$

$$\leqslant \frac{1}{n-1}\sum_{k=1}^{k_0-1}\mathscr{E}|\gamma_k| + \frac{n-k_0}{n-1}\epsilon$$

$$\leqslant \frac{1}{n-1}\sum_{k=1}^{k_0-1}\mathscr{E}|\gamma_k| + \epsilon.$$

It follows that

$$\mathscr{E}\xi_n \to 0. \qquad (2.54)$$

From (2.31), one has that $|W_k(M)| \leqslant M$ for all k. Then (2.48) gives that $\mathscr{E}\gamma_k \leqslant 2M$, so that (2.53) implies that

$$\xi_n \leqslant 2M. \qquad (2.55)$$

Hence

$$\mathscr{E}(Y_0\xi_n) = \int_{(Y_0>c)} (Y_0\xi_n)\,\mathrm{d}\hat{P}$$

$$+ \int_{(Y_0\leqslant c)} (Y_0\xi_n)\,\mathrm{d}\hat{P} \leqslant 2M\int_{(Y_0>c)} Y_0\,\mathrm{d}\hat{P} + c\mathscr{E}\xi_n.$$

Thus the fact that $\mathscr{E}Y_0 = 1$ and (2.54) imply that

$$\mathscr{E}(Y_0\xi_n) \to 0. \qquad (2.56)$$

Therefore, on account of (2.53) and (2.56),

$$\frac{1}{n}\sum_{k=1}^{n-1}\mathscr{E}\,|Y_0\gamma_k| \leqslant \frac{1}{n-1}\sum_{k=1}^{n-1}\mathscr{E}\,|Y_0\gamma_k| = \mathscr{E}(Y_0\xi_n) \to 0.$$

This establishes (2.50) and hence the lemma itself. ∎

LEMMA 2.6A   Referring to (2.34), one has

$$\frac{t^4}{n^2}\theta_n(t)\sum_{k=1}^{n-1}\phi_{k+1}(t)\,\mathscr{E}\left\{\exp{(itZ_{k-1})}\left[\sum_{j=0}^{k-1}W_j(M)\right]\right\} \to 0.$$

Proof Applying arguments similar to the ones used in the proof of the previous lemma, one has that

$$\mathscr{E}\left\{\exp{(itZ_{k-1})}\left[\sum_{j=0}^{k-1}W_j(M)\right]\right\} = \mathscr{E}\left\{\exp{(itZ_0)}\left[\sum_{j=-(k-1)}^{0}W_j(M)\right]\right\}$$

$$= k\mathscr{E}[\exp{(itZ_0)}\,\gamma_k] + k\mathscr{E}W_0(M),$$

where γ_k is given by (2.48).

Hence, proceeding as in (2.49), we obtain

$$\left|\frac{t^4}{n^2}\theta_n(t)\sum_{k=1}^{n-1}\phi_{k+1}(t)\,\mathscr{E}\left\{\exp{(itZ_{k-1})}\left[\sum_{j=0}^{k-1}W_j(M)\right]\right\}\right|$$

$$\leqslant \frac{t^4}{n}\sum_{k=1}^{n-1}\mathscr{E}\,|\gamma_k| + t^4\,|\mathscr{E}W_0(M)|.$$

The right-hand side of this last inequality converges to zero by means of (2.46) (for $k = 0$), (2.53) and (2.54). The proof of the lemma is completed. ∎

LEMMA 2.7A   Referring to (2.34), one has

$$\frac{t^2}{n}\sum_{k=1}^{n-1}\phi_{k+1}(t)\,\mathscr{E}\left\{\exp{(itZ_{k-1})}\left[\sum_{j=0}^{k-1}W_j(M)\right]R_n(t,X_k)\right\} \to 0.$$

Proof Once again, stationarity yields

$$\mathscr{E}\left\{\exp{(itZ_{k-1})}\left[\sum_{j=0}^{k-1}W_j(M)\right]R_n(t,X_k)\right\}$$

$$= \mathscr{E}\left\{\exp{(itZ_0)}\left[\sum_{j=-(k-1)}^{0}W_j(M)\right]R_n(t,X_1)\right\}$$

$$= k\mathscr{E}\left\{\exp{(itZ_0)}\left[\frac{1}{k}\sum_{j=-(k-1)}^{0}W_j(M) - \mathscr{E}W_0(M)\right]R_n(t,X_1)\right\}$$

$$+ k\mathscr{E}[\exp{(itZ_0)}\,R_n(t,X_1)]\,\mathscr{E}W_0(M).$$

Therefore (2.40) and (2.48) imply that

$$\left| \frac{t^2}{n} \sum_{k=1}^{n-1} \phi_{k+1}(t)\, \mathscr{E}\left\{ \exp\left(\mathrm{it}Z_{k-1}\right) \left[\sum_{j=0}^{k-1} W_j(M) \right] R_n(t, X_k) \right\} \right|$$

$$\leqslant t^2 \sum_{k=1}^{n-1} \mathscr{E}[|\gamma_k|\, |R_n(t, X_1)|] + \frac{t^2}{n} n^2 (\mathscr{E}\, |R_n(t, X_1)|)\, |\mathscr{E}W_0(M)|$$

$$\leqslant \frac{t^4}{n} \sum_{k=1}^{n-1} \mathscr{E}(|\gamma_k|\, X_1^2) + t^4 (\mathscr{E}X_1^2)\, |\mathscr{E}W_0(M)|$$

$$= \frac{t^4}{n} \sum_{k=1}^{n-1} \mathscr{E}(|\gamma_k|\, X_1^2) + t^4 |\mathscr{E}W_0(M)|. \tag{2.57}$$

Since $\mathscr{E}W_0(M) \to 0$ as $M \to \infty$ by (2.46), it follows that in order to prove (2.57) it suffices to show that

$$\frac{1}{n} \sum_{k=1}^{n-1} \mathscr{E}(|\gamma_k|\, X_1^2) \to 0. \tag{2.58}$$

But on account of (2.53),

$$\frac{1}{n} \sum_{k=1}^{n-1} \mathscr{E}(|\gamma_k|\, X_1^2) \leqslant \mathscr{E}(\xi_n X_1^2) \tag{2.59}$$

and $\mathscr{E}(\xi_n X_1^2) = \displaystyle\int_{(X_1^2 > c)} (\xi_n X_1^2)\, \mathrm{d}\hat{P} + \int_{(X_1^2 \leqslant c)} (\xi_n X_1^2)\, \mathrm{d}\hat{P}$

$$\leqslant 2M \int_{(X_1^2 > c)} X_1^2\, \mathrm{d}\hat{P} + c\mathscr{E}\xi_n$$

by means of (2.55). Now the fact that $\mathscr{E}X_1^2 = 1$ and $\mathscr{E}\xi_n \to 0$ by (2.54), implies that $\mathscr{E}(\xi_n X_1^2) \to 0$ and therefore (2.58) by way of (2.59). This completes the proof of the lemma. ∎

Finally, the following lemma completes the proof of the theorem.

LEMMA 2.8A   Referring to (2.34), one has

$$\frac{t^2}{n} \phi_n(t)\, \mathscr{E}\left\{ \exp\left(\mathrm{it}Z_{n-1}\right) \left[\sum_{j=0}^{n-1} W_j(M) \right] \right\} \to 0.$$

Proof Utilizing stationarity and (2.48), one has

$$\mathscr{E}\left\{\exp(itZ_{n-1})\left[\sum_{j=0}^{n-1}W_j(M)\right]\right\}$$

$$= \mathscr{E}\left\{\exp(itZ_0)\left[\sum_{j=-(n-1)}^{0}W_j(M)\right]\right\}$$

$$= n\mathscr{E}\left\{\exp(itZ_0)\left[\frac{1}{n}\sum_{j=-(n-1)}^{0}\mathscr{E}W_j(M) - \mathscr{E}W_0(M)\right]\right\}$$

$$\qquad\qquad + n\mathscr{E}\exp(itZ_0)\,\mathscr{E}W_0(M)$$

$$= n\mathscr{E}[\exp(itZ_0)\,\gamma_n] + n\mathscr{E}\exp(itZ_0)\,\mathscr{E}W_0(M).$$

Hence

$$\left|\frac{t^2}{n}\phi_n(t)\,\mathscr{E}\left\{\exp(itZ_{n-1})\left[\sum_{j=0}^{n-1}W_j(M)\right]\right\}\right| \leqslant t^2\mathscr{E}\,|\gamma_n| + t^2\,|\mathscr{E}W_0(M)|.$$

Then (2.46) and (2.52) complete the proof of the lemma and therefore that of the theorem itself.❚

Exercise

Referring to Remark 2.1 A (v), consider the stationary and ergodic coordinate process $\{X_n\}$, $n \geqslant 1$, defined on the probability space $(R^\infty, \mathscr{B}^\infty, \hat{P})$ and set $(R^0_{-\infty}, \mathscr{B}^0_{-\infty}) = \prod_{j=-\infty}^{0}(R_j, \mathscr{B}_j)$, where $(R_j, \mathscr{B}_j) = (R, \mathscr{B})$. Also set $(R^\infty_{-\infty}, \mathscr{B}^\infty_{-\infty}) = (R^0_{-\infty}, \mathscr{B}^0_{-\infty}) \times (R^\infty, \mathscr{B}^\infty)$ and let $\{X_n\}$, $n = 0, \pm 1, ...$, be the coordinate process defined on $(R^\infty_{-\infty}, \mathscr{B}^\infty_{-\infty})$. On $\mathscr{B}^0_{-\infty}$, define the probability measure $\hat{P}_0$ by

$$\hat{P}_0(X_n \leqslant x_n, ..., X_{n+k} \leqslant x_{n+k})$$

$$= P(X_1 \leqslant x_n, ..., X_{k+1} \leqslant x_{n+k}) \quad (n \leqslant 0, \, k > 0),$$

and let $\hat{P}_1$ be defined on $\mathscr{B}^\infty_{-\infty}$ by $\hat{P}_1 = \hat{P}_0 \times \hat{P}$. Then for the process $\{X_n\}$, $n = 0, \pm 1, ...$, show that

   (i) It is stationary,

   (ii) $\mathscr{L}[(X_1, X_2, ...)|\hat{P}_1] = \mathscr{L}[(X_1, X_2, ...)|\hat{P}]$,

   (iii) It is ergodic.

(See also Proposition 6.5 in Breiman [1], p. 105, and Billingsley [3], p. 206.)

3   A theorem employed in Chapter 5

THEOREM 3.1A   Let C_1, C_2 be $k \times k$ positive definite, symmetric matrices. Then $C_1 - C_2^{-1}$ is positive semi-definite if and only if $(h'C_2h)^{\frac{1}{2}} \geqslant v'h/(v'C_1v)^{\frac{1}{2}}$ for all non-zero $h, v \in R^k$.

The proof of this theorem is based on the following lemma which can be found in Rao [3], p. 48.

LEMMA 3.1A   Let A be a positive definite $k \times k$ matrix and let $u \in R^k$. Then

$$\sup \left[\frac{(u'x)^2}{x'Ax}; x \in R^k \right] = u'A^{-1}u$$

and the supremum is attained at $x^* = A^{-1}u$.

Proof of Theorem 3.1A   The proof is by contradiction. Suppose that $C_1 - C_2^{-1}$ is positive semi-definite and that there exist non-zero, $h, v \in R^k$ such that $(h'C_2h)^{\frac{1}{2}} < v'h/(v'C_1v)^{\frac{1}{2}}$. Hence

$$v'C_1v < (v'h)^2/h'C_2h.$$

By Lemma 3.1A,

$$\sup \left[\frac{(v'h)^2}{h'C_2h}; h \in R^k \right] = v'C_2^{-1}v,$$

so that $v'C_1v < v'C_2^{-1}v$, or equivalently, $v'(C_1 - C_2^{-1})v < 0$. However, this contradicts the assumption that $C_1 - C_2^{-1}$ is positive semi-definite and therefore $(h'C_2h)^{\frac{1}{2}} \geqslant v'h/(v'C_1v)^{\frac{1}{2}}$ for all non-zero $h, v \in R^k$.

Now suppose that $(h'C_2h)^{\frac{1}{2}} \geqslant v'h/(v'C_1v)^{\frac{1}{2}}$ for all non-zero $h, v \in R^k$ and let $C_1 - C_2^{-1}$ be not positive semi-definite. Then there exists $0 \neq x \in R^k$ such that $x'(C_1 - C_2^{-1})x < 0$, or equivalently, $x'C_1x < x'C_2^{-1}x$. Utilizing our assumption with $v = x$, we obtain $(h'C_2h)^{\frac{1}{2}} \geqslant x'h/(x'C_1x)^{\frac{1}{2}}$ and hence $x'C_1x \geqslant (x'h)^2/h'C_2h$. Thus

$$x'C_1x \geqslant \sup \left[\frac{(x'h)^2}{h'C_2h}; h \in R^k \right] = x'C_2^{-1}x$$

(by Lemma 3.1A), so that $x'C_1x \geqslant x'C_2^{-1}x$. However, this contradicts the previously established inequality $x'C_1x < x'C_2^{-1}x$. Therefore $C_1 - C_2^{-1}$ is positive semi-definite and the proof of the theorem is completed. ∎

To this theorem, there is the following immediate corollary.

COROLLARY 3.1A   The conclusion of the theorem remains valid if the inequality in the theorem is true for all non-zero rationals $h, v \in R^k$.

4   Some theorems employed in Chapter 6

In the present section, as in the previous ones, we gather together some of the theorems used in the various proofs of Chapter 6 for easy reference.

THEOREM 4.1A   If Γ is a $k \times k$ positive definite, symmetric matrix, there exists a non-singular matrix M such that $M'M = \Gamma$.

Proof   From the assumptions it follows that there exists an orthogonal matrix P such that $P'\Gamma P = D$, where D is a diagonal matrix with (diagonal) elements $d_j, j = 1, ..., k$ (the characteristic roots of Γ). From the orthogonality of P it follows that $P^{-1} = P'$. Therefore $P'\Gamma P = D$ gives that $\Gamma = PDP'$. Let now D_0 be the diagonal matrix with (diagonal) elements $\sqrt{d_j}$ $(j = 1, ..., k)$. Then, clearly, $D_0 D_0 = D$. Therefore

$$\Gamma = PDP' = PD_0 D_0 P' = PD_0 D_0' P' = (PD_0)(PD_0)'.$$

Since P, D_0 are non-singular, so is PD_0. Therefore, by setting $(PD_0)' = M$, we obtain $\Gamma = M'M$, as was to be seen. ∎

We now wish to establish Theorem 4.2A below along with its Corollary 4.2A. For this purpose, some notation and auxiliary results are needed. The proofs of some of these results are presented herein and the reader is referred to other sources for the omitted proofs.

To start with, let T be an open subset of R^k and let V be a k-dimensional random vector whose distribution $Q_{t, V}$ is absolutely continuous with respect to $Q_{t_0, V}$ (to be shortened to Q_0) for some $t_0 \in T$ and has a probability density of the standard exponential type, namely,

$$[dQ_{t, V}/dQ_0] = C(t) \exp t'v \quad (v \in R^k, t \in T). \tag{4.1}$$

We shall assume that if P is any probability measure on T, then the indefinite integral of $C(t)$ is a σ-finite measure; i.e.

$$\text{if} \quad \mu_P(B) = \int_B C(t)\,\mathrm{d}P, \quad \text{then } \mu_P \text{ is } \sigma\text{-finite for every } P. \quad (4.2)$$

In the family (4.1), suppose we are interested in testing the hypothesis $H\colon t = t_0$ against the alternative $A\colon t \neq t_0$ on the basis of the random vector V. Let $\bar{A} = \{a_0, a_1\}$ be the *action space*, where a_0 stands for accepting H and a_1 for rejecting H, and let ϕ be a test function defined on R^k. Then if $V = v$, $\phi(v)$ means rejecting H with probability $\phi(v)$ and accepting H with probability $1 - \phi(v)$. Furthermore, let L be a bounded *loss function* defined on $T \times \bar{A}$ into R. Then if $V = v$, the loss incurred is given by

$$L(t, a_1)\,\phi(v) + L(t, a_0)\,[1 - \phi(v)].$$

The *risk* $R_\phi(t)$ corresponding to ϕ, i.e. the expected loss under $t \in T$ when the test ϕ is employed, is given by

$$R_\phi(t) = L(t, a_1)\,\mathscr{E}_t^!\phi(V) + L(t, a_0)\,[1 - \mathscr{E}_t\phi(V)]$$

and is easily seen that

$$R_\phi(t) = L(t, a_0) + [L(t, a_1) - L(t, a_0)]\beta_\phi(t), \quad (4.3)$$

where
$$\beta_\phi(t) = \mathscr{E}_t\phi(V) = \int \phi(v)\,\mathrm{d}Q_{t,\,V}.$$

In all that follows, the loss function L is taken to be as specified below

$$\left. \begin{array}{ll} L(t_0, a_0) = 0, & L(t_0, a_1) = 1 \\ t \neq t_0, \quad L(t, a_0) = 1, & L(t, a_1) = 0. \end{array} \right\} \quad (4.4)$$

and for

For such a choice of the loss function L, the corresponding risk, relative to the test ϕ, becomes as follows

$$R_\phi(t_0) = \beta_\phi(t_0), \quad \text{and for} \quad t \neq t_0, \quad R_\phi(t) = 1 - \beta_\phi(t). \quad (4.5)$$

Now for a fixed test ψ (see proof of Theorem 4.2A for its specification), consider $R_\psi(t)$ as a function of t and define the following (bounded) loss function

$$L^*(t, a) = L(t, a) - R_\psi(t).$$

Then for any test ϕ the corresponding risk R_ϕ^*, associated with L^*, is given by

$$R_\phi^*(t) = L^*(t, a_0) + [L^*(t, a_1) - L^*(t, a_0)]\beta_\phi(t), \qquad (4.6)$$

according to (4.2), and is easily seen that

$$R_\phi^*(t) = R_\phi(t) - R_\psi(t). \qquad (4.7)$$

For the above choice of L, one has then that

$$R_\phi^*(t_0) = \beta_\phi(t_0) - R_\psi(t_0),$$

and for $\qquad t \neq t_0, \quad R_\phi^*(t) = 1 - \beta_\phi(t) - R_\psi(t). \qquad (4.8)$

Let W be a probability distribution over T. Then the *Bayes* (or *global*) *risk* $r(\phi, W)$, corresponding to the test ϕ and the prior distribution W, is given by

$$r(\phi, W) = \int_T R_\phi \, dW. \qquad (4.9)$$

A *Bayes test*, with respect to the prior distribution W, is any test which minimizes $r(\phi, W)$, as ϕ varies over the class of all tests. A *minimax test* (associated with the risk $R_\phi(t)$) is any test within the class of all tests which minimizes the following quantity $\sup[R_\phi(t); t \in T]$.

The first result in this section is to the effect that a Bayes test with respect to a prior distribution is essentially the indicator of the complement of a closed, convex set. More precisely, one has the following result.

LEMMA 4.1A   In connection with the family given by (4.1), suppose we are interested in testing the hypothesis $H: t = t_0$ against the alternative $A: t \neq t_0$ on the basis of V. Then any Bayes test ϕ_W, with respect to the prior distribution W, is given by

$$\phi_W(v) = \begin{cases} 1 & \text{if} \quad v \in C_W^c, \\ 0 & \text{if} \quad v \in C_W, \end{cases} \qquad (4.10)$$

where C_W is a closed, convex set in R^k. The test ϕ_W may be defined in an arbitrary (measurable) way on the boundary ∂C_W of the set C_W.

Proof The lemma is established along the lines of the proof presented by Birnbaum [1]. Let w_0 be the weight assigned to $\{t_0\}$. Then one has, by means of (4.10) and (4.5), that

$$r(\phi, W) = \int_T R_\phi \, dW = \int_{\{t_0\}} R_\phi \, dW + \int_{(t \neq t_0)} R_\phi \, dW = w_0 \beta_\phi(t_0)$$

$$+ \int_{(t \neq t_0)} (1 - \beta_\phi) \, dW$$

$$= (1 - w_0) + 2 w_0 \beta_\phi(t_0) - \int_T \beta_\phi \, dW.$$

But

$$\int_T \beta_\phi \, dW = \int_T \left(\int \phi \, dQ_{t,V} \right) dW = \int_T \left[\int \phi(v) \, C(t) \exp t'v \, dQ_0 \right] dW$$

$$= \int \left[\int_T C(t) \exp t'v \, dW \right] \phi(v) \, dQ_0,$$

where the interchange of the order of integration is valid by Fubini's theorem; also $\beta_\phi(t_0) = \int \phi(v) \, dQ_0$, so that

$$r(\phi, W) = (1 - w_0) + \int \left[2 w_0 - \int_T C(t) \exp t'v \, dW \right] \phi(v) \, dQ_0.$$

It follows that the Bayes risk $r(\phi, W)$ is minimized by the test ϕ_W which is defined as follows

$$\phi_W(v) = \begin{cases} 1 & \text{if } v \in A_1 = A_1(W) \\ & \quad = \left\{ v \in R^k; \, 2 w_0 < \int_T C(t) \exp t'v \, dW \right\} \\ 0 & \text{if } v \in A_2 = A_2(W) \\ & \quad = \left\{ v \in R^k; \, 2 w_0 > \int_T C(t) \exp t'v \, dW \right\}; \end{cases}$$

$$(4.11)$$

ϕ_W may be arbitrary (but measurable) on the set

$$A_3 = A_3(W) = \left\{ v \in R^k; \, 2 w_0 = \int_T C(t) \exp t'v \, dW \right\}.$$

In order for this test to be of the form (4.10), it remains for us to show that the set $A_2 \cup A_3$ is a closed, convex set and that A_3 is its boundary. To this end, set

$$C_W = A_2 \cup A_3 = \left\{ v \in R^k; \int_T C(t) \exp t'v \, dW \leqslant 2 w_0 \right\}$$

and consider the inequality

$$\exp\left[\lambda u_1 + (1-\lambda)u_2\right] \leqslant \lambda \exp u_1 + (1-\lambda)\exp u_2 \quad (0 \leqslant \lambda \leqslant 1).$$

Let $v_1, v_2 \in C_W$ and set $u_1 = t'v_1, u_2 = t'v_2$. Then

$$\int_T C(t)\exp t'[\lambda v_1 + (1-\lambda)v_2]\,\mathrm{d}W$$

$$= \int_T C(t)\exp\left[\lambda t'v_1 + (1-\lambda)t'v_2\right]\mathrm{d}W$$

$$\leqslant \lambda \int_T C(t)\exp t'v_1\,\mathrm{d}W + (1-\lambda)\int_T C(t)\exp t'v_2\,\mathrm{d}W$$

$$\leqslant \lambda 2w_0 + (1-\lambda)\,2w_0 = 2w_0,$$

so that $\lambda v_1 + (1-\lambda)v_2 \in C_W$; i.e. C_W is convex. Next, by means of (4.2), one has

$$\int_T C(t)\exp t'v\,\mathrm{d}W = \int_T \exp t'v\,\mathrm{d}\mu_W = \int_T \exp v't\,\mathrm{d}\mu_W$$

and $\int_T \exp v't\,\mathrm{d}\mu_W$ is continuous (in v) by Theorem 9 on p. 52 in Lehmann [3]. Therefore the set C_W is also closed. Finally, to show that $A_3 = \partial C_W$. In the first place, $\partial C_W \subseteq A_3$. In fact, if $v_0 \in \partial C_W$, then $v_0 \in C_W$ since C_W is closed and therefore $\partial C_W \subseteq C_W$. Thus $v_0 \in A_2 \cup A_3$. However, $v_0 \notin A_2$, provided, of course, that $A_2 \neq \varnothing$. This is so, because A_2 is open, being the inverse image of $(0, 2w_0)$ under the continuous function $\int_T \exp v't\,\mathrm{d}\mu_W$, and therefore if $v_0 \in A_2$, there would exist a neighbourhood of v_0 lying entirely in A_2. Hence $v_0 \in A_3$. The same conclusion holds if $A_2 = \varnothing$. This shows that $\partial C_W \subseteq A_3$. Next, we assert that $A_3 \subseteq \partial C_W$. For a contradiction, suppose that $v_0 \in A_3$ whereas $v_0 \notin \partial C_W$. Then $v_0 \in C_W$ and $v_0 \notin \partial C_W$ which implies that $v_0 \in C_W^0$. Of course, if $C_W^0 = \varnothing$, we already have a contradiction. Thus, suppose that $C_W^0 \neq \varnothing$. Then every (sufficiently small) neighbourhood of v_0 lies entirely in C_W. Actually, we shall show that every such neighbourhood lies entirely in A_3. To see this, suppose that there is a $v_1 \in A_2$ in such a neighbourhood and consider the line segment connecting v_0 and v_1. Let v_2 be on this line segment, opposite to v_1

with respect to v_0 and within the neighbourhood in question. Then for some $0 < \lambda < 1$, one has $v_0 = \lambda v_1 + (1-\lambda) v_2$ and therefore

$$2w_0 = \int_T C(t) \exp t' v_0 \, dW$$

$$\leqslant \lambda \int_T C(t) \exp t' v_1 \, dW + (1-\lambda) \int_T C(t) \exp t' v_2 \, dW$$

$$< \lambda 2w_0 + (1-\lambda) 2w_0 = 2w_0,$$

a contradiction. Thus $v_0 \in A_3$ and every (sufficiently small) neighbourhood of v_0 lies entirely in A_3. Furthermore, this is true for all $0 < w_0 \leqslant 1$. (For $w_0 = 0$, $A_2 = A_3 = \varnothing$ and the theorem is trivially true.) However, this cannot be true because

$$A_3^0(W) = \varnothing$$

for any probability distribution W in T. This is true for $w_0 = 1$ because then A_3 is the hyperplane $t_0' v = \log [2/C(t_0)]$. Thus suppose that $0 < w_0 < 1$ and let v_1, v_2 be two points in a neighbourhood of v_0 lying in A_3 such that v_0, v_1, v_2 lie on a line segment, $v_0 = \lambda v_1 + (1-\lambda) v_2$ for some $0 < \lambda < 1$ and v_1, v_2 lie on the two different halfspaces into which R^k is divided by the hyperplane $t_0' v = t_0' v_0$. This last restriction implies that $t_0' v_1 \neq t_0' v_2$ and therefore

$$\exp t_0' v_0$$

$$= \exp \{t_0' [\lambda v_1 + (1-\lambda) v_2]\}$$

$$= \exp [\lambda(t_0' v_1) + (1-\lambda)(t_0' v_2)] < \lambda \exp t_0' v_1 + (1-\lambda) \exp t_0' v_2.$$

Now $\qquad \exp t' v_0 \leqslant \lambda \exp t' v_1 + (1-\lambda) \exp t' v_2 \quad (t \in T)$

and hence

$$\int_{T-\{t_0\}} C(t) \exp t' v_0 \, dW \leqslant \lambda \int_{T-\{t_0\}} C(t) \exp t' v_1 \, dW$$
$$+ (1-\lambda) \int_{T-\{t_0\}} C(t) \exp t' v_2 \, dW,$$

whereas

$$C(t_0) \exp (t_0' v_0) < \lambda C(t_0) \exp (t_0' v_1) + (1-\lambda) C(t_0) \exp (t_0' v_2).$$

Adding up the last two inequalities, one obtains

$$2w_0 < \lambda 2w_0 + (1-\lambda)\,2w_0 = 2w_0,$$

a contradiction. The proof of the lemma is completed. ∎

The following result will also be needed in the sequel.

LEMMA 4.2 A (*weak compactness theorem*)   Given any sequence of tests $\{\phi_n\}$ there is a subsequence $\{\phi_m\}$ and a test ϕ such that

$$\int \phi_m g \, dQ_0 \to \int \phi g \, dQ_0$$

for every real-valued, integrable function g defined on R^k.

Proof   It can be found in Lehmann [3], pp. 354–6. ∎

LEMMA 4.3 A   If $\{\phi_m\}$ is a sequence of tests converging weakly to the test ϕ (relative to Q_0), then

$$R_{\phi_m}^*(t) \to R_\phi^*(t) \quad (t \in T).$$

Proof   By (4.6),

$$R_{\phi_m}^*(t) = L^*(t, a_0) + [L^*(t, a_1) - L^*(t, a_0)]\beta_{\phi_m}(t), \quad (4.12)$$

whereas

$$\beta_{\phi_m}(t) = \int \phi_m(v)\,dQ_{t,V} = \int \phi_m(v)\,C(t)\exp t'v\,dQ_0. \quad (4.13)$$

Now $C(t)\exp t'v$ is Q_0-integrable and therefore the assumption of weak convergence implies that

$$\int \phi_m(v)\,C(t)\exp t'v\,dQ_0 \to \int \phi(v)\,C(t)\exp t'v\,dQ_0$$

$$= \int \phi(v)\,dQ_{t,V} = \beta_\phi(t). \quad (4.14)$$

By (4.12), (4.13) and (4.14), it follows then that $R_{\phi_m}(t) \to R_\phi^*(t)$ for every $t \in T$, as was to be seen. ∎

In Lemma 4.1 A, it was shown that any Bayes test with respect to the prior distribution W was given by (4.10). The next objective in this section is that of showing that the weak limit of a sequence of Bayes tests is again a Bayes test. To this end, the following notions and results will be necessary. Let

$$S_r = S(0, r) = \{z \in R^k;\ \|z\| \leqslant r\}$$

and for any closed subset, A, of S_r, define its δ-*neighbourhood*, $N_\delta(A)$, as follows

$$N_\delta(A) = \{z \in R^k;\ (z-u)'\,(z-u) < \delta \text{ for some } u \in A\} \quad (\delta > 0).$$

For any two closed subsets A and B of S_r, define the numbers δ_1 and δ_2 as follows

$$\delta_1 = \inf\{\delta > 0;\ A \subset N_\delta(B)\} \quad \text{and} \quad \delta_2 = \inf\{\delta > 0;\ B \subset N_\delta(A)\}.$$

Then set $d(A,B) = \delta_1 + \delta_2$. It can be shown that the function d defined as above on the Cartesian product of the class of all closed subsets of S_r with itself is, in fact, a metric. For the proof of it, the reader is referred to Eggleston [1], p. 60. The distance $d(A,B)$ between A and B is known as the *Hausdorff distance*.

In terms of the metric just defined, one can formulate the following standard result on convex sets.

LEMMA 4.4 A (*Blaschke selection theorem*)   Given any sequence of closed, convex subsets $\{C_n\}$ of S_r, there exists a subsequence $\{C_\nu\}$ and a closed, non-void, convex subset C of S_r such that $d(C_\nu, C) \to 0$.

Proof   It can be found in Eggleston [1], pp. 64–7.]

In the sequel, a generalized version of this lemma, due to Mathes and Truax [1], p. 684, will be needed. Namely,

LEMMA 4.5 A   Let $\{C_m\}$ be a sequence of closed, convex sets not necessarily subsets of a bounded set. Then there is a subsequence $\{C_\nu\}$ and a closed, convex set C (which may be empty) such that

$$d(C_\nu \cap S_j,\ C \cap S_j) \to 0 \quad \text{for} \quad j = r, r+1, \ldots, \quad \text{some} \quad r > 0.$$

Proof   First suppose that for a sufficiently large r there are infinitely many ms, forming a subsequence $\{C_{q_0}\}$ such that $C_{q_0} \cap S_r \neq \varnothing$. For a fixed r as above, consider the sequence of closed, convex subsets $\{C_{q_0} \cap S_r\}$ of S_r. Then by Blaschke selection theorem, there exists a subsequence $\{q_1\} \subseteq \{q_0\}$ and a non-empty, closed, convex subset C^r of S_r such that $d(C_{q_1} \cap S_r, C^r) \to 0$. Next, consider the sequence $\{C_{q_1} \cap S_{r+1}\}$. Then, as above, there exists a subsequence $\{q_2\} \subseteq \{q_1\}$ and a non-empty, closed, convex subset C^{r+1} of S_{r+1} such that $d(C_{q_2} \cap S_{r+1}, C^{r+1}) \to 0$. It is easily seen that

$C^{r+1} \cap S_r = C^r$ (see also Exercise 1 (i)). By induction, one has that there exists a subsequence $\{q_{i+1}\} \subseteq \{q_i\}$ and a non-empty, closed, convex subset C^{r+i} of S_{r+i} such that $d(C_{q_{i+1}} \cap S_{r+1}, C^{r+i}) \to 0$, and that
$$C^{r+i} \cap S_{r+i-1} = C^{r+i-1}, i = 1, 2, \ldots.$$
Now consider the sequences $\{C_{q_i}\}$, $i = 0, 1, \ldots$ and let $\{C_\nu\}$ be the diagonal sequence. Then, clearly, $d(C_\nu \cap S_j, C^j) \to 0$ for each $j = r, r+1, \ldots$. Set $C = \bigcup_{\alpha=r}^{\infty} C^\alpha$. Then C is non-empty, closed and convex (see also Exercise 1 (ii)). Furthermore, since
$$C^{r+i} \cap S_{r+i-1} = C^{r+i-1}, i = 1, 2, \ldots,$$
it follows that
$$C \cap S_j = \bigcup_{\alpha=r}^{\infty} (C^\alpha \cap S_j) = C^j, j = r, r+1, \ldots.$$
Therefore $d(C_\nu \cap S_j, C \cap S_j) \to 0$, $j = r, r+1, \ldots$, as was to be seen. Next, suppose that for every positive r, $C_m \cap S_r \neq \varnothing$ for only finitely many ms. In this case, take $C = \varnothing$. One then has that the lemma is trivially true with $\{\nu\} = \{m\}$. The proof of the lemma is completed.∎

Now the lemma just proved has the following important corollary which is also contained in a result in the reference last cited. Namely,

COROLLARY 4.1A Let $\{\phi_m\}$ be a sequence of Bayes tests with respect to some prior distribution W, i.e. tests of the form (4.10), and suppose that this sequence converges weakly to the test ϕ (relative to Q_0). Then ϕ is the indicator of the complement of a closed, convex set in R^k a.s. $[Q_0]$ (and hence a.s. $[Q_{t,V}]$ for all $t \in T$). However, ϕ may not correspond to any prior distribution W on the parameter set T.

Proof Let $\{C_m\}$ be the sequence of closed, convex sets corresponding to $\{\phi_m\}$ by (4.10). Then, according to Lemma 4.5A, there exists a subsequence $\{C_\nu\}$ and a closed, convex set C such that $d(C_\nu \cap S_j, C \cap S_j) \to 0$ for $j = r, r+1, \ldots$, some positive r. There are two cases to consider, namely, C is non-empty and C is empty. First, suppose that $C \neq \varnothing$. Let A be a compact subset of R^k disjoint from C. Then A is closed and bounded and there exists a δ-neighbourhood, $N_\delta(A)$, such that $N_\delta(A) \cap C = \varnothing$ (see also Exercise 2(i)). Choose j sufficiently large, so that $N_\delta(A) \subset S_j$. Then by the fact that $d(C_\nu \cap S_j, C \cap S_j) \to 0$, one has that $C_\nu \cap A = \varnothing$, or equivalently, $A \subseteq C_\nu^c$ for all sufficiently large ν (see also Exercise 2(ii)). Now $\{\phi_\nu\} \subseteq \{\phi_m\}$ and hence $\{\phi_\nu\}$ converges weakly to ϕ. That is,

$$\int \phi_\nu g \, dQ_0 \to \int \phi g \, dQ_0 \qquad (4.15)$$

for all real-valued, integrable functions g defined on R^k. Let B be an arbitrary measurable subset of A and take $g = I_B$. Then relation (4.15) gives $\displaystyle\int_B \phi_\nu \, dQ_0 \to \int_B \phi \, dQ_0$. But $B \subseteq C_\nu^c$ and $\phi_\nu = 1$ on C_ν^c. Thus

$$\int_B \phi_\nu \, dQ_0 = \int_B dQ_0, \quad \text{so that} \quad \int_B \phi \, dQ_0 = \int_B dQ_0$$

for every measurable subset B of A. It follows that $\phi = 1$ a.s. $[Q_0]$ on A.

Now C^c can be covered by a denumerable number of compact sets A_i, $i = 1, 2, \ldots$ and applying the previous argument to each one of these sets, one concludes that $\phi = 1$ a.s. $[Q_0]$ on C^c.

A similar argument shows that $\phi = 0$ a.s. $[Q_0]$ on C. If C is empty, one has that $\phi = 1$ a.s. $[Q_0]$ on $C^c = R^k$ from the previous result. This completes the proof of the corollary.▌

The following definition will be needed in the sequel.

DEFINITION 4.1A A class of tests is said to be *essentially complete* for testing the hypothesis $H: t = t_0$ against the alternative $A: t \neq t_0$, if for any test ψ not in the class there is a test ϕ in the class for which $\beta_\psi(t_0) \geqslant \beta_\phi(t_0)$ and $\beta_\psi(t) \leqslant \beta_\phi(t)$ for $t \neq t_0$.

Now consider the class of tests ϕ of the following form

$$\phi(z) = \begin{cases} 1 & \text{if } z \in C^c, \\ 0 & \text{if } z \in C^0, \end{cases} \qquad (4.16)$$

where C is a closed, convex set in R^k. The test may be defined in an arbitrary (but measurable) manner on the boundary ∂C of C. The tests given by (4.10) are also of the form (4.16). However, in the following, we will be interested in tests of the form (4.16) which may not correspond to any prior distribution W on the parameter set.

We are now in a position to show that the class of tests of the form (4.16) is an essentially complete class. The following lemma will facilitate the proof of this fact.

LEMMA 4.6A   Let $\{t_n\}$ be a dense sequence in T with $t_1 = t_0$ (and distinct terms). Then for each $j = 1, 2, ...,$ there exists a test ϕ_j of the form (4.10) such that

$$R_{\phi_j}^*(t_i) \leqslant 0 \quad (i = 1, ..., j, j = 1, 2, ...). \tag{4.17}$$

Proof   For each j, consider the *risk set* $C_j(R^*)$ in R^j associated with the risk R^*; i.e.

$$C_j(R^*) = \{z = (x_1, ..., x_j)' \in R^j;$$

$$x_i = R_\phi^*(t_i), i = 1, ..., j \text{ for some test } \phi\}.$$

Then the set $C_j(R^*)$ is convex. In fact, by setting

$$\phi_\lambda = \lambda \phi_1 + (1 - \lambda) \phi_2$$

for any two tests ϕ_1, ϕ_2 and $0 \leqslant \lambda \leqslant 1$, one has, by means of (4.6),

$$
\begin{aligned}
R_{\phi_\lambda}^*(t) &= L^*(t, a_0) \\
&\quad + [L^*(t, a_1) - L^*(t, a_0)] \mathscr{E}_t[\lambda \phi_1(V) + (1 - \lambda) \phi_2(V)] \\
&= \lambda \{ L^*(t, a_0) + [L^*(t, a_1) - L^*(t, a_0)] \mathscr{E}_t \phi_1(V) \} \\
&\quad + (1 - \lambda) \{ L^*(t, a_0) + [L^*(t, a_1) - L^*(t, a_0)] \mathscr{E}_t \phi_2(V) \} \\
&= \lambda R_{\phi_1}^*(t) + (1 - \lambda) R_{\phi_2}^*(t).
\end{aligned}
$$

Next, $C_j(R^*)$ is closed. In fact, let $R_{\phi_n}^*(t) \to u(t)$. To show that $u(t) = R_\phi^*(t)$ for some test ϕ. By the weak compactness theorem, there is a subsequence $\{\phi_m\} \subseteq \{\phi_n\}$ which converges weakly to a test ϕ. But then $R_{\phi_n}^*(t) \to R_\phi^*(t)$ by Lemma 4.3 A. Therefore $R_{\phi_n}^*(t) \to R_\phi^*(t)$, as was to be shown.

Finally, for each $j = 1, 2, ...,$ the risk set $C_j(R^*)$ is bounded, as is immediate from the boundedness of L^*.

By taking $\phi = c$ and varying c in $[0, 1]$, one sees that the risk set $C_j(R^*)$ has points on all sides of the main diagonal, so that the main diagonal intersects $C_j(R^*)$. It is then clear that the minimax test ϕ_j, say, corresponds to that point of $C_j(R^*)$ on the main diagonal with the smallest coordinates. We shall show now that ϕ_j is also a Bayes test with respect to some prior distribution. To this end, let $d_j = \inf\{\delta \in R; \delta(1, \ldots, 1)' \in C_j(R^*)\}$ and define the set $C(d_j)$ by

$$C(d_j) = \{z = (x_1, \ldots, x_j)' \in R^j; x_i < d_j, i = 1, \ldots, j\}.$$

Then, clearly, $C(d_j)$ is a convex set and $C(d_j)$ and $C_j(R^*)$ are disjoint. Then by the separating hyperplane theorem (see, e.g. Theorem 2 in Ferguson [1], p. 73), there is a hyperplane $w'z = a$ such that

$$w'z \geqslant a \text{ for } z \in C_j(R^*), \quad w'z < a \text{ for } z \in C(d_j)$$

and $\qquad\qquad w'z = a \text{ for } z = d_j(1, \ldots, 1)'.$

Now if $w = (y_1, \ldots, y_j)'$, then $y_i \geqslant 0, i = 1, \ldots, j$ because if $y_i < 0$ for some i, then by letting $x_i \to -\infty$ in $w'z < a$ and keeping the other coordinates fixed, we would have $w'z \to \infty$ which contradicts the fact that $w'z < a$ for $z \in C(d_j)$. Next, replacing the constant a by $a/\|w\| = \bar{a}$ and also setting $\bar{w} = w/\|w\|$, we have that $\bar{w} = (\bar{y}_1, \ldots, \bar{y}_j)'$, where $\bar{y}_i = y_i/\|w\|$, $i = 1, \ldots, j$, is a probability distribution over the set $\{t_1, \ldots, t_j\}$. So the test ϕ_j, corresponding to the point $z_j = d_j(1, \ldots, 1)'$ of $C_j(R^*)$, has the property that $\bar{w}'z_j = \bar{a}$ and $\bar{w}'z \geqslant \bar{a}$ for $z \in C_j(R^*)$. Therefore ϕ_j is also Bayes with respect to the distribution $\bar{w} = (\bar{y}_1, \ldots, \bar{y}_j)'$ and hence ϕ_j is of the form (4.10). Furthermore, from (4.7), it follows that $R_\psi^*(t) = 0$ for all $t \in T$, and so, in particular, $R_\psi^*(t_i) = 0, i = 1, \ldots, j$. Since ϕ_j is a minimax test, it follows that $R_{\phi_j}^*(t_i) \leqslant 0, i = 1, \ldots, j$, which is what (4.17) asserts. $\blacksquare$

We finally have the following result.

THEOREM 4.2A   For testing the hypothesis $H: t = t_0$ against the alternative $A: t \neq t_0$ in the family (4.1), the class of tests of the form (4.16) is essentially complete.

Proof   Let ψ be a test which does not belong to the class just described (this is the specification of the test ψ which was employed in the definition of the risk $R_\phi^*(t)$). We shall show that there exists a test ϕ in the class under consideration such that

$$\beta_\psi(t_0) \geqslant \beta_\phi(t_0) \quad \text{and} \quad \beta_\psi(t) \leqslant \beta_\phi(t) \quad \text{for} \quad t \neq t_0. \quad (4.18)$$

Consider the sequence of tests $\{\phi_j\}$ of Lemma 4.6 A. Then, by the weak compactness theorem, the sequence $\{\phi_j\}$ contains a subsequence $\{\phi_m\}$ which converges weakly to a test ϕ and ϕ is equal to a test of the form (4.16) a.s. $[Q_{t,\,V}]$ for all $t \in T$, by Corollary 4.1 A. Therefore, without loss of generality, we may assume that ϕ itself is of the form (4.16). According to Lemma 4.3 A,

$$R^*_{\phi_m}(t_i) \to R^*_{\phi}(t_i) \quad (i = 1, 2, \ldots).$$

On the other hand, $R^*_{\phi_m}(t_i) \leqslant 0$ for all i and m. Therefore

$$R^*_{\phi}(t_i) \leqslant 0 \quad (i = 1, 2, \ldots). \tag{4.19}$$

In particular, $R^*_{\phi}(t_0) \leqslant 0$, or equivalently, $\beta_{\psi}(t_0) \geqslant \beta_{\phi}(t_0)$, by means of (4.5) and (4.8). This is the first of the inequalities in (4.18). Once more, by means of (4.5) and (4.8), inequality (4.19) is equivalent to

$$\beta_{\phi}(t_i) \geqslant \beta_{\psi}(t_i) \quad (i = 2, 3, \ldots). \tag{4.20}$$

It remains for us to show that (4.20) is true for all $t \in T$, $t \neq t_0$. But this is an immediate consequence of (4.20), by the fact that the sequence $\{t_n\}$ was chosen to be dense in T, and the continuity of the power function of any test, by Theorem 9 in Lehmann [3], pp. 52–4 (this theorem applies because of our assumption in (4.2)). The proof is completed. ∎

In connection with Theorem 4.2 A, it should be mentioned that this result is also stated in Chibisov [1] (see Theorem 1.2, p. 92) but no proof is presented there.

Now instead of the family of probability densities in (4.1), suppose that we have the following family

$$[dQ^*_t / dQ^*_0] = C(t) \exp{(t' \Sigma v)} \quad (v \in R^k, t \in T), \tag{4.21}$$

where Σ is a positive definite, symmetric matrix.

Consider the transformation $\lambda = \Sigma t$, so that $t' \Sigma = \lambda'$. Then $\lambda = \lambda_0$ if and only if $t = t_0 (= \Sigma^{-1} \lambda_0)$, since Σ is positive definite. Setting $Q^*_{\Sigma^{-1} \lambda} = Q^{**}_{\lambda}$, the family (4.21) becomes

$$[dQ^{**}_{\lambda} / dQ^*_0] = C^*(\lambda) \exp{(\lambda' v)} \quad (v \in R^k, \lambda \in T^*), \tag{4.22}$$

where $C^*(\lambda) = C(\Sigma^{-1} \lambda)$ and $T^* = \Sigma T$, the image of T under the transformation $\lambda = \Sigma t$. In connection with the family (4.22), the

hypothesis testing problem $H: t = t_0$ against $A: t \neq t_0$ becomes
$H^*: \lambda = \lambda_0$ against $A^*: \lambda \neq \lambda_0$. By (4.2) and the fact that (4.22)
is of the same form as (4.1), it follows that Theorem 4.2 A is still
true. Thus we have the following corollary to Theorem 4.2 A.

COROLLARY 4.2 A For testing the hypothesis $H: t = t_0$
against the alternative $A: t \neq t_0$ in the family (4.21), the class of
tests of the form (4.16) is essentially complete.

REMARK 4.1 A The significance of Theorem 4.2 A (and also
Corollary 4.2 A) in the present context is the following. For testing
the hypothesis $H: t = t_0$ against the alternative $A: t \neq t_0$, if ψ is
a test of whatever size but not of the form (4.16), there exists
a test of that form with no larger size and no smaller power than
that of ψ. Consequently, whatever criterion is proposed in terms
of power for specifying a class of tests, it is always possible to
restrict ourselves to a subclass of the class of all tests of the form
(4.16).

The first theorem stated below has been established in R. R.
Rao [1] (see Theorem 4.2), where the reader is also referred for its
proof. It provides a very useful uniform version of weak convergence of measures.

THEOREM 4.3 A For $n = 1, 2, ...,$ let μ_n and μ be measures
in R^k such that $\mu(\partial C) = 0$ for every $C \in \mathscr{C}$, where $\mathscr{C}$ is the class of
all measurable convex sets in R^k. Then $\mu_n \Rightarrow \mu$ if and only if

$$\sup [|\mu_n(C) - \mu(C)| ; C \in \mathscr{C}] \to 0.$$

In particular, the conclusion is true if μ is absolutely continuous
with respect to Lebesgue measure in R^k.

We finally formulate the following theorem established by
Wald [2] (see Proposition II, pp. 445–8). This theorem may also
be shown on the basis of some general results obtained by
Lehmann [2]. An explicit proof of it based on invariance considerations has been given by Soms [1].

THEOREM 4.4 A Let Z be a k-dimensional random vector
distributed as $N(\theta, \Sigma)$ with unknown mean θ and known non-
singular covariance matrix Σ. Then for testing the hypothesis

$H: \theta = \theta_0$ on the basis of a single observation on each of the components of Z, the critical region given by

$$(z - \theta_0)' \Sigma^{-1}(z - \theta_0) \geqslant d$$

has uniformly best average power with respect to the surfaces $E_{n,c}$ defined in (1.11) of Chapter 6 and the weight function given in (1.9) of the same chapter.

Exercises

1. Referring to the proof of Lemma 4.5 A, show that

   (i) $C^{r+1} \cap S_r = C^r$;

   (ii) $C = \bigcup\limits_{\alpha=r}^{\infty} C^\alpha$ is a non-empty, closed and convex set.

2. Referring to the proof of Corollary 4.1 A, show that

   (i) There exists a δ-neighbourhood, $N_\delta(A)$, of A such that

   $$N_\delta(A) \cap C = \varnothing \,;$$

   (ii) $A \subseteq C_\nu^c$ for all sufficiently large ν.

Bibliography

(referred to in the text)

[1] ANDERSON, T. W. *An Introduction to Multivariate Analysis.* New York (1958).

[1] BAHADUR, R. R. On Fisher's bound for asymptotic variances. *Ann. Math. Statist.* **35** (1964), 1545–52.

[1] BICKEL, P. Personal communication.

[1] BILLINGSLEY, P. *Statistical Inference for Markov Processes.* Univ. of Chicago Press (1961).

[2] BILLINGSLEY, P. The Lindeberg–Lévy theorem for martingales. *Proc. Amer. Math. Soc.* **12** (1961), 788–92.

[3] BILLINGSLEY, P. *Ergodic Theory and Information.* New York (1965).

[4] BILLINGSLEY, P. *Convergence of Probability Measures.* New York (1968).

[1] BIRNBAUM, A. Characterizations of complete classes of tests of some multiparametric hypotheses, with applications to likelihood ratio tests. *Ann. Math. Statist.* **26** (1955), 21–36.

[1] BREIMAN, L. *Probability.* Reading, Mass. (1968).

[1] CHIBISOV, D. M. A theorem on admissible tests and its application to an asymptotic problem of testing hypotheses. *Theor. Probability Appl.* **12** (1967), 90–103.

[1] CRAMÉR, H. *Mathematical Methods of Statistics.* Princeton (1946).

[1] DOOB, J. L. *Stochastic Processes.* New York (1953).

[1] EGGLESTON, H. G. *Convexity.* Cambridge University Press (1966).

[1] FERGUSON, T. S. *Mathematical Statistics.* New York (1967).

[1] GÄNSSLER, P. Note on minimum contrast estimates for Markov processes. *Mathematisches Institut der Universität zu Köln* (1969).

[1] HÁJEK, J. & ŠIDÁK, Z. *Theory of Rank Tests.* New York (1967).

[2] HÁJEK, J. A characterization of limiting distributions of regular estimates. *Z. Wahrscheinlichkeitstheorie und Verw. Gebiete,* **14** (1970), 323–30.

[3] HÁJEK, J. Local asymptotic minimax and admissibility in estimation. To appear in the *Proceedings of the Sixth Berkeley Symposium on Mathematical Statistics and Probability.*

[1] IBRAGIMOV, I. A. A central limit theorem for a class of dependent random variables. *Theor. Probability Appl.* **8** (1963), 83–9.

[1] JOHNSON, R. A. & ROUSSAS, G. G. Asymptotically most powerful tests in Markov processes. *Ann. Math. Statist.* **40** (1969), 1207–15.

[2] JOHNSON, R. A. & ROUSSAS, G. G. Asymptotically optimal tests in Markov processes. *Ann. Math. Statist.* **41** (1970), 918–38.

[1] KAUFMAN, S. Asymptotic efficiency of the maximum likelihood estimator. *Ann. Inst. Statist. Math.* **18** (1966), 155–78.

[1] KENDALL, M. G. & STUART, A. *The Advanced Theory of Statistics.* 2nd ed. New York (1967).

[1] KRAFT, C. Some conditions for consistency and uniform consistency of statistical procedures. *Univ. California Publ. Statist.* **2** (1955), 125–241.

[1] LECAM, L. On some asymptotic properties of the maximum likelihood estimates and related Bayes' estimates. *Univ. California Publ. Statist.* **1** (1953), 277–330.

[2] LECAM, L. On the asymptotic theory of estimation and testing hypothesis. *Proceedings of the Third Berkeley Symposium on Mathematical Statistics and Probability,* **1** (1956), 129–56.

[3] LECAM, L. Les propriétés asymptotiques des solutions de Bayes. *Publ. Inst. Statist. Univ. Paris,* **7** (1958), 17–35.

[4] LECAM, L. Locally asymptotically normal families of distributions. *Univ. California Publ. Statist.* **3** (1960), 37–98.

[5] LeCam, L. Likelihood functions for large numbers of independent observations. In *Research Papers in Statistics.* (F. N. David, editor.) New York (1966).

[6] LcCam, L. On the assumptions used to prove asymptotic normality of maximum likelihood estimates. *Ann. Math. Statist.* **41** (1970), 802–28.

[7] LeCam, L. Limits of experiments and a theorem of J. Hájek. To appear in the *Proceedings of the Sixth Berkeley Symposium on Mathematical Statistics and Probability.*

[8] LeCam, L. *Théorie asymptotic de la décision Statistique.* University of Montreal Press, Montreal (1970).

[1] Lehmann, E. L. Some comments on large sample tests. *Proceedings of the First Berkeley Symposium on Mathematical Statistics and Probability* (1949), 451–7.

[2] Lehmann, E. L. Optimum invariant tests. *Ann. Math. Statist.* **30** (1959), 881–4.

[3] Lehmann, E. L. *Testing Statistical Hypotheses.* New York (1959).

[1] Lind, B. & Roussas, G. G. A remark on quadratic mean differentiability. *Technical Report no. 272* (1971). University of Wisconsin, Madison.

[1] Loève, M. *Probability Theory.* 3rd ed. Princeton (1963).

[1] Mathes, T. K. & Truax, D. R. Tests of composite hypotheses for the multivariate exponential family. *Ann. Math. Statist.* **38** (1967), 681–97.

[1] Neveu, J. *Mathematical Foundations of the Calculus of Probability.* San Francisco (1965).

[1] Neyman, J. Optimal asymptotic tests of composite statistical hypotheses. In *Probability and Statistics.* (Ulf Grenander, editor.) New York (1959).

[1] Pfanzagl, J. On the asymptotic efficiency of median unbiased estimates. *Ann. Math. Statist.* **41** (1970), 1550–9.

[1] Philippou, A. Tests of hypotheses for the logarithmic, logistic and Cauchy distributions. Master's essay (1971). University of Wisconsin, Madison.

[1] Pratt, J. W. On interchanging limits and integrals. *Ann. Math. Statist.* **31** (1960), 74–7.

[1] RAO, C. R. Large sample tests of statistical hypotheses concerning several parameters with applications to problems of estimation. *Proc. Cambridge Philos. Soc.* **44** (1948), 50–7.

[2] RAO, C. R. Criterion for estimation in large samples. *Sankhyā Ser. A.* **25** (1963), 189–206.

[3] RAO, C. R. *Linear Statistical Inference and Its Applications.* New York (1965).

[1] RAO, R. R. Relations between weak and uniform convergence of measures with applications. *Ann. Math. Statist.* **33** (1962), 659–80.

[1] ROUSSAS, G. G. Asymptotic inference in Markov processes. *Ann. Math. Statist.* **36** (1965), 978–92.

[2] ROUSSAS, G. G. Asymptotic inference in Markov processes, II. *Bull. Soc. Math. Grèce,* **6** (1965), 37–50.

[3] ROUSSAS, G. G. Extension to Markov processes of a result by A. Wald about the consistency of the maximum likelihood estimate. *Z. Wahrscheinlichkeitstherorie und Verw. Gebiete,* **4** (1965), 69–73.

[4] ROUSSAS, G. G. Asymptotic normality of the maximum likelihood estimate in Markov processes. *Metrika,* **14** (1968), 62–70.

[5] ROUSSAS, G. G. Some applications of the asymptotic distribution of the likelihood functions to the asymptotic efficiency of estimates. *Z. Wahrscheinlichkeitstheorie und Verw. Gebiete,* **10** (1968), 252–60.

[6] ROUSSAS, G. G. On the concept of contiguity and related theorems. *Aarhus Universitet, Preprint Series* (1968–9), no. 33.

[7] ROUSSAS, G. G. The usage of the concept of contiguity in discussing some statistical problems. *Aarhus Universitet, Preprint Series* (1968–9), no. 34.

[8] ROUSSAS, G. G. & SOMS, A. A remark on some characterizations of contiguity. *Technical Report no. 255* (1970). University of Wisconsin, Madison.

[9] ROUSSAS, G. G. & SOMS, A. On the exponential approximation of a family of probability measures and a representation theorem of Hájek. *Technical Report no. 234* (1971). University of Wisconsin, Madison.

[1]   SCHMETTERER, L. On the asymptotic efficiency of estimates.
      In *Research Papers in Statistics*. (F. N. David, editor.) New
      York (1966).
[1]   SOMS, A. Personal communication.
[1]   WALD, A. Asymptotically most powerful tests of statistical
      hypotheses. *Ann. Math. Statist.* **12** (1941), 1–19.
[2]   WALD, A. Tests of statistical hypotheses concerning several
      parameters when the number of observations is large.
      Trans. Amer. Math. Soc. (1943), 426–82.
[1]   WEISS, L. & WOLFOWITZ, J. Generalized maximum
      likelihood estimators. *Theor. Probability Appl.* **11** (1966),
      58–81.
[2]   WEISS, L. & WOLFOWITZ, J. Maximum probability esti-
      mators. *Ann. Inst. Statist. Math.* **19** (1967), 193–206.
[1]   WOLFOWITZ, J. Asymptotic efficiency of the maximum
      likelihood estimator. *Theor. Probability Appl.* **10** (1965),
      247–60.

Some other references of related interest

[1]   BARTOO, J. B. & PURI, P. S. On optimal asymptotic tests
      of composite hypotheses. *Ann. Math. Statist.* **38** (1967),
      1845–52.
[1]   BÜHLER, W. J. & PURI, P. S. On optimal asymptotic tests
      of composite hypotheses with several constraints. *Z. Wahr-
      scheinlichkeitstheorie und Verw. Gebiete,* **5** (1966), 71–88.
[1]   INAGAKI, NOBUO. On the limiting distribution of a sequence
      of estimators with uniformity property. *Ann. Inst. Statist.
      Math.* **22** (1970), 1–13.
[1]   LIND, B. Quadratic mean differentiability and some hypo-
      theses testing problems for Markov processes. *Ph.D. Thesis*
      (1972), University of Wisconsin, Madison.

[2] LIND, B. & ROUSSAS, G. G. A remark on quadratic mean differentiability. *Technical Report no. 272.* University of Wisconsin, Madison. Also *Ann. Math. Statist.* **42** (1971) (Abstract no. 71T–88) and *Ann. Math. Statist.* **43** (1972).

[3] LIND, B. & ROUSSAS, G. G. Cramér-type conditions and quadratic mean differentiability. *Technical Report no. 273.* University of Wisconsin, Madison. Also *Ann. Math. Statist.* **42** (1971) (Abstract no. 71T–87).

[1] MICHEL, R. & PFANZAGL, J. Asymptotic normality. *Mathematisches Institut der Universität zu Köln* (1970).

[1] NOCTURNE, D. Asymptotic efficiency of the maximum likelihood estimators for the parameters of certain stochastic processes. *Ph.D Thesis, Technical Report no. 105,* Department of Operations Research, Cornell University.

[1] PHILIPPOU, A. & ROUSSAS, G. G. Asymptotic distribution of the likelihood function in the independent not identically distributed case. *Technical Report no.* 291. University of Wisconsin, Madison. Also *Ann. Math. Statist.* **42** (1971) (Abstract no. 71T–85).

[2] PHILIPPOU, A. & ROUSSAS, G. G. Exponential approximation and asymptotically optimal tests in the independent not identically distributed case. *Technical Report no.* 305. University of Wisconsin, Madison.

[1] ROUSSAS, G. G. Multi-parameter asymptotically optimal tests for Markov processes. *Aarhus Universitet, Preprint Series* (1968–9), no. 42.

[1] WEISS, L. & WOLFOWITZ, J. Generalized maximum likelihood estimators in a particular case. *Theor. Probability Appl.* **13** (1968), 622–7.

[2] WEISS, L. & WOLFOWITZ, J. Asymptotically minimax tests of composite hypotheses. *Z. Wahrscheinlichkeitstheorie und Verw. Gebiete,* **14** (1969), 161–8.

[3] WEISS, L. & WOLFOWITZ, J. Maximum probability estimators and asymptotic sufficiency. *Ann. Inst. Statist. Math.* **22** (1970), 225–44.

Index

action space, 226
Anderson, T W., 94
application of
  Berry–Esseen theorem, 156
  composition theorem, 138
  dominated convergence theorem, 166, 200, 204
  ergodic theorem, 55, 61
  Fatou's lemma, 160
  Fubini's theorem, 228
  Helly–Bray lemma, 75, 110
  Helly–Bray theorem, 3, 108
  Hölder inequality, 59, 70, 205
  Kolmogorov inequality, 62
  L_r-convergence theorem, 76
  Minkowski inequality, 63, 134
  Neyman–Pearson fundamental lemma, 30, 101, 103, 150, 154, 155, 163
  normal convergence criterion, 92
  Polya's lemma, 74, 88, 89, 92, 97, 115, 122, 159
  Scheffé's lemma, 204
  separating hyperplane theorem, 236
  Slutsky theorems, 65
  Tulcea theorem, 162
  weak compactness theorem, 136, 231, 235
assumptions, 45, 46, 87, 197
asymptotic distribution, 35, 41, 42, 53, 54, 85
asymptotic efficiency, 128
asymptotic efficiency, classical approach, 157, 158, 160
asymptotic efficiency in Wolfowitz sense (W-efficiency), 128, 130, 131, 141, 147, 158
asymptotic level of significance, 108, 116, 170, 171, 172, 188, 189, 191
asymptotic normality of, 52
asymptotic squared error, lower bound for, 159

asymptotically locally most powerful tests, 104, 105
asymptotically locally most powerful unbiased tests, 121, 126, 127
asymptotically most stringent tests, 191
asymptotically unbiased tests, 108
asymptotically uniformly most powerful tests, 99
asymptotically uniformly most powerful unbiased tests, 108

Bahadur, R. R., 129, 133, 157, 159, 160
Bayes risk, 227, 228
Bayes test, 227, 231, 233, 236
Bickel, P., 136
Billingsley, P., 3, 157, 199, 202, 203, 205, 223
Birnbaum, A., 228
Blaschke selection theorem, 232
Breiman, L., 206, 207, 210, 212, 223

central limit theorem for martingales, 205
characterization of
  contiguity, 8, 11, 17, 31, 32, 33
  relative compactness, 6
  tightness, 6
  weak convergence, 3, 199
Chibisov, D. M., 176, 237
classical efficiency, the multiparameter case, 160
contiguity
  characterization of, 8, 11, 17, 31, 32, 33
  consequences of, 33, 37
  definition of, 7
  transitivity of, 8, 25
continuity set, 199
continuous convergence, 131–2
Cramér, H., 156

derivative in q.m., 44, 45, 46, 49
differentiability in q.m., 43, 44
differentially (asymptotically) equivalent, 80
differentially (asymptotically) sufficient, 81
Doob, J. L., 43, 45, 46, 50
double exponential density, 49

Eggleston, H. G., 232
envelope power function, 190, 191
essentially complete class, 234, 236, 238
examples, 9, 11, 47, 48, 49, 50, 87, 88, 90, 92, 106, 107
exponential approximation, 68, 116, 123
exponential family, 42, 53, 67, 68, 69, 83, 84, 116

Ferguson, T. S., 121, 236
Fisher's information matrix (number), 42, 128, 130, 155.

Gänssler, P., 157
Gaussian process, 50

Hájek, J. A., 39, 68, 136, 157
Hausdorff distance, 232
Hodges, J. L., Jr, 129

Ibragimov, I. A., 205

Johnson, R. A., 47, 68, 105

Kaufman, S., 131
Kendall, M. G., 100
Krafft, O., 100
Kraft, C., 39

LeCam, L., 1, 33, 39, 41, 68, 98, 129, 136, 157
Lehmann, E. L., 100, 119, 125, 229, 231, 237, 238
Lind, B., 41
Loève, M., 3, 42, 60, 62, 63, 65, 74, 75, 76, 88, 92, 108, 110, 112, 133, 138, 156, 166, 204
log-likelihood
    asymptotic distribution of, 41, 42, 53, 54, 85
    asymptotic expansion of, 41, 52, 53, 85

L_1-norm, 6
loss function, 226

main theorems, proof of, 63
Markov case, 105, 107
Markov processes, 41, 43, 50, 53, 86
Markov property, 60
Mathes, T. K., 232
maximum likelihood estimate, 42, 105, 107, 128, 130, 157
median-unbiasedness, 154
metrically transitive, 50
minimax test, 227, 235-6
most stringent test, 191
multiparameter asymptotically optimal tests, 167

Neveu, J., 162
non-local alternatives, 196

Pfanzagl, J., 131, 148, 154
Philippou, A., 41
Pratt, J. W., 133, 134

Rao, C. R., 65, 69, 74, 88, 107, 157, 161, 185, 224
Rao, R. R., 238
relation of
    absolute continuity and contiguity, 9, 10
    continuous and uniform convergence, 133
    convergence in L_1-norm and contiguity, 8, 9, 12
    relative compactness and contiguity, 11, 13. 14, 16, 17, 31, 33
    relative compactness and tightness, 4, 202
    tightness and contiguity, 10, 12, 16, 31, 33
    W-efficiency and classical efficiency, 158
relative compactness, 4
risk (global), 226, 227
risk, set, 235
Roussas, G. G., 33, 41, 47, 68, 105, 131, 136, 145, 157, 160, 166

Schmetterer, L., 131, 135, 141, 145
Šidák, Z., 39
singular part of, 10
Soms, A., 33, 136, 238

stationarity, 55, 63
statistical applications, 2, 33, 35, 54, 84, 85
statistical interpretation, 29
Stuart, A., 100
superefficient estimate, 129

testing hypothesis (es), 30, 31, 81, 82, 86, 96, 99, 103, 107, 123, 170, 171, 172, 191, 197, 226
tightness, 4
Truax, D. R., 232

truncated version(s) of, 67, 70

uniformly integrable, 74
uniformly most powerful tests, 106

Vitali's theorem, 204

Wald, A., 99, 105, 108, 150, 168, 238
weak convergence, 2
weight function, 168
Weiss, L., 157
Wolfowitz, J., 128, 129, 130, 131, 157